# Remote Sensing Image Processing and Application

# Remote Sensing Image Processing and Application

Guest Editors

**Weitao Chen**
**Ailong Ma**
**Guohua Wu**

MDPI

Basel • Beijing • Wuhan • Barcelona • Belgrade • Novi Sad • Cluj • Manchester

*Guest Editors*
Weitao Chen
School of Computer Science
China University of Geosciences
Wuhan
China

Ailong Ma
The State Key Laboratory of Information Engineering in Surveying, Mapping, and Remote Sensing (LIESMARS)
Wuhan University
Wuhan
China

Guohua Wu
School of Automation
Central South University
Changsha
China

*Editorial Office*
MDPI AG
Grosspeteranlage 5
4052 Basel, Switzerland

This is a reprint of the Special Issue, published open access by the journal *Applied Sciences* (ISSN 2076-3417), freely accessible at: https://www.mdpi.com/journal/applsci/special_issues/Remote_Sensing_Image.

For citation purposes, cite each article independently as indicated on the article page online and as indicated below:

| |
|---|
| Lastname, A.A.; Lastname, B.B. Article Title. *Journal Name* **Year**, *Volume Number*, Page Range. |

**ISBN 978-3-7258-7274-9 (Hbk)**
**ISBN 978-3-7258-7275-6 (PDF)**
**https://doi.org/10.3390/books978-3-7258-7275-6**

# Contents

# About the Editors

**Weitao Chen**

Weitao Chen, born in October 1980, is a Professor and Ph.D. Supervisor at the School of Computer Science, China University of Geosciences (Wuhan). He concurrently serves as the Deputy Director of the Key Laboratory of Geological Survey and Evaluation of the Ministry of Education and has been selected as a leading scientific and technological talent of the Ministry of Natural Resources. He has long been engaged in interdisciplinary research between artificial intelligence and geospatial temporal information engineering, with his main research directions including geoscience remote sensing, spatiotemporal computing, and intelligent services. Professor Chen Weitao has published over 100 academic papers, including more than 70 SCI-indexed papers as the first/corresponding author. He has authored 5 monographs, obtained over 50 national invention patents, and has received multiple awards, including the China Patent Excellence Award and the Hubei Provincial Science and Technology Progress Third Prize.

**Ailong Ma**

Ailong Ma is an Associate Professor and Ph.D. Supervisor at the State Key Laboratory of Information Engineering in Surveying, Mapping and Remote Sensing (LIESMARS), Wuhan University. He received his Ph.D. degree in Photogrammetry and Remote Sensing from Wuhan University in 2017. His research primarily focuses on theoretical methods for disaster remote sensing information extraction and assessment, as well as the processing and application of multimodal remote sensing imagery.

**Guohua Wu**

Guohua Wu, born in 1986, is a Professor and Ph.D. Supervisor at the School of Automation, Central South University. He completed his Bachelor's, Master's, and Doctoral degrees at the National University of Defense Technology and was a joint training Ph.D. student at the University of Alberta, Canada. He was selected for the National "Ten Thousand Talents Program" Young Top-notch Talent and the Hunan Outstanding Youth Fund and has been named among Stanford University's World's Top 2% Most-Cited Scientists for five consecutive years. Professor Wu Guohua's main research directions include intelligent optimization and decision-making, and aerospace resource scheduling. He has published over 100 high-level papers, led more than 40 national-level projects, and has received numerous honors, including the First Prize of Natural Science from the China Simulation Federation and the Second Prize of Natural Science from Hunan Province.

# Preface

The rapid development and wide application of remote sensing technology have made it one of the most important research fields today. This Special Issue aims to gather a series of the latest research findings in remote sensing image processing, exploring new insights, developments, current challenges, and future prospects in this field. Advances in remote sensing image processing technology have provided strong support for several critical areas, including environmental monitoring, disaster management, urban planning and management, and agriculture and resource management.

The motivation for compiling this Special Issue stems from the enormous potential of remote sensing technology in practical applications. Through advanced sensors such as hyperspectral, multispectral, and LiDAR, scientists can obtain more accurate and richer data on the Earth's surface. This data not only supports scientific research in atmospheric science, geology, and ecology but also drives innovations in sensor technology, data processing, and analysis algorithms. We hope that this Special Issue will provide researchers, engineers, and practitioners in related fields with the latest technological advancements and application cases, promoting exchange and collaboration between academia and industry.

The target audience for this Special Issue includes scientists, engineers, policymakers, and students involved in remote sensing technology and application research. The successful publication of this Special Issue would not have been possible without the contributions of numerous authors and peer reviewers. We extend our heartfelt thanks to them. Special thanks to the institutions and individuals who provided assistance and support during the research process. Your efforts and wisdom were key to the successful publication of this Special Issue.

We hope that this Special Issue will provide readers with valuable reference materials, inspire more innovative research and applications, and contribute to the development and application of remote sensing technology. With continuous technological advancements and interdisciplinary cooperation, remote sensing technology will undoubtedly continue to drive the development of earth sciences and related application fields in the future.

**Weitao Chen, Ailong Ma, and Guohua Wu**
*Guest Editors*

*applied sciences*

MDPI

*Article*

# A Novel Global-Local Feature Aggregation Framework for Semantic Segmentation of Large-Format High-Resolution Remote Sensing Images

**Shanshan Wang [1], Zhiqi Zuo [2], Shuhao Yan [1], Weimin Zeng [3] and Shiyan Pang [3,*]**

1 China Aero Geophysical Survey and Remote Sensing Center for Natural Resources, Beijing 100083, China; wangshanshan@mail.cgs.gov.cn (S.W.); yanshuhao@mail.cgs.gov.cn (S.Y.)
2 College of Informatics, Huazhong Agricultural University, Wuhan 430070, China; zuo668@mail.hzau.edu.cn
3 Faculty of Artificial Intelligence in Education, Central China Normal University, 152 Luoyu Road, Wuhan 430079, China; zwm18370831109@mails.ccnu.edu.cn
* Correspondence: pangsy@ccnu.edu.cn; Tel.: +86-13618669171

**Abstract:** In high-resolution remote sensing images, there are areas with weak textures such as large building roofs, which occupy a large number of pixels in the image. These areas pose a challenge for traditional semantic segmentation networks to obtain ideal results. Common strategies like downsampling, patch cropping, and cascade models often sacrifice fine details or global context, resulting in limited accuracy. To address these issues, a novel semantic segmentation framework has been designed specifically for large-format high-resolution remote sensing images by aggregating global and local features in this paper. The framework consists of two branches: one branch deals with low-resolution downsampled images to capture global features, while the other branch focuses on cropped patches to extract high-resolution local details. Also, this paper introduces a feature aggregation module based on the Transformer structure, which effectively aggregates global and local information. Additionally, to save GPU memory usage, a novel three-step training method has been developed. Extensive experiments on two public datasets demonstrate the effectiveness of the proposed approach, with an IoU of 90.83% on the AIDS dataset and 90.30% on the WBDS dataset, surpassing state-of-the-art methods such as DANet, DeepLab v3+, U-Net, ViT, TransUNet, CMTFNet, and UANet.

**Keywords:** high-resolution remote sensing images; optical large-format images; semantic segmentation; transformer; building extraction

## 1. Introduction

Remote sensing images are widely used in economic construction, national defense construction, and people's daily life. Semantic segmentation of remote sensing images is an important aspect of remote sensing image processing, which plays a crucial role in urban planning, disaster relief, traffic management, and climate modeling. Semantic segmentation of remote sensing images refers to the process of assigning specific class labels to individual pixels in an image, enabling a comprehensive understanding of the image content [1]. The classified objects include buildings, roads, vegetation, etc.

Deep learning-based semantic segmentation algorithms have become the mainstream for remote sensing image processing. According to the difference in network structure, existing networks fall into three main categories: convolutional neural network(CNN)-based, Transformer-based, and hybrid CNN-Transformer models.

In terms of CNN-based semantic segmentation, classical convolutional neural networks include FCN [2], U-Net [3], DeepLab [4–6], PSPNet [7], etc. FCN [2] represents the inaugural convolutional neural network to attain image semantic segmentation. It eliminates the fully connected layer from the conventional classification network and effectively transforms the image classification network into an image segmentation network. This is achieved by reinstating the resolution of the feature map through deconvolution. U-Net [3] is a symmetrical network structure of encoder and decoder, which makes full use of the multi-scale image convolution features generated by the encoder through skip connections. DeepLab uses atrous convolution to solve the problem of spatial resolution degradation in the downsampling process of traditional CNN networks. To enhance the DeepLab networks, many researchers have incorporated depthwise separable convolution, parallel atrous separable convolution, and symmetric encoder-decoder structures. This integration has resulted in the development of more advanced variants. PSPNet [7] designs a new method of multi-scale feature calculation and usage, namely using four pooling layers of different sizes to generate feature maps of different levels and then obtaining image features that fuse multi-scale information through feature map aggregation, so as to improve the global information mining ability. Inspired by the idea of feature map accumulation in ResNet networks [8], DenseNet [9] improves feature reuse through denser connections and slows down gradient vanishing. In addition to the network structure, many attempts have also been made to improve the semantic segmentation of remote sensing images. These methods include the use of the attention mechanisms to enhance feature representation [10–16], generative adversarial network to enable cross-domain semantic segmentation [17–19], a self-updating CNN model [20] and a progressive edge guidance network [21] to incorporate geographic knowledge into CNN models, a semantic category balance-aware involved anti-interference network named SCBANet [22] to handle category imbalance issue, a high-order semantic decoupling network (HSDN) to disentangle features [23], a uncertainty-aware network (UANet) [24] to facilitate level-by-level feature refinement.

However, since the convolution operation is only performed within a certain range around the center point, the range of the receptive field is limited, and there is a lack of global understanding of the image itself. To make full use of the context information of images, some researchers have introduced the Transformer [25] structure with global modeling capabilities into the field of image processing, and a series of image semantic segmentation networks based on Transformer have emerged, such as Vision Transformer (ViT) [26], Swin Transformer [27], etc. ViT [26] is the first semantic segmentation network based on Transformer structure. It first splits and serializes the entire image, introduces the image positional information by adding position embeddings, and then uses the Transformer structure to process the resulting sequence of vectors to obtain more powerful image features, thereby improving the accuracy of image semantic segmentation. Since ViT splits the entire images into fixed-size patches, the result is rough when the patch size is large, while the memory consumption and computation amount are large when the patch size is small. To this end, some researchers have proposed a Swin Transformer [27] with a hierarchical design through shifted windows, which limits the self-attention computation to local windows through a sliding window operation, thereby significantly reducing computational complexity while retaining the details of the image.

Semantic segmentation networks utilizing Transformers necessitate substantial training data and computational resources. Despite numerous scholarly efforts to enhance and optimize these networks, exemplified by innovations like Swin Transformer [27], TopFormer [28] and SiamixFormer [29], they can only mitigate training complexity to a certain degree. To this end, some researchers have designed semantic segmentation networks that combine CNNs and Transformers, such as TransUNet. TransUNet [30] is one of the most classic hybrid networks, which improves the classic U-Net network by adding a series of Transformer units at the end of the encoder to globally model deep features with global context information, thereby significantly improving the accuracy of image segmentation. Since then, Wu et al. [31] introduced the CNN and Multiscale Transformer Fusion Network (CMTFNet), an encoder-decoder architecture that integrates CNN and transformer techniques to extract local details and merge multiscale global context for precise high-resolution remote-sensing image segmentation. With transformer as the backbone network, STransFuse [32], Cmt [33] and CMFNet [34] have emerged. STransFuse [32] incorporates swim transformer coding branches alongside CNN coding, extracting feature representations at various scales through hierarchical stages. Cmt [33] is a transformer-based hybrid network that substitutes convolution for the multilayer perceptron in the transformer, enhancing model accuracy without sacrificing speed. CMFNet [34] is a crossmodal multiscale fusion network that utilizes transformer architecture to capture long-range dependencies across multiscale convolutional feature maps of remote sensing data from diverse modalities, with the goal of enhancing semantic segmentation.

Deep learning-based semantic segmentation has achieved good results in the field of natural images and remote sensing images. However, classical semantic segmentation algorithms are mainly focus on processing small-sized images. Compared with natural images, remote sensing images have high resolution and large format, and there are many large-sized objects with weak-texture areas such as large building roofs, waters, and vegetation. Small-sized images (e.g., 256 × 256) in these areas contain less information and often have low segmentation accuracy in these areas. Meanwhile, due to the limitations of graphics processing unit (GPU) memory, large-format high-resolution images (e.g., 2048 × 2048) cannot be processed directly. Typical solutions such as downsampling, patch cropping, and the use of cascade models often entail a trade-off where they sacrifice intricate details or comprehensive contextual information, leading to accuracy constraints. To address this issue, we propose a novel global-local feature aggregation framework for the semantic segmentation of large-format high-resolution remote sensing images. This framework extracts global information and local details through two branches, which are then aggregated to effectively resolve category determination in weakly textured areas with fewer spatial features, such as water bodies and large building rooftops. Moreover, to optimize GPU memory usage during training, we introduce a novel three-step training method that divides the extensive semantic segmentation network into three parts, thereby reducing the reliance on graphics cards during the model training process.

The main contributions of this paper are summarized as follows:

- A new semantic segmentation framework is designed for large-format high-resolution remote sensing images. In this framework, two branches are used to extract global and local features respectively, and then a global-local feature aggregation module is designed based on the transformer global attention mechanism, which effectively improves the accuracy of semantic segmentation through the fusion of global and local information. Comprehensive comparisons with the state of the arts on two public datasets and ablation studies are conducted to verify its effectiveness.
- To reduce the large memory occupation that occurs during large-format image processing, a new model training method based on step-by-step training and gradient accumulation is devised. This method divides the entire network into three parts and trains them in three steps by designing a separate classification head for each part. Extensive experiments demonstrate that this method effectively reduces the memory requirements of the GPU during model training.
- This framework can effectively integrate the mainstream semantic segmentation networks and its strong scalability has been validated using four mainstream backbone networks, including U-Net, ViT, TransUNet, and ConvNeXt V2.

In the following sections, we first present the methodology of our framework. We will then describe the experimental results and analysis, followed by the conclusion.

## 2. Methodology

### 2.1. Overview

The network structure of our proposed method is depicted in Figure 1 and comprises five key components: global feature extraction module, feature coding module, feature aggregation module, information decoding module, and lightweight convolution module.

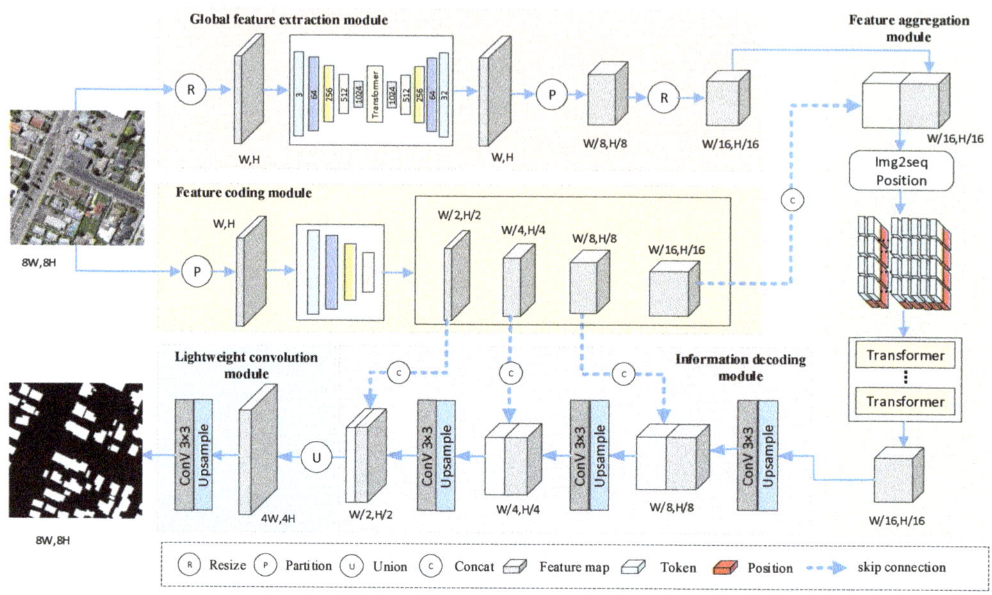


**Figure 1.** Network structure of the proposed method.

Here's a brief overview of the functions of each module:

- **Global feature extraction module**: This module's primary function is to extract global contextual information from large-format images. It takes a downsampled low-resolution image as input and produces global semantic features with a consistent size as the output. The training dataset for this module includes low-resolution images and their corresponding labels. In the subsequent sections of this paper, global semantic features refer to the features output by this module.
- **Feature coding module**: The feature coding module processes cropped small image patches. It generates multi-scale convolution features from the input, with the highest and smallest features being used for fusion with the output feature from the global feature extraction module, while other multi-scale features are utilized for skip connections. In the subsequent sections of this paper, local semantic features refer to the features output by this module.
- **Feature aggregation module**: This module is responsible for combining the global features produced by the global feature extraction module with the local features generated by the feature coding module. Initially, these two features are concatenated, and then a series of Transformer units process the concatenated features to obtain new aggregated features that incorporate both global and local information.
- **Information decoding module**: The information decoding module's primary purpose is to generate a prediction map consistent with the original image size. It takes as input the new features produced by the feature aggregation module and the multi-scale shallow features generated by the feature coding module. The feature map gradually returns to its original size through a series of upsampling, convolution, and skip connection operations, resulting in a prediction map consistent with the original image size.
- **Lightweight convolution module**: This module primarily stitches the feature map at the original resolution. This stitching corresponds to the slicing operation in the feature coding module, which is used to create a feature map of the same size as the original image. Subsequently, the prediction map with the same size as the original image is restored through a simple combination of lightweight convolution.

### 2.2. Global Feature Extraction Module

The global feature extraction module is primarily used to extract global semantic features from large-format input images. In this paper, we first downscale large-format images into low-resolution small-size images and also downscale labels corresponding to large-format images to small-size labels. Then, we select the appropriate semantic segmentation model to extract global semantic features. In this paper, we have chosen the TransUNet [30], a classic semantic segmentation network, which is primarily composed of three components: an encoder, an attention module, and a decoder. The encoder utilizes the ResNet-50 network to generate multi-scale feature maps. The height and width of these multi-scale feature maps are 1/2, 1/4, 1/8, and 1/16 of the small-size image's dimensions, respectively. The attention module consists of 12 transformer units connected in series to process the encoder's final output feature map, which has a height and width that is 1/16 of the small-size image. This process yields an enhanced feature map that is rich in contextual information. The encoder consists of a series of convolutional upsampling and skip connection units. It employs a U-Net-like strategy to progressively restore the feature map layer by layer. The final output of the TransUNet is a pixel-by-pixel semantic feature map that matches the size of the low-resolution small-size image.

Subsequently, to fuse with the local features output by the subsequent feature coding module, we divided the global semantic features into 64 blocks, which we referred to as a partition operation. Finally, to fuse with the highest-level features output by the feature encoding module, we further resized each block of the global semantic features to the same size as the highest-level features output by the feature encoding module, e.g., 1/16 of the small-size image's dimensions.

The equation for the global feature extraction module is summarized as follows:

$$x_g = Resize(Partition(TransUNet(Resize(x)))), \tag{1}$$

where $x$ denotes the large-format input image, Resize represents the operation of downscaling the image, *TransUnet* represents the network structure used by the global feature extraction module in this paper, *Partition* stands for partition operation, and $x_g$ represents the global feature of the final output.

It should be noted that to better utilize the global semantic information extracted by the semantic information extraction module, high-dimensional semantic features are used as input to the subsequent feature aggregation module, instead of the semantic segmentation results. To alleviate the computational load and GPU memory usage during the training of large-format image semantic segmentation, we use the downsampling image and corresponding label to train the model in advance. During the training of the entire proposed network, the parameters of the global feature extraction module are fixed and do not participate in the backpropagation and gradient update, greatly reducing the GPU memory consumption by the network.

### 2.3. Feature Coding Module

The feature coding module crops the large-format image into a series of small-size image patches, which are then processed by a convolutional neural network to extract the multi-scale convolutional features of each patch.

The input of the feature coding module is the cropped image patches (e.g., a large-format image is divided into 64 patches), and ResNet-50 is used as the network structure. The output is multi-scale feature maps with dimensions of 1/2, 1/4, 1/8, and 1/16 of the original image, respectively. The equation for the feature coding module is described as follows:

$$\left[x_l^1, x_l^2, x_l^3, x_l^4\right] = ResNet(Partition(x)), \tag{2}$$

where $x$ represents the large-format input image, *Partition* stands for partition operation, ResNet represents the network structure adopted by the feature coding module in this paper, and $[x_l^1, x_l^2, x_l^3, x_l^4]$ represents the final output of four local features at different scales.

### 2.4. Feature Aggregation Module

The feature aggregation module primarily serves to integrate both global and local features, resulting in a new feature that encompasses local and global information. The global features are derived from the global feature extraction module, while the local features are obtained from the feature coding module. The module comprises 6 transformer units within the network architecture, with the detailed computation described as follows:

$$x_{gl} = Transformers\left(Concat\left(x_l^4,\ x_g\right)\right), \tag{3}$$

where $x_g$ represents the global feature output of the global feature extraction module, $x_l^4$ represents the last layer of local features output by the feature coding module, *Concat* stands for feature concatenation operation, which connects two sets of features together along the channel dimension to form a new, larger set of features, *Transformers* represents the network structure adopted by the feature aggregation module in this paper, which consists of 6 multi-head transformer units, and $x_{gl}$ represents the new features after feature aggregation. The transformer unit used here is the same as the transformer unit of TransUNet [30] in Section 2.2.

### 2.5. Information Decoding Module

The information decoding module consists of two parts of input data: high-order features output by the feature aggregation module and multi-scale features output by the feature coding module. It decodes high-order features output by the feature aggregation module layer by layer and outputs feature maps with a consistent size of the original image. This module is mainly composed of convolutional upsampling blocks and skip connection blocks. The convolutional upsampling block consists of convolution and upsampling operations, which are mainly used to increase the size of the feature map. The skip connection block is implemented by the concatenation operation. It is mainly used to connect the multi-scale features output by the feature coding module and improve the detailed information of the feature map by introducing low-level features. The specific calculation process is as follows:

$$y_1 = Concat\left(x_l^3, Conv\_up\left(x_{gl}\right)\right), \tag{4}$$

$$y_2 = Concat\left(x_l^2, Conv\_up(y_1)\right), \tag{5}$$

$$y_{patch} = Concat\left(x_l^1, Conv\_up(y_2)\right), \tag{6}$$

where $x_{gl}$ represents the new features after feature aggregation, $x_l^1, x_l^2, x_l^3$ represents the multi-scale shallow features output by the feature coding module, which is used to gradually restore the feature map to its original size, *Conv_up* represents convolution and upsampling operations, *Concat* stands for feature concatenation operation, $y_{patch}$ represents the new feature after information decoding, and the size is the same as the image patch.

### 2.6. Lightweight Convolution Module

The lightweight convolution module is primarily used to restore the prediction map to the same size as the original image and improve the segmentation result at the tile boundary of the image. Since convolution of large-format feature maps consumes a significant amount of GPU memory, this module consists of only a few simple convolutional layers. Specifically, this module combines the small patch-size feature map into a large-format feature map that is consistent with the original large-format image through the stitching operation. Finally, it obtains the final prediction map through the convolution operation. The calculation process is as follows:

$$y = Conv\left(Union\left(y_{patch}\right)\right), \tag{7}$$

where *Union* represents the feature stitching operation, which corresponds to the *Partition* operation of the feature coding module, and *Conv* represents the convolution operation.

### 2.7. Training Process

Training our model for large-format images demands significant GPU memory and computational power, necessitating high-performance GPUs. It is difficult to use regular GPU for direct training. To address this challenge, we optimize the training process by breaking it down into three distinct steps, as shown in Figure 2. Details are as follows.

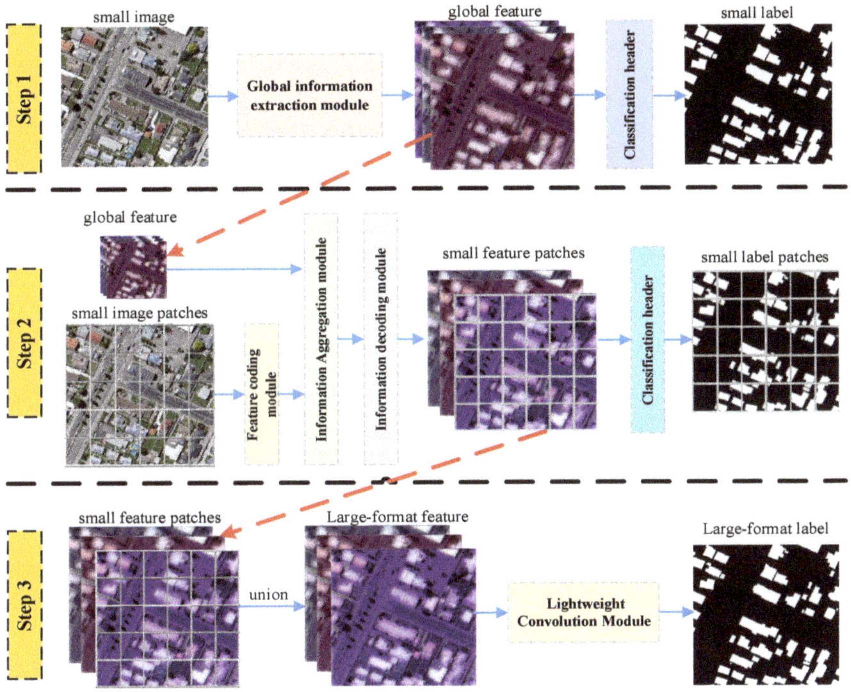


**Figure 2.** Training process.

Step 1: Training the Global Feature Extraction Module

In the first step, we train the global feature extraction module using downsampled images and their corresponding labels. The training process aligns with that of a traditional semantic segmentation network.

Step 2: Training the Feature Encoding, Feature Aggregation, and Information Decoding Modules

Building upon the fixed parameters of the global feature extraction module, we proceed to train the following three modules: the feature encoding module, the feature aggregation module, and the information decoding module. To facilitate this training, we introduce modifications to the information decoding module, incorporating a classification head. The classification head serves the dual purpose of generating classification results and computing the loss. In this step, we scale the input large-sized image down to a smaller $256 \times 256$ image, and leverage the global feature extraction model to extract global features. Subsequently, these three modules are trained, with inputs comprising global feature information, small image patches, and their corresponding labels, while the output yields prediction results determined by the classification head.

During the training process, we encountered high GPU memory consumption. To mitigate this issue, we implemented a gradient-accumulation training approach, following these specific steps: (1) Dividing the input image patches, corresponding labels, and global features into several small batches. (2) Performing backward propagation for each small batch, but refraining from updating the model parameters. (3) Updating the model parameters only after all batches have been processed.

Step 3: Training the Lightweight Convolution Module

In the final step, we train the lightweight convolution module based on the fixed parameters of the global feature extraction module, feature encoding module, feature aggregation module, and information decoding module. We begin by employing the global feature extraction module to extract global features from the downsampled image. Subsequently, we utilize the feature encoding module, feature aggregation module, and information decoding module to obtain feature maps with the same resolution as the original image. Finally, these feature maps are seamlessly stitched together into a large feature map mirroring the original image's dimensions. This unified feature map is then fed into the lightweight convolution module for parameter training.

This partitioned training approach optimizes GPU memory usage and computational efficiency, making it feasible to train our model for large-format images on GPUs with more constrained resources.

## 3. Experimental Results and Analysis

### 3.1. Datasets

To evaluate the effectiveness of our proposed method, we used two datasets for validation, both sourced from the WHU Building Dataset [35], available at http://gpcv.whu.edu.cn/data/building_dataset.html (accessed on 2 March 2023). To differentiate between them, we have named them the Aerial Imagery Dataset (AIDS) and the Building Change Detection Dataset (WBDS). Both datasets consist of large-area images and corresponding building labels in Shapefile format. During the process of use, we convert the vector data in Shapefile format into a raster binary image that has the same resolution as the image, with buildings represented in white and the background in black. Due to GPU memory constraints, we were unable to process these large-area images directly. Therefore, we cropped the area images to produce relatively large-format images (e.g., 2048 × 2048). The size of the cropped large-format image is 2048 × 2048, which is eight times larger than the common image size of 256 × 256. For large-format labels, the cropping processing is exactly the same as the original image. Prior to cropping, the datasets were divided into training, validation, and test sets in a 6:2:2 ratio.

**AIDS Dataset**: Collected in New Zealand, the AIDS dataset comprises more than 22,000 individual buildings. The original area image is a large image that measures 1,560,159 pixels × 517,909 pixels with a spatial resolution of 0.075 m. Cropping the entire area image resulted in 12,940 large-format images with sizes of 2048 × 2048. The dataset was divided into 7764 training sets, 2588 validation sets, and 2588 test sets according to the 6:2:2 ratio.

**WBDS Dataset**: Covering a substantial portion of Christchurch, New Zealand, the WBDS dataset includes bi-temporal images from 2012 and 2016, each accompanied by corresponding building label data. The original image size is 32,507 pixels × 15,345 pixels with a spatial resolution of 0.2 m. After cropping, 630 large-format images with sizes of 2048 × 2048 were generated. The dataset was divided into 378 training sets, 126 validation sets, and 126 test sets in a 6:2:2 ratio.

### 3.2. Implementation Details

**Data Preprocessing**: To enhance the network's robustness and avoid overfitting, we employed data augmentation techniques such as random rotation, mirror flipping, and adjustments to image color, saturation, and contrast.

**Training Details**: Our experiments were conducted using the PyTorch framework with CUDA version 11.0 on an Ubuntu 20 environment. We utilized an NVIDIA A40 with 46GB of memory to accelerate model training. The AdamW optimizer was chosen with an initial learning rate of 0.0001, a decay rate of $1 \times 10^{-8}$, $\beta 1 = 0.9$, and $\beta 2 = 0.999$. The batch size varies slightly across different steps: it is 4 for steps 1 and 2, and 2 for step 3. The epoch is set to 200. Details are shown in Table 1. In our large-format image semantic segmentation task, we chose binary cross-entropy as the loss function. The global feature extraction module and feature coding module were initialized with ResNet50 on ImageNet-1k as the initialization parameters. During network training, to conserve GPU memory, the parameters of the global feature extraction module remained fixed and were not updated. We employed a gradient accumulation strategy for image patch processing.

**Table 1.** Hyperparameters.

| Items | Settings |
| --- | --- |
| Optimizer | AdamW optimizer |
| Initial learning rate | 0.0001 |
| Decay rate | $1 \times 10^{-8}$ |
| Momentum | $\beta 1 = 0.9$, $\beta 2 = 0.999$ |
| Batch size | 4 for steps 1 and 2, and 2 for step 3 |
| Epoch | 200 |

**Metrics**: To evaluate our method's performance, we used four common metrics: Intersection over Union (IoU), precision, recall, and F1-score. These metrics are calculated based on true positives (TP), false positives (FP), true negatives (TN), and false negatives (FN) as follows:

$$\mathrm{IoU} = \mathrm{TP}/(\mathrm{TP} + \mathrm{FP} + \mathrm{FN}), \tag{8}$$

$$\mathrm{Precison} = \mathrm{TP}/(\mathrm{TP} + \mathrm{FP}), \tag{9}$$

$$\mathrm{Recall} = \mathrm{TP}/(\mathrm{TP} + \mathrm{FN}), \tag{10}$$

$$\mathrm{F1-score} = 2 \times (\mathrm{Precison} \times \mathrm{Recall})/(\mathrm{Precison} + \mathrm{Recall}) \tag{11}$$

### 3.3. Evaluation of GPU Memory Usage

To evaluate GPU memory usage, we monitored GPU memory consumption during the model training process. We used a single NVIDIA A40 graphics card with 46GB of memory for the experiments. When training the entire model directly, GPU memory consumption exceeded 46 GB, rendering training infeasible on a single A40. Consequently, we employed a step-by-step training strategy, consisting of three steps:

- Step 1: Training the global feature extraction module with a batch size of 4, resulting in GPU memory consumption of less than 20 GB.
- Step 2: Training the feature encoding module, feature aggregation module, and information decoding module with a batch size of 1 and GPU memory consumption of 43 GB. Using the gradient accumulation strategy with a batch size of 4 for small image groups reduced GPU memory usage to a maximum of 16 GB.
- Step 3: Training the lightweight convolution module with a batch size of 2 and GPU memory consumption of 26 GB.

This three-step training strategy substantially reduced GPU memory usage and eased the hardware requirements for model training.

### 3.4. Ablation Study

To further assess the effectiveness of our proposed framework, we conducted an ablation study. All experimental data and parameter settings remained consistent throughout the study. The ablation study was designed as follows:

- Assessment of the Proposed Framework's Effectiveness: We evaluated the performance improvement achieved by incorporating four popular networks (i.e., U-Net, ViT, TransUNet and ConvNeXt V2) as global feature extraction modules in our proposed framework.

- Evaluation of Different Modules' Effectiveness: Using TransUNet as the global feature extraction module, we added feature aggregation module and lightweight convolution module sequentially to form three cases, aiming to verify the role and effectiveness of these modules.

### 3.4.1. Assessment of the Proposed Framework's Effectiveness

To gauge the effectiveness of our framework, we employed four mainstream networks (i.e., U-Net, ViT, TransUNet and ConvNeXt V2) as global feature extraction modules and compared their performance improvements within our framework on the AIDS and WBDS datasets. Table 2 presents the performance improvements of the proposed framework with different global feature extraction modules.

**Table 2.** Performance improvements of the proposed framework with different global feature extraction modules.

| Datasets | Network | Direct Stitching | | Ours | | Improvement | |
|---|---|---|---|---|---|---|---|
| | | IoU | F1 | IoU | F1 | IoU | F1 |
| AIDS | ViT | 84.29 | 91.51 | -- | -- | -- | -- |
| | U-Net | 83.53 | 91.19 | 90.00 | 94.73 | 6.47 | 3.54 |
| | TransUNet | 87.11 | 93.12 | 90.12 | 94.80 | 3.01 | 1.68 |
| | ConvNeXt V2 | 87.17 | 93.14 | 89.98 | 94.73 | 2.81 | 1.59 |
| WBDS | ViT | 83.78 | 91.56 | 91.74 | 95.69 | 7.96 | 4.13 |
| | U-Net | 85.17 | 91.99 | 91.17 | 95.38 | 6.00 | 3.39 |
| | TransUNet | 87.89 | 93.60 | 92.07 | 95.87 | 4.18 | 2.27 |
| | ConvNeXt V2 | 87.81 | 93.51 | 89.65 | 94.54 | 1.84 | 1.03 |

Table 2 reveals that TransUNet outperforms U-Net, ViT and ConvNeXt V2 on both the AIDS and WBDS datasets. This demonstrates that TransUNet, chosen as our global feature extraction module, is effective. Moreover, adopting our proposed framework yields performance improvements over direct stitching results of U-Net, ViT, TransUNet and ConvNeXt V2, highlighting the effectiveness of our framework in utilizing the global characteristics of large-format images.

### 3.4.2. Evaluation of Different Modules' Effectiveness

To assess the individual contributions of each module, we examined the results on the WBDS dataset when various combinations of modules were employed. The different module combinations were as follows:

- **Baseline**: The TransUNet network is employed as the baseline for processing small-sized image patches, and the outcomes from these smaller images are subsequently stitched together directly to produce large-format results.
- **+Global Feature Extraction Module (GFEM)**: TransUNet serves as the global feature extraction module, ResNet-50 acts as the feature coding model, and concatenation operations are employed to replace the feature aggregation module in the global-local feature aggregation process.
- **+Feature Aggregation Module (FAM)**: We added the feature aggregation module to integrate global and local features, without employing lightweight convolution. The output was the prediction result of image patches, and we stitched these predictions to obtain large-format predictions.
- **+Lightweight Convolution Module (LCM)**: The lightweight convolution module was introduced to process large-format feature maps stitched from small patches. The small patch feature maps were produced by the feature aggregation module.

Table 3 showcases the results of these different module combinations. Table 3 demonstrates that, using TransUNet as the baseline, the model's performance is incrementally enhanced as the global feature extraction module, the feature aggregation module, and the lightweight convolution module are incorporated one by one. Compared to the baseline TransUNet, the IoU metrics increased by 0.78%, 0.91%, and 2.49% respectively. This demonstrates the effectiveness of the modules within the network structure designed in this paper.

**Table 3.** Results of different module combinations on the WBDS dataset. The items with "√" indicate the options that have been adopted. The boldfaced values indicate the optimal result.

| TransUNet | +GFEM | +FAM | +LCM | IoU | F1 |
|---|---|---|---|---|---|
| √ | | | | 87.89 | 93.60 |
| | √ | | | 88.67 | 93.99 |
| | √ | √ | | 89.58 | 94.52 |
| | √ | √ | √ | **92.07** | **95.87** |

To further illustrate the efficacy of the modules used in the ablation study, we selected several images from the dataset to visualize the semantic segmentation outcomes of different module combinations, as depicted in Figure 3. It is evident that the addition of the global information extraction module leads to a slight improvement in performance. However, without advanced feature aggregation module, the enhancement is not substantial. Upon incorporating the feature aggregation module, there is a full integration of global and local information, resulting in a significantly improved outcome. The inclusion of the lightweight convolution module effectively enhances the edge areas of buildings. Hence, this further affirms the validity of the modules presented in this paper.

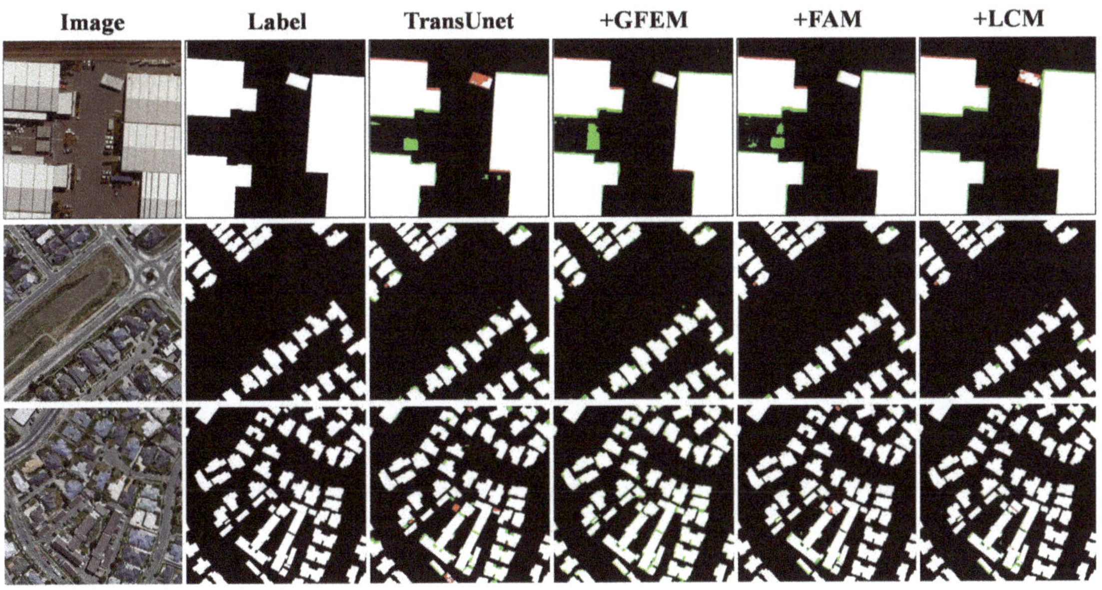


**Figure 3.** Visualization of ablation study on the WBDS dataset. white represents the building area, black represents the background, red represents missed detection, and green represents error detection.

### *3.5. Comparisons with the State of the Arts*

To further validate the effectiveness of the proposed method, we conducted comparative experiments on AIDS and WBDS datasets, including DANet [10], DeepLab v3+ [6], U-Net [3], ViT [26], TransUNet [30], CMTFNet [31] and UANet [24]. To make the experimental results more convincing, except for the ViT network, we selected ResNet50 as the feature extraction backbone network for all networks during the experiment. ResNet50 loaded the pretrained model on ImageNet-1k and ViT loaded the pretrained model of ViT-B on ImageNet21k as the initialization parameters. Table 4 displays the comparisons on the AIDS and WBDS datasets, with the best results indicated in bold.

**Table 4.** Comparisons of different models on the AIDS and WBDS datasets. The boldfaced values indicate the optimal result.

| Model | Computational Efficiency | | AIDS | | | | WBDS | | | |
|---|---|---|---|---|---|---|---|---|---|---|
| | Params ($\times 10^6$) | FLOPs ($\times 10^9$) | IoU | precision | Recall | F1 | IoU | precision | Recall | F1 |
| DANet | 46.19 | 125.14 | 79.68 | 88.25 | 89.14 | 88.69 | 77.95 | 94.07 | 81.98 | 87.61 |
| DeepLab v3+ | 40.35 | 17.36 | 82.98 | **96.73** | 85.37 | 90.70 | 85.12 | 95.99 | 88.25 | 91.96 |
| U-Net | 34.53 | 65.52 | 83.53 | 96.10 | 86.47 | 91.03 | 85.17 | 93.55 | 90.43 | 91.97 |
| ViT | 87.80 | 24.28 | 84.29 | 96.08 | 87.30 | 91.48 | 83.78 | 94.79 | 87.82 | 91.17 |
| TransUNet | 93.23 | 32.23 | 87.11 | 96.20 | 90.22 | 93.11 | 87.89 | 94.05 | 93.07 | 93.56 |
| CMTFNet | 30.07 | 8.56 | 88.05 | 94.26 | 93.03 | 93.64 | 86.21 | 90.12 | **95.21** | 92.59 |
| UANet | **26.73** | **7.45** | 88.83 | 94.90 | 93.28 | 94.08 | 88.50 | 92.73 | 95.09 | 93.90 |
| Ours | 93.23<br>30.28 | 32.23<br>19.93 | 90.12 | 95.67 | 93.95 | 94.80 | **92.07** | **96.65** | 95.11 | **95.87** |
| Ours (gradient-accumulation) | 93.23<br>30.28 | 32.23<br>19.93 | **90.83** | 95.58 | **94.80** | **95.19** | 90.30 | 95.80 | 94.02 | 94.90 |

Table 4 illustrates that the early CNN-based semantic segmentation networks (such as U-Net and DeepLab v3+) are not satisfactory due to the absence of attention mechanisms. Results from the ViT network, relying solely on self-attention mechanisms, also proved unsatisfactory. However, subsequent networks combining CNNs with attention mechanisms (such as TransUNet, CMTFNet and UANet) notably improved results by integrating various attention modules into CNNs. Among them, despite its relatively simple structure, TransUNet exhibited stable segmentation results. Compared with the state of the arts, the proposed method achieves superior semantic segmentation results on both datasets due to the introduction of large-format global features. Additionally, implementing gradient accumulation strategies introduced some fluctuations in accuracy, but still yielded favorable outcomes. This is primarily because our framework effectively combines the global semantic features from large-format images with the local details of image patches. This integration allows for a more precise handling and extraction of objects across various scales, which notably enhances the accuracy and reliability of semantic segmentation in high-resolution remote sensing images.

To further verify the computational efficiency of the proposed method, we have detailed the training parameters and Floating Point of Operations (FLOPs) for different models on the AIDS dataset in Table 4. Our network is divided into two parts for statistical purposes: global semantic feature extraction (i.e., Part I) and other components (i.e., Part II). Part I is TransUNet, which is mainly responsible for extracting global semantic features. Part II consists of our four modules: feature coding, feature aggregation, information decoding, and lightweight convolution. As indicated in Table 4, our supplementary processing modules for large-format images (i.e., Part II) impose only a modest increase in computational load, thereby preserving computational efficiency.

To further demonstrate the effectiveness of the proposed method, we selected several images from two datasets to visualize the semantic segmentation results of buildings. The results are shown in Figure 4. It can be observed that the segmentation results of the comparison method in some building areas are not ideal, and there are problems such as blurred and incoherent boundaries. The proposed method achieves the best detection results, especially in large building areas, which significantly reduces error detection.

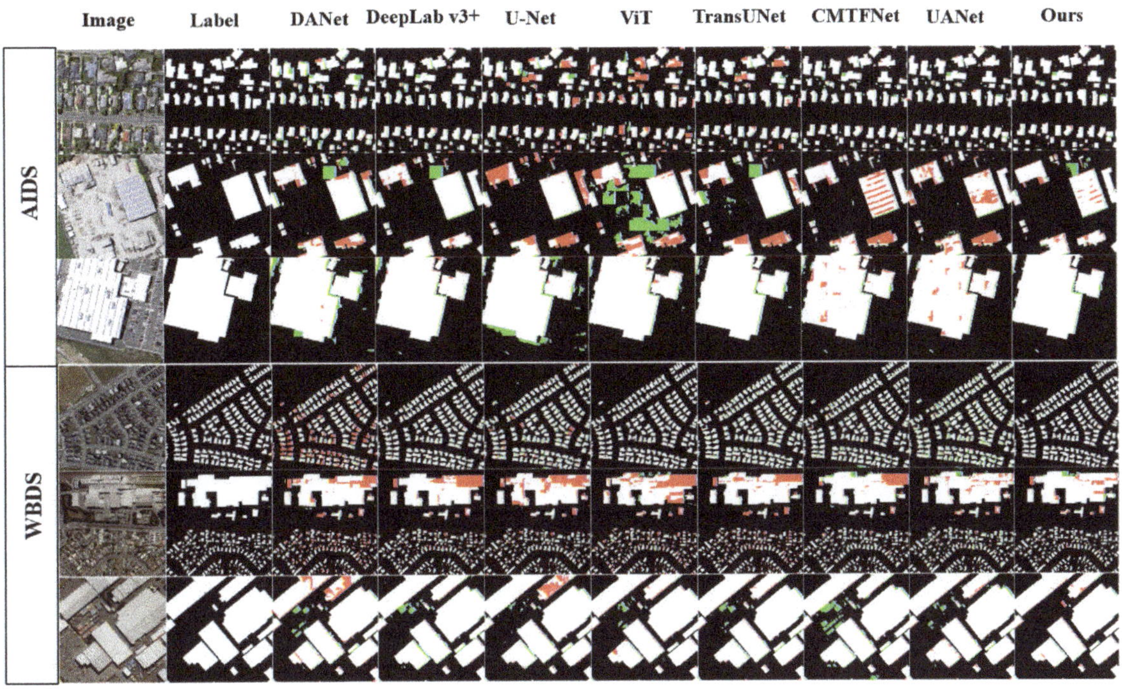


**Figure 4.** Comparisons with the state of the arts on the AIDS and WBDS datasets. white represents the building area, black represents the background, red represents missed detection, and green represents error detection.

## 4. Conclusions

In this paper, we have introduced a novel gloabl-local feature aggregation framework for the semantic segmentation of large-format high-resolution remote sensing images. This framework effectively combines the advantages of local information from cropped small-size images and global information from downsampled large-format images. Furthermore, we have implemented strategies such as step-by-step training and gradient accumulation, resulting in a significant reduction in GPU memory consumption.

Extensive experiments on two public datasets have demonstrated that our framework effectively enhances the accuracy and reliability of semantic segmentation for large-format high-resolution remote sensing images. We have also conducted thorough comparisons with state-of-the-art models, including DANet, DeepLab v3+, U-Net, ViT, TransUNet, CMTFNet and UANet, further showcasing the effectiveness of our proposed method. Additionally, our framework exhibits strong scalability, making it adaptable to various mainstream semantic segmentation networks such as U-Net, ViT, TransUNet and ConvNeXt V2.

In our future work, we will explore semantic segmentation models that are tightly integrated with large models and integrate them into our large-format image processing frameworks to meet the needs of more diverse remote sensing image scenarios.

**Author Contributions:** Conceptualization, S.W., Z.Z. and S.P.; methodology, S.W., Z.Z. and S.P.; validation, Z.Z., S.Y. and W.Z.; writing—original draft preparation, S.W., Z.Z. and W.Z.; writing—review and editing, S.W., S.Y. and S.P.; visualization, S.Y. and W.Z.; supervision, S.W; project administration, S.W.; funding acquisition, S.W. and S.P. All authors have read and agreed to the published version of the manuscript.

**Funding:** This research was supported in part by National Key Research and Development Program of China under Grant 2021YFC3000400, Ministry of Education of the People's Republic of China under Grant 22YJC880058, Knowledge Innovation Program of Wuhan-Shuguang Project under Grant 2022010801020281, University-Industry Collaborative Education Program under Grant 230806008021539 and the Fundamental Research Funds for the Central Universities under Grant CCNU22QN011 and CCNU22QN019.

**Data Availability Statement:** The original data presented in the study are openly available at http://gpcv.whu.edu.cn/data/building_dataset.html, accessed on 13 July 2024.

**Conflicts of Interest:** The authors declare no conflict of interest.

## Abbreviations

| | |
|---|---|
| AIDS | Aerial Imagery DataSet |
| CMFNet | Crossmodal Multiscale Fusion Network |
| Cmt | Convolutional neural networks Meet vision Transformers |
| CMTFNet | Crossmodal Multiscale Fusion Network |
| DANet | Dual Attention Network |
| DenseNet | Densely Connected Convolutional Network |
| DRDG | Depth-assisted ResiDualGAN |
| FCN | Fully Convolutional Network |
| FN | false negative |
| FP | false positive |
| GLF-Net | A Semantic Segmentation Model Fusing Global and Local Features for High-Resolution Remote Sensing Images |
| FLOPs | Floating Point Operations |
| GPU | Graphics Processing Unit |
| HSDN | High-order Semantic Decoupling Network |
| IoU | Intersection over Union |
| LANet | Local Attention Network |
| MACU-Net | Multiscale skip connected and Asymmetric-Convolution-based U-Net |
| MLDANets | MultiLevel Deformable Attention-aggregated Network |
| PEG-Net | Progressive Edge Guidance Network |
| PSPNet | Pyramid Scene Parsing Network |
| ResNet | Residual Network |
| SCBANet | Semantic Category Balance-Aware involved anti-interference Network |
| SLU-CNN | Self-Learning-Update CNN |
| SPGAN | Semantic-Preserving Generative Adversarial Network |
| SSCNet | Spectral-Spatial Cooperation Network |
| STransFuse | Fusing Swin Transformer and Convolutional Neural Network for Remote Sensing Image Semantic Segmentation |
| TN | true negative |
| TopFormer | Token Pyramid Transformer for Mobile Semantic Segmentation |
| TP | true positive |
| UANet | Uncertainty-Aware Network |
| ViT | Vision Transformer |
| WFE | Wavelet Feature Enhancement |

## References

1. Camps-Valls, G.; Tuia, D.; Bruzzone, L.; Benediktsson, J.A. Advances in Hyperspectral Image Classification: Earth Monitoring with Statistical Learning Methods. *IEEE Signal Process. Mag.* **2014**, *31*, 45–54. [CrossRef]
2. Shelhamer, E.; Long, J.; Darrell, T. Fully Convolutional Networks for Semantic Segmentation. *IEEE Trans. Pattern Anal. Mach. Intell.* **2017**, *39*, 640–651. [CrossRef] [PubMed]
3. Ronneberger, O.; Fischer, P.; Brox, T. U-Net: Convolutional Networks for Biomedical Image Segmentation. In Proceedings of the Medical Image Computing and Computer-Assisted Intervention—MICCAI 2015, Cham, Switzerland, 5–9 October 2015; pp. 234–241.

4. Chen, L.C.; Papandreou, G.; Kokkinos, I.; Murphy, K.; Yuille, A.L. DeepLab: Semantic Image Segmentation with Deep Convolutional Nets, Atrous Convolution, and Fully Connected CRFs. *IEEE Trans. Pattern Anal. Mach. Intell.* **2018**, *40*, 834–848. [CrossRef] [PubMed]
5. Chen, L.-C.; Zhu, Y.; Papandreou, G.; Schroff, F.; Adam, H. Encoder-Decoder with Atrous Separable Convolution for Semantic Image Segmentation. In Proceedings of the Computer Vision—ECCV 2018, Munich, Germany, 8–14 September 2018; pp. 833–851.
6. Chen, L.-C.; Papandreou, G.; Schroff, F.; Adam, H. Rethinking Atrous Convolution for Semantic Image Segmentation. *arXiv* **2017**, arXiv:1706.05587.
7. Zhao, H.; Shi, J.; Qi, X.; Wang, X.; Jia, J. Pyramid Scene Parsing Network. In Proceedings of the 2017 IEEE Conference on Computer Vision and Pattern Recognition (CVPR), Honolulu, HI, USA, 21–26 July 2017; pp. 6230–6239.
8. He, K.; Zhang, X.; Ren, S.; Sun, J. Deep Residual Learning for Image Recognition. In Proceedings of the 2016 IEEE Conference on Computer Vision and Pattern Recognition (CVPR), Las Vegas, NV, USA, 27–30 June 2016; pp. 770–778.
9. Huang, G.; Liu, Z.; Maaten, L.V.D.; Weinberger, K.Q. Densely Connected Convolutional Networks. In Proceedings of the 2017 IEEE Conference on Computer Vision and Pattern Recognition (CVPR), Honolulu, HI, USA, 21–26 July 2017; pp. 2261–2269.
10. Fu, J.; Liu, J.; Tian, H.; Li, Y.; Bao, Y.; Fang, Z.; Lu, H. Dual Attention Network for Scene Segmentation. In Proceedings of the 2019 IEEE/CVF Conference on Computer Vision and Pattern Recognition (CVPR), Long Beach, CA, USA, 15–20 June 2019; pp. 3141–3149.
11. Ding, L.; Tang, H.; Bruzzone, L. LANet: Local Attention Embedding to Improve the Semantic Segmentation of Remote Sensing Images. *IEEE Trans. Geosci. Remote Sens.* **2021**, *59*, 426–435. [CrossRef]
12. Song, W.; Zhou, X.; Zhang, S.; Wu, Y.; Zhang, P. GLF-Net: A Semantic Segmentation Model Fusing Global and Local Features for High-Resolution Remote Sensing Images. *Remote Sens.* **2023**, *15*, 4649. [CrossRef]
13. Li, Y.; Liu, Z.; Yang, J.; Zhang, H. Wavelet Transform Feature Enhancement for Semantic Segmentation of Remote Sensing Images. *Remote Sens.* **2023**, *15*, 5644. [CrossRef]
14. Zhang, X.; Yu, W.; Pun, M.O. Multilevel Deformable Attention-Aggregated Networks for Change Detection in Bitemporal Remote Sensing Imagery. *IEEE Trans. Geosci. Remote Sens.* **2022**, *60*, 1–18. [CrossRef]
15. Li, X.; Xu, F.; Yong, X.; Chen, D.; Xia, R.; Ye, B.; Gao, H.; Chen, Z.; Lyu, X. SSCNet: A Spectrum-Space Collaborative Network for Semantic Segmentation of Remote Sensing Images. *Remote Sens.* **2023**, *15*, 5610. [CrossRef]
16. Li, R.; Duan, C.; Zheng, S.; Zhang, C.; Atkinson, P.M. MACU-Net for Semantic Segmentation of Fine-Resolution Remotely Sensed Images. *IEEE Geosci. Remote Sens. Lett.* **2022**, *19*, 1–5. [CrossRef]
17. Zhao, Y.; Guo, P.; Gao, H.; Chen, X. Depth-Assisted ResiDualGAN for Cross-Domain Aerial Images Semantic Segmentation. *IEEE Geosci. Remote Sens. Lett.* **2023**, *20*, 1–5. [CrossRef]
18. Xi, Z.; Meng, Y.; Chen, J.; Deng, Y.; Liu, D.; Kong, Y.; Yue, A. Learning to Adapt Adversarial Perturbation Consistency for Domain Adaptive Semantic Segmentation of Remote Sensing Images. *Remote Sens.* **2023**, *15*, 5498. [CrossRef]
19. Li, Y.; Shi, T.; Zhang, Y.; Ma, J. SPGAN-DA: Semantic-Preserved Generative Adversarial Network for Domain Adaptive Remote Sensing Image Semantic Segmentation. *IEEE Trans. Geosci. Remote Sens.* **2023**, *61*, 1–17. [CrossRef]
20. Zheng, C.; Hu, C.; Chen, Y.; Li, J. A Self-Learning-Update CNN Model for Semantic Segmentation of Remote Sensing Images. *IEEE Geosci. Remote Sens. Lett.* **2023**, *20*, 1–5. [CrossRef]
21. Pan, S.; Tao, Y.; Nie, C.; Chong, Y. PEGNet: Progressive Edge Guidance Network for Semantic Segmentation of Remote Sensing Images. *IEEE Geosci. Remote Sens. Lett.* **2021**, *18*, 637–641. [CrossRef]
22. Nie, J.; Wang, Z.; Liang, X.; Yang, C.; Zheng, C.; Wei, Z. Semantic Category Balance-Aware Involved Anti-Interference Network for Remote Sensing Semantic Segmentation. *IEEE Trans. Geosci. Remote Sens.* **2023**, *61*, 1–12. [CrossRef]
23. Zheng, C.; Nie, J.; Wang, Z.; Song, N.; Wang, J.; Wei, Z. High-Order Semantic Decoupling Network for Remote Sensing Image Semantic Segmentation. *IEEE Trans. Geosci. Remote Sens.* **2023**, *61*, 1–15. [CrossRef]
24. Li, J.; He, W.; Cao, W.; Zhang, L.; Zhang, H. UANet: An Uncertainty-Aware Network for Building Extraction From Remote Sensing Images. *IEEE Trans. Geosci. Remote Sens.* **2024**, *62*, 1–13. [CrossRef]
25. Vaswani, A.; Shazeer, N.; Parmar, N.; Uszkoreit, J.; Jones, L.; Gomez, A.N.; Kaiser, Ł.; Polosukhin, I. Attention is all you need. In Proceedings of the Proceedings of the 31st International Conference on Neural Information Processing Systems, Long Beach, CA, USA, 4–9 December 2017; pp. 6000–6010.
26. Dosovitskiy, A.; Beyer, L.; Kolesnikov, A.; Weissenborn, D.; Zhai, X.; Unterthiner, T.; Dehghani, M.; Minderer, M.; Heigold, G.; Gelly, S.; et al. An Image is Worth 16x16 Words: Transformers for Image Recognition at Scale. *arXiv* **2020**, arXiv:2010.11929.
27. Liu, Z.; Lin, Y.; Cao, Y.; Hu, H.; Wei, Y.; Zhang, Z.; Lin, S.; Guo, B. Swin Transformer: Hierarchical Vision Transformer using Shifted Windows. In Proceedings of the 2021 IEEE/CVF International Conference on Computer Vision (ICCV), Virtual, 10–17 October 2021; pp. 9992–10002.
28. Zhang, W.; Huang, Z.; Luo, G.; Chen, T.; Wang, X.; Liu, W.; Yu, G.; Shen, C. TopFormer: Token Pyramid Transformer for Mobile Semantic Segmentation. In Proceedings of the 2022 IEEE/CVF Conference on Computer Vision and Pattern Recognition (CVPR), New Orleans, LA, USA, 18–24 June 2022; pp. 12073–12083.
29. Mohammadian, A.; Ghaderi, F. SiamixFormer: A fully-transformer Siamese network with temporal Fusion for accurate building detection and change detection in bi-temporal remote sensing images. *Int. J. Remote Sens.* **2023**, *44*, 3660–3678. [CrossRef]
30. Chen, J.; Lu, Y.; Yu, Q.; Luo, X.; Zhou, Y. TransUNet: Transformers Make Strong Encoders for Medical Image Segmentation. *arXiv* **2021**, arXiv:2102.04306.

31. Wu, H.; Huang, P.; Zhang, M.; Tang, W.; Yu, X. CMTFNet: CNN and Multiscale Transformer Fusion Network for Remote-Sensing Image Semantic Segmentation. *IEEE Trans. Geosci. Remote Sens.* **2023**, *61*, 1–12. [CrossRef]
32. Gao, L.; Liu, H.; Yang, M.; Chen, L.; Wan, Y.; Xiao, Z.; Qian, Y. STransFuse: Fusing Swin Transformer and Convolutional Neural Network for Remote Sensing Image Semantic Segmentation. *IEEE J. Sel. Top. Appl. Earth Observ. Remote Sens.* **2021**, *14*, 10990–11003. [CrossRef]
33. Guo, J.; Han, K.; Wu, H.; Tang, Y.; Chen, X.; Wang, Y.; Xu, C. CMT: Convolutional Neural Networks Meet Vision Transformers. In Proceedings of the 2022 IEEE/CVF Conference on Computer Vision and Pattern Recognition (CVPR), New Orleans, LA, USA, 18–24 June 2022; pp. 12165–12175.
34. Ma, X.; Zhang, X.; Pun, M.O. A Crossmodal Multiscale Fusion Network for Semantic Segmentation of Remote Sensing Data. *IEEE J. Sel. Top. Appl. Earth Obs. Remote Sens.* **2022**, *15*, 3463–3474. [CrossRef]
35. Ji, S.; Wei, S.; Lu, M. Fully Convolutional Networks for Multisource Building Extraction From an Open Aerial and Satellite Imagery Data Set. *IEEE Trans. Geosci. Remote Sens.* **2019**, *57*, 574–586. [CrossRef]

applied sciences

MDPI

*Article*

# Detection of Ocean Internal Waves Based on Modified Deep Convolutional Generative Adversarial Network and WaveNet in Moderate Resolution Imaging Spectroradiometer Images

**Zhongyi Jiang [1], Xing Gao [1], Lin Shi [1], Ning Li [1] and Ling Zou [1,2,*]**

1 School of Computer and Artificial Intelligence, Changzhou University, Changzhou 213164, China; jiangzhongyi2008@hotmail.com (Z.J.); gaoxingtt@outlook.com (X.G.); slcczu@cczu.edu.cn (L.S.); lncczu@cczu.edu.cn (N.L.)
2 School of Microelectronics and Control Engineering, Changzhou University, Changzhou 213164, China
* Correspondence: zouling@cczu.edu.cn

**Abstract:** The generation and propagation of internal waves in the ocean are a common phenomenon that plays a pivotal role in the transport of mass, momentum, and energy, as well as in global climate change. Internal waves serve as a critical component of oceanic processes, contributing to the redistribution of heat and nutrients in the ocean, which, in turn, has implications for global climate regulation. However, the automatic identification of internal waves in oceanic regions from remote sensing images has presented a significant challenge. In this research paper, we address this challenge by designing a data augmentation approach grounded in a modified deep convolutional generative adversarial network (DCGAN) to enrich MODIS remote sensing image data for the automated detection of internal waves in the ocean. Utilizing t-distributed stochastic neighbor embedding (t-SNE) technology, we demonstrate that the feature distribution of the images produced by the modified DCGAN closely resembles that of the original images. By using t-SNE dimensionality reduction technology to map high-dimensional remote sensing data into a two-dimensional space, we can better understand, visualize, and analyze the quality of data generated by the modified DCGAN. The images generated by the modified DCGAN not only expand the dataset's size but also exhibit diverse characteristics, enhancing the model's generalization performance. Furthermore, we have developed a deep neural network named "WaveNet," which incorporates a channel-wise attention mechanism to effectively handle complex remote sensing images, resulting in high classification accuracy and robustness. It is important to note that this study has limitations, such as the reliance on specific remote sensing data sources and the need for further validation across various oceanic regions. These limitations are essential to consider in the broader context of oceanic research and remote sensing applications. We initially pre-train WaveNet using the EuroSAT remote sensing dataset and subsequently employ it to identify internal waves in MODIS remote sensing images. Experiments show the highest average recognition accuracy achieved is an impressive 98.625%. When compared to traditional data augmentation training sets, utilizing the training set generated by the modified DCGAN leads to a 5.437% enhancement in WaveNet's recognition rate.

**Keywords:** MODIS; internal waves; classification; DCGAN; transfer learning; deep neural network; attention

## 1. Introduction

### 1.1. Internal Waves

Internal waves are a phenomenon arising from variations in temperature and salinity within the ocean, typically occurring in regions with density stratification [1]. These waves can exhibit significant amplitudes, exceeding 100 m, and travel distances spanning tens to hundreds of kilometers [2], rendering their detection challenging. Consequently, internal waves have emerged as a prominent focus of research within the field of oceanography.

The generation and propagation of internal waves are pervasive phenomena in the ocean, playing a crucial role in the transport of mass, momentum, and energy within oceanic systems. Breaking oceanic internal waves induces turbulent mixing, which in turn facilitates the vertical transport of water, heat, and other crucial climatic tracers within the ocean. This process holds significant importance as it actively influences the circulation patterns and the distribution of heat and carbon in the climate system [3], contributing to global climate change [4]. As a result, they can exert substantial influence on the safety and efficiency of marine engineering, oceanic communications, and oil exploration [5], and contribute to broader environmental factors, including their role in global climate change. Therefore, the study of internal waves in the ocean, particularly through the automatic recognition of these waves in remote sensing imagery, holds immense academic and practical significance.

With the continuous advancement of remote sensing technology, the exploration of internal waves in the ocean has shifted away from traditional field observations. Instead, researchers have embraced the use of remote sensing imagery, presenting a novel approach to this study. Among the various remote sensing instruments, the moderate resolution imaging spectroradiometer (MODIS) stands out as one of the most vital and distinctive tools currently available. It is integrated into platforms like Terra and Aqua and represents a state-of-the-art "all-in-one" optical remote sensing device in today's world.

MODIS boasts an impressive array of data with 36 bands, offering varying spatial resolutions of 250 m, 500 m, and 1000 m. Its scanning capacity spans an impressive 2330 km [6]. On average, local data can be acquired on a daily basis, and these data are readily accessible, making them the premier data source for global monitoring purposes. Nonetheless, the data provided by MODIS images are not only extensive in quantity but also substantial in size. Given that internal waves within the ocean occupy only a small fraction of the image area, the initial challenge lies in identifying and isolating images containing internal waves from the vast MODIS dataset before embarking on the process of detection and characterization.

### 1.2. Challenges and Research Objectives

To understand the characteristics of internal waves in MODIS satellite imagery, it is important to consider their distinctive optical signatures. When examining internal waves in the ocean through MODIS satellite imagery, they are frequently observed to be more prevalent in the vicinity of optical glare areas. This phenomenon can be attributed to the fact that signals received in these regions primarily result from sunlight reflecting off the sea surface. The presence of internal waves induces alterations in the sea surface roughness, leading to variations in the gradient of the sea surface at various scales. These variations, in turn, modulate the intensity of reflected sunlight received by remote sensors. Consequently, internal waves manifest as stripes with varying degrees of brightness in optical images. In visible satellite images, the divergent regions within the glare area typically appear as dark stripes, while the convergent regions appear as bright stripes. Outside the glare area, the divergent areas are characterized by bright stripes, while the convergent areas exhibit dark stripes. For a visual representation of how internal waves are depicted in MODIS imagery, please consult Figure 1.

The classification of remote sensing images depicting oceanic internal waves presents two significant challenges. Firstly, owing to the high resolution and abundant spatial and semantic information within remote sensing images, traditional machine learning methods like SVM, KNN, and decision trees struggle to effectively capture the intricate features inherent to these images. Secondly, the task of cropping and annotating remote sensing images is exceptionally labor-intensive, rendering it challenging for researchers to amass a substantial quantity of labeled image datasets for remote sensing data.

This study aims to address these challenges by developing an innovative approach to automatically detect and classify internal waves in MODIS satellite imagery, leveraging advanced deep learning techniques and data augmentation methods. By doing so, we aim to contribute to the field of oceanography and remote sensing by providing a more efficient and accurate means of studying and monitoring internal waves in the ocean, ultimately enhancing our understanding of their role in oceanic and global climate processes.

**Figure 1.** Oceanic internal waves in MODIS.

### 1.3. Contributions

The primary objective of this paper is to introduce an automated deep learning approach for the identification of internal waves within MODIS images.

1. We devised a modified DCGAN specifically tailored for data augmentation of MODIS remote sensing images.
2. We developed WaveNet, which incorporates a channel-wise attention mechanism, with the purpose of identifying internal waves.
3. We established a transfer learning methodology for the pre-training of WaveNet.

### 1.4. Paper Structure

The structure of this paper is organized as follows: Section 2 presents an overview of related work. The proposed approach is detailed in Section 3. Section 4 outlines the experimental methodology and reports the obtained results. Finally, in Section 5, conclusions and discuss possible avenues for future research are presented.

## 2. Related Work

Previous classification models, such as neural networks and support vector machines (SVMs) [7], typically featured a network structure with either one hidden layer node or none at all, earning them the designation of "shallow" classification models. However, these shallow classification models are categorized under shallow learning and possess limited capability to extract deeper features from constrained datasets, thereby constraining their overall model generalization ability [8].

In contrast, deep learning, as a novel machine learning paradigm, aspires to emulate the analytical learning capabilities of the human brain. By leveraging substantial volumes of training data and employing deep models with multiple hidden layers, deep learning can uncover more valuable features, consequently enhancing classification accuracy. Unlike shallow learning, deep learning architectures are characterized by their depth, typically consisting of more than three layers of hidden nodes. This depth enables them to explore deeper and more abstract features, thereby acquiring more precise feature information and ultimately affording superior generalization capabilities. In recent years, deep learning has achieved remarkable success in various image classification applications, prompting an increasing number of researchers to apply it to the domain of remote sensing image processing.

Deep learning has been integrated into the classification of hyperspectral data to harness the wealth of spectral information within hyperspectral images [1]. Notably, Haut et al. [9] introduced an innovative classification model that leverages both spectral and spatial information present in hyperspectral data. This approach effectively mitigates the issue of rapid overfitting and accuracy degradation typically encountered when using convolutional neural networks (CNNs) with limited training data.

Bao et al. [10] employed the faster R-CNN framework, incorporating convolutional neural network features, to detect oceanic internal waves. Their efforts resulted in an impressive recognition rate of 94.78%. On a related note, Yu et al. [11] harnessed the lightweight convolutional neural network MobileNetv2 to extract deep and abstract image features. By combining feature fusion with bilinear pooling, they achieved higher accuracy in the realm of remote sensing image classification while utilizing fewer parameters and computational resources, surpassing other state-of-the-art methods.

In scenarios involving small-sample datasets, Li et al. [12] introduced a novel fault-tolerant deep learning approach known as RSSC-ETDL for remote sensing image scene classification. This method effectively mitigates the adverse effects stemming from inaccurately labeled datasets.

In recent years, researchers have also proposed many excellent models. In 2022, Zheng et al. [13] proposed a stripe segmentation algorithm based on SegNet for synthetic aperture radar (SAR) images. This method effectively identifies the presence of oceanic internal waves in SAR images and accurately locates both light and dark stripes associated with these waves. Also using SAR images, Tao et al. [14] construct a comprehensive dataset of 390 Sentinel-1 synthetic aperture radar (SAR) images, spanning multiple oceanic regions. These images are used to develop a machine learning model achieving high precision and recall when applied to detect internal waves (IW) across different scales and propagation directions in SAR imagery. Also in that year, Serebryany et al. [15] conducted an analysis using a collection of optical multispectral satellite images, including Sentinel-2 and Landsat-8 data, in conjunction with sea-truth data to identify internal wave features within the Black Sea.

Deep learning models typically necessitate multiple iterations of data analysis and processing, often involving substantial amounts of data [16]. Although the above work also has high performance in identifying ocean internal waves, it either requires a large number of high-quality datasets as data support, or the network model has room for improvement. However, the challenges associated with image cropping, annotation, and the acquisition of rare remote sensing images can present substantial obstacles for researchers when striving to compile extensive remote sensing image datasets during the data collection phase. Therefore, there is a special need in the field for a method that can greatly increase the data volume of remote sensing datasets, thereby effectively reducing the cost of annotation, and at the same time have a very high recognition rate of ocean internal waves.

This paper conducts classification on full-space images acquired through the Moderate Resolution Imaging Spectroradiometer (MODIS), encompassing four distinct categories: ocean scenes, clouds, terrestrial landscapes, and ocean waves. To address the challenge of limited dataset availability, the author employs an enhanced deep convolutional generative adversarial network (DCGAN) to substantially augment the data within each category sample. Additionally, a novel residual network is designed, taking into consideration the channel information of deep features, termed "WaveNet," to enable automated detection of internal ocean waves in MODIS images through an end-to-end approach.

## 3. Methods and Data

This section will introduce in detail the collection of remote sensing data, the construction of datasets, the modified DCGAN model structure and the WaveNet model structure. The entire workflow diagram is shown in Figure 2.

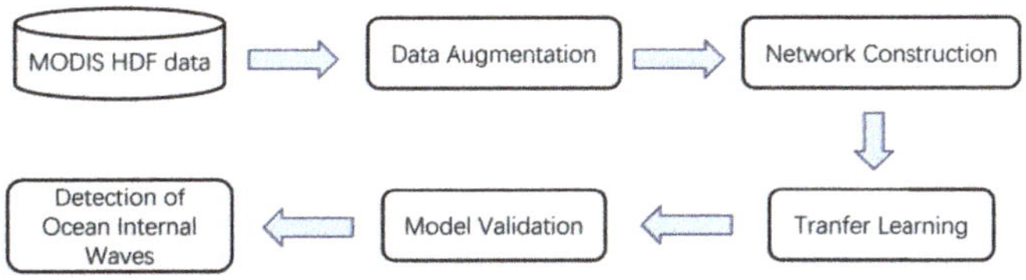


**Figure 2.** Flow chart of internal waves detection framework.

### *3.1. Data Augmentation*

#### 3.1.1. MODIS Images

First, obtain the HDF format data of MOD02QKM from the official website of the National Aeronautics and Space Administration (NASA) (https://ladsweb.modaps.eosdis.nasa.gov, accessed on 1 October 2022) [17]. Research [18] shows that internal ocean waves in the South China Sea occur frequently in summer, but less frequently in other seasons. Therefore, the MODIS data collection time used in this article is from 1 June to 31 August every year, and is conducted in the northern South China Sea. In order to reflect the data enhancement work, this article only collected a total of 1217 pieces of data from 2019 to 2022. Then, use ENVI Classic 5.3 professional softwareto read the HDF format file in Earth View 250M Reflective Solar Bands Scaled Integers format into BSQ format, then save it into IMG image format, and apply histogram equalization operation to improve the brightness of the image, where the maximum resolution. The rate is 5416 × 8120. The image is then split into smaller sub-images, each 64 × 64 pixels in size. Finally, experts who have studied remote sensing for many years divided these sub-images into different categories, including 700 images each of internal waves, clouds, oceans, and land. Figure 3 shows some classification results. The above process finally provides data support for this study, allowing us to study the existence and characteristics of internal waves in the ocean.

Clouds

Ocean

Land

Ocean internal waves

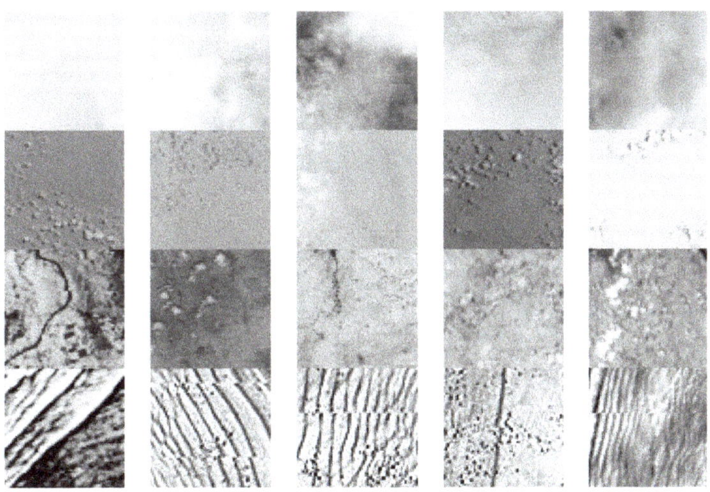

**Figure 3.** Partial data samples of clouds, ocean, land, and oceanic internal waves.

#### 3.1.2. Data Augmentation

Currently, data augmentation techniques can be categorized into two main groups: traditional data augmentation methods and image generation algorithms based on generative adversarial networks (GANs), which are relatively new and capable of generating images with similar features to the original dataset but different from them. This approach significantly enhances dataset diversity, thereby improving the model's generalizability [19].

In order to compare the effects of these two data enhancement methods on training models, we applied these two methods to the original data and conducted experimental comparisons. First, we applied ten traditional data augmentation methods to each class of training samples. These methods include color truncation, min–max normalization, standard normalization, flipping, sharpening, Gaussian filtering, random erasing, random brightness transformation, random contrast transformation, and uniform noise. For each augmentation, we randomly selected 130 images, resulting in a total of 1300 augmented samples for each class of training data. When combined with the original dataset, this created a total of 2000 training samples for each class, constituting our Training Set 1.

Subsequently, we applied the GAN to data augmentation. Generative adversarial networks (GANs) are a type of deep learning model employed for generating synthetic data, including images [20]. This method leverages two neural networks: a generator and a discriminator. The generator's objective is to produce synthetic data that closely resembles real data by learning the distribution of real data [21]. Conversely, the discriminator is tasked with distinguishing between real and synthetic data. These two networks engage in a competitive process during training, leading to the continual improvement of the generator's ability to produce realistic synthetic data and the discriminator's ability to effectively differentiate between real and synthetic data.

The objective function $V(D;G)$ for the GAN is as follows:

$$min_G max_D V(D,G) = E_{X\sim P_{data(x)}}[logD(x)]E_{z\sim P_z(z)}[log(1-D(G(z)))] \quad (1)$$

In the equations provided, where x represents a real sample, $D(x)$ signifies the probability assigned by the discriminator networks for classifying $x$ as a real sample. $G(z)$ corresponds to a sample generated from noise $z$ by the generator network $G$, and $D(G(z))$ indicates the probability assigned by the discriminator network $D$ for classifying $G(z)$ as a real sample.

Deep convolutional generative adversarial networks (DCGAN) [22] represent a variant of GANs designed to transform noise into images. They excel at generating images that fall within the same category as those present in the training set. DCGANs combine convolutional neural networks (CNNs) with unsupervised learning in the context of supervised learning, finding widespread applications in image generation.

In order to facilitate the generation of remote sensing images with dimensions of 64 × 64 pixels, we have made modifications to the DCGAN network architecture, as depicted in Figure 4. These modifications include the use of convolutional neural networks as both the generator and discriminator, the implementation of batch normalization for expedited training, the utilization of the LeakyReLU activation function to overcome the limitations of ReLU, the elimination of fully connected layers to prevent overfitting, the adoption of the Adam optimizer, and the capacity to generate high-quality images.

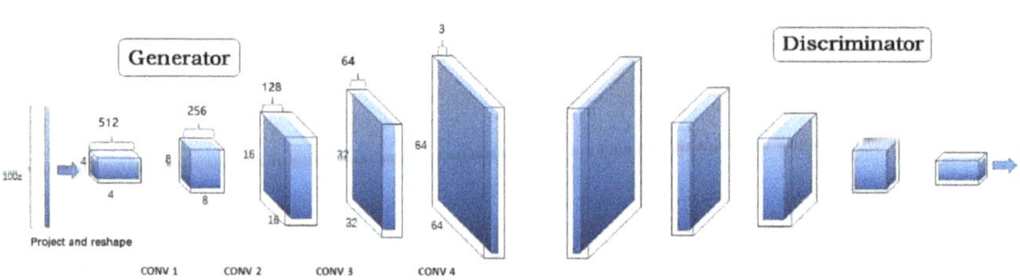


**Figure 4.** Modified DCGAN network structure.

We applied the modified DCGAN data augmentation to clouds, land, ocean, and internal waves, with parameters listed in Table 1.

**Table 1.** Experimental parameters for the modified DCGAN network.

| Original Data Volume | Batch Size | Learning Rate | Training Epochs | Optimizer | Exponential Decay Rate for the First Moment Is Estimated in the Optimizer | Exponential Decay Rate for the Second-Moment Estimates in the Optimizer |
|---|---|---|---|---|---|---|
| 700 | 16 | 0.0005 | 1000 | Adam | 0.5 | 0.999 |

The modified DCGAN network is trained using the original dataset, with the addition of a dropout layer at the end of each layer in the generator. This dropout layer serves to reduce the model's reliance on specific input features, enhance its generalization ability, and mitigate overfitting. The dropout parameter is set to 0.5, indicating that 50% of the neurons are randomly deactivated during each training iteration.

The training process for the modified DCGAN network has demonstrated success, as evidenced by the stabilization of loss functions for both the generator G and the discriminator D after 500 rounds. This suggests that the model has effectively learned the underlying data features and can generate images that closely resemble those in the original dataset.

Following this, the generator G is employed to generate 1300 images for each category, including ocean internal waves, clouds, land, and ocean. A selection of these generated images is presented in Figure 5. To further enhance the dataset, these generated images were combined with the original dataset, resulting in the creation of a new data augmentation training set referred to as Training Set 2.

The generated images, whether depicting clouds, oceans, land, or internal waves, convincingly simulate various real-world scenarios. These images effectively capture the texture of cloud layers, the topography of the land, the undulations of the ocean's surface, and the oscillations of internal waves. Consequently, our approach provides an effective means to expand existing remote sensing image datasets, with potential applications extending to various other applications within the realm of remote sensing. The dataset consisting of original data, traditional data augmentation, and the modified DCGAN-generated data are shown in Table 2.

In Part 4, we will compare the effects of two data enhancement methods in model training.

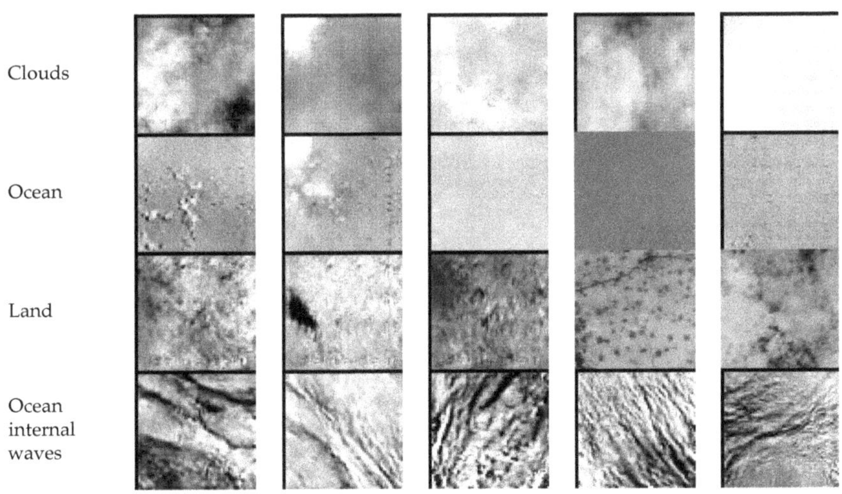


**Figure 5.** Some of the images generated by the modified DCGAN network.

**Table 2.** Number of training and testing sets.

| Unit: Images | Original Data Volume | Dataset Size for Traditional Data Augmentation | Dataset Size Generated by the Modified DCGAN | Training Set 1 | Training Set 2 | Test Set |
|---|---|---|---|---|---|---|
| Cloud | 700 | 1300 | 1300 | 2000 | 2000 | 400 |
| Land | 700 | 1300 | 1300 | 2000 | 2000 | 400 |
| Ocean | 700 | 1300 | 1300 | 2000 | 2000 | 400 |
| Oceanic internal waves | 700 | 1300 | 1300 | 2000 | 2000 | 400 |

### 3.2. *Construction of the WaveNet Network Model*

This article provides a detailed introduction to a residual convolutional neural network called "WaveNet," enhanced by a channel attention mechanism. WaveNet is designed to effectively process complex remote sensing images while achieving high classification accuracy and robustness. It achieves greater network depth by sequentially combining convolutional layers and pooling layers, enabling autonomous learning and the capture of essential features in remote sensing images. Simultaneously, it employs a channel attention mechanism to assign varying weights to channels within the feature map. These learned features are subsequently consolidated through a fully connected layer to produce the final classification result. The experimental findings presented in Section 4 unequivocally demonstrate the outstanding performance of WaveNet in diverse image classification tasks. Consequently, this method holds immense potential for widespread application in remote sensing image processing and is poised to contribute significantly to advancements in the field of remote sensing image classification.

#### 3.2.1. Residual Block

The residual block structure in WaveNet is illustrated in Figure 6. The size of the input feature map is $C/2 \times H \times W$, where C represents the number of channels in the feature map, and H and W represent the height and width of the feature map, respectively. Between every two convolutional layers, there is a batch normalization layer and a rectified linear unit (ReLU) activation function [23]. The use of batch normalization ensures that the input distribution of each neuron remains consistent, which can accelerate the convergence speed of the network and avoid the issues of gradient vanishing and exploding, thereby improving the generalization performance of the model. ReLU is a commonly used activation function in deep learning. ReLU is defined as follows:

$$f(x) = max(0, x) \quad (2)$$

In short, for input $x$, if $x$ is greater than zero, then output $x$, otherwise output zero. The advantage of the ReLU activation function is its simplicity and non-linearity. Compared with traditional activation functions (such as sigmoid or tanh), ReLU is more stable for the back propagation of gradients and helps alleviate the vanishing gradient problem. In addition, ReLU introduces non-linearity, allowing the neural network to learn more complex functions.

In the architecture of the residual block within WaveNet, the initial convolutional layer plays a pivotal role. Its primary function is to double the number of channels in the feature map while halving its spatial size. This process is crucial for maintaining consistent feature map sizes during the initial addition operation. To achieve this, we employ a $1 \times 1$ convolutional layer that processes the input data, augmenting dimensionality along the channel axis and aligning it with the dimensions of other components within the residual block [24].

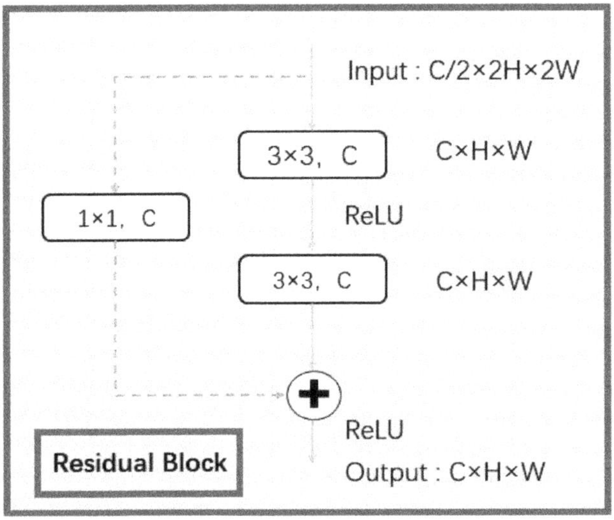


**Figure 6.** Residual block structure in WaveNet.

Following this dimension alignment, a 3 × 3 convolutional layer is utilized for feature extraction. Importantly, the resulting feature map maintains the same spatial dimensions as the input feature map. Subsequently, the output feature map from the initial convolutional layer undergoes further feature extraction via another 3 × 3 convolutional layer. An element-wise addition operation is then applied to the output feature map of the second convolutional layer, effectively creating a residual connection. This connection enables the network to learn the difference between the input and output, which enhances network optimization and training.

After the residual connection, we apply a ReLU activation function to the feature map, effectively setting all negative values to zero. This introduces non-linear features and amplifies the network's representational capacity. Through this sequence of operations, the residual block efficiently extracts and propagates essential feature information, thereby enhancing the network's overall performance and its ability to learn complex patterns within remote sensing images.

The mapping relationship in the residual block can be succinctly represented as follows:

$$y = F(x, W_i) + W_s x \tag{3}$$

In this context, we utilize the symbols x and y to represent the input and output feature maps of the residual block, respectively. The primary objective of the residual block is to acquire knowledge of the residual mapping function $F(x, W_i)$, with $W_i$ denoting the set of parameters involved in the learning process. To facilitate the integration of shortcut connections and ensure dimension alignment, a 1 × 1 convolutional layer is introduced, with parameters $W_s$ responsible for managing dimension adjustments. These pivotal steps within the residual blocks of WaveNet effectively facilitate the extraction of essential features and maintain stable gradient flow during training. Consequently, this simplifies the training of deep neural networks by addressing challenges related to vanishing and exploding gradients.

3.2.2. SE Residual Block

The squeeze and excitation (SE) residual block, illustrated in Figure 7, represents an essential architectural component within WaveNet. It combines the benefits of a traditional residual block with a channel-wise attention mechanism. This integration enhances the model's capability to capture and leverage crucial global information across feature channels, facilitating the recognition of important patterns and relationships within the data. The SE residual block plays a pivotal role in enhancing the overall performance of the WaveNet model.

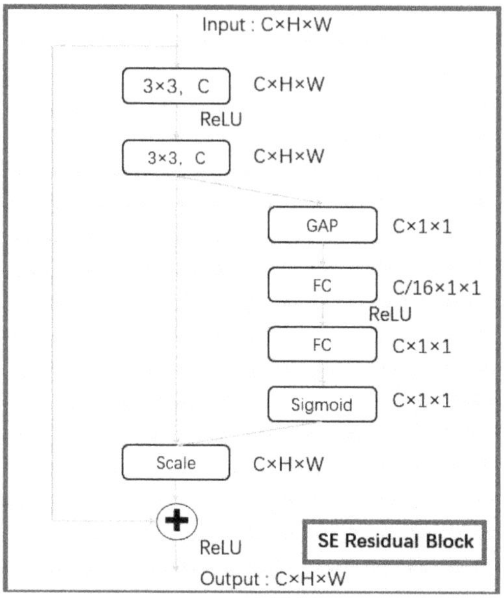


**Figure 7.** The squeeze and excitation attention mechanism in the residual block (SE residual block) structure.

The initial size of the input feature map $X_0$ of the SE residual block is C × H × W. and the output feature map X after two layers of convolution operations will be used as the input of global average pooling. We will perform feature compression on the feature map X along the spatial dimension by applying global average pooling. This operation transforms each two-dimensional feature channel into a single scalar value. Each scalar, in a sense, possesses a global receptive field and shares the same dimensionality as the number of input feature channels. These scalars represent the global response distribution across feature channels and enable layers closer to the input to access global information. Consequently, feature compression transforms the size of the feature map from C × H × W to C × 1 × 1. Following feature compression, the resulting compressed feature vector serves as the input to the channel-wise attention mechanism. This integrated mechanism effectively allows the model to capture and leverage global information across feature channels, thereby enhancing its capacity to recognize important patterns and relationships within the data.

The formula of global average pooling of the SE residual block can be expressed as follows:

$$F_c(X) = \frac{1}{H \times W} \sum_{i=1}^{H} \sum_{j=1}^{W} X(i,j) \tag{4}$$

In the equation, $F_c$ represents the compressed feature map, and $X$ represents the output feature map of the last convolutional layer.

The subsequent two fully connected layers are used to model the correlations between channels and output weights equal to the number of input feature channels. Firstly, in the first fully connected layer, we reduce the dimensionality of the channel features to 1/16 of the original size. Then, a ReLU activation function is applied for non-linear transformation, followed by another fully connected layer to increase the dimensionality back to the dimension of the original feature channels. This design, compared to using only one fully connected layer, offers greater non-linear capability, enabling better fitting of complex correlations between channels, while also reducing the number of parameters and computational complexity. For the fully connected layer input FC, the output can be expressed as:

$$F_o = (ReLU(F_c \times w_1 + b_1)) \times w_2 + b_2 \tag{5}$$

In the equation, $F_o$ represents the final output of the fully connected layer, and $w_1$ and $b_1$ denote the weight and bias of the first fully connected layer, respectively. Similarly, $w_2$ and $b_2$ represent the weight and bias of the second fully connected layer.

The output from the final fully connected layer undergoes a non-linear transformation facilitated by a sigmoid activation function. The formula of the sigmoid function is as follows.

$$Sigmoid(x) = \frac{1}{1+e^{-x}} \tag{6}$$

The sigmoid function generates a normalized weight for each channel within the range of 0 to 1. These normalized weights are then applied in an element-wise multiplication operation with the original channel features, thus finalizing the re-scaling of the original features along the channel dimension. This re-scaling process allows the network to assign weights to individual features based on their importance for the given classification task. Channels that are more relevant to the task receive higher weights, while the influence of less relevant channels is suppressed. By employing this approach, the network effectively harnesses the inter-channel correlations, elevates its feature representation capacity, and enhances overall classification performance. This mechanism ensures that the network pays greater attention to the most informative features, thereby improving its ability to recognize complex patterns and make accurate predictions. This process can be represented as follows:

$$\acute{X} = X \odot Sigmoid(F_o) \tag{7}$$

$\acute{X}$ represents the feature map after re-scaling. X denotes the output feature map of the last convolutional layer. $F_o$ represents the output of the last fully connected layer.

For the SE residual block, the size of the final output feature map Y is also C × H × W, which can be expressed by the following formula:

$$Y = ReLU(X_0 + X') \tag{8}$$

Among them, $X_0$ represents the input feature map of the SE residual block, $\acute{X}$ represents the feature map after re-scaling.

#### 3.2.3. WaveNet

The WaveNet network architecture is composed of several key components, including convolutional layers with a 3 × 3 kernel size, max-pooling layers with a 2 × 2 size, three residual blocks, three channel-wise attention mechanism residual blocks, and a global average pooling layer. The final layer is a fully connected layer that utilizes softmax transformation to derive the probability distribution for each sample across different classes.

For input remote sensing images with dimensions of 3 × 64 × 64, each layer of the WaveNet network has specific input and output feature map sizes, as illustrated in Table 3.

To manage computational complexity effectively, WaveNet initially employs a sequence of 3 × 3 convolutional layers and 2 × 2 max-pooling layers to reduce the feature map size. However, in the first residual block, the feature map size remains unchanged to preserve crucial information, while the number of channels is doubled. In the subsequent three residual blocks, the channel count is doubled, but the height and width are halved by the first convolutional layer.

The final layer in the network is a fully connected layer that incorporates softmax transformation. The last layer in the network is a fully connected layer containing a softmax transformation. This layer converts the final output of the network into a probability distribution between 0 and 1. The formula of the softmax function is as follows:

$$p_i = \frac{e^{z_i}}{\sum_{j=1}^{K} e^{z_j}} \quad (9)$$

Among them, $p_i$ represents the probability of the $i$-th category, $z_i$ is the $i$-th element of the input vector $z$, and $K$ is the total number of categories, which is 4 in this article.

The softmax function has specific advantages in classification tasks because it can calculate the predicted probability for each class, which is very useful for the training and interpretation of neural networks. The softmax function will emphasize the element with the largest value in the input vector so that its corresponding category probability is the highest, thereby classifying, and ultimately determines the final classification result.

**Table 3.** Size changes of feature map.

| Network Layer | Input Feature Map Size | Output Feature Map Size |
|---|---|---|
| 3 × 3 Conv | 3 × 64 × 64 | 64 × 64 × 64 |
| 2 × 2 Max Pooling | 64 × 64 × 64 | 64 × 32 × 32 |
| Residual Block | 64 × 32 × 32 | 128 × 32 × 32 |
| SE Residual Block | 128 × 32 × 32 | 128 × 32 × 32 |
| Residual Block | 128 × 32 × 32 | 256 × 16 × 16 |
| SE Residual Block | 256 × 16 × 16 | 256 × 16 × 16 |
| Residual Block | 256 × 16 × 16 | 512 × 8 × 8 |
| SE Residual Block | 512 × 8 × 8 | 512 × 8 × 8 |
| Global Average Pooling | 512 × 8 × 8 | 512 × 1 × 1 |
| Fully Connected | 512 | 4 |

### 3.3. *Transfer Learning*

In the domain of deep learning models, training models with numerous hidden layers and parameters often demand a substantial amount of high-quality labeled data, which can result in significant time and computational costs. However, when dealing with satellite imagery of oceanic internal waves, data availability is frequently limited, presenting a challenge in training an effective deep learning model with a small dataset. To address this limitation, transfer learning has emerged as a valuable approach in deep learning model training, alleviating the data requirement.

Transfer learning involves leveraging a pre-trained model, which is then fine-tuned for a new task. The fundamental concept of transfer learning is to initially train the network model parameters using large-scale datasets and subsequently fine-tune them for the specific image recognition task at hand. This approach greatly assists the classifier in performing image recognition tasks, even when confronted with limited data resources.

The EuroSAT remote sensing dataset, as described in reference [25], comprises 10 distinct scene categories, including agricultural land, forest, herbaceous vegetation, highways, industrial areas, pastures, permanent crops, residential areas, rivers, and lakes, with a total of 3000 samples for each category. The primary objective of this study is to evaluate the performance of employing the pre-trained WaveNet network for classification tasks using the EuroSAT remote sensing dataset. Detailed parameter settings for this study are provided in Table 4. These settings cover various aspects of the model and training process, offering a comprehensive overview of the experimental configuration and methodology employed in the classification task using the EuroSAT dataset.

**Table 4.** Pre-trained network parameters.

| Batch Size | Learning Rate | Training Epochs | Optimizer |
|---|---|---|---|
| 32 | 0.001 | 500 | Adam |

During the fine-tuning stage, the parameters of the WaveNet network model are initialized with pre-trained parameters, and the output of the last fully connected layer is adjusted to 4 in order to perform classification on the MODIS image dataset. Utilizing a pre-training strategy instead of initializing the network parameters with random weights allows the WaveNet model to initially learn rich texture features from the EuroSAT dataset. This approach not only eliminates the need to train the model from scratch but also helps overcome potential overfitting issues, thereby enhancing the model's performance in situations with limited data. By leveraging pre-training, the model can benefit from the learned representations, enabling it to generalize more effectively and achieve higher performance even when the available data are limited.

## 4. Experiments and Results

In this paper, we will use PyTorch as a deep learning framework to build network models. We also utilize graphics processing units (GPUs) to improve computing speed and training efficiency during network training.

In the experimental phase, we will first analyze the quality of data generated based on the modified DCGAN through the t-SNE dimensionality reduction method. Then, compare the training effects of Datasets 1 and 2 obtained by using two different data enhancement methods on the WaveNet network. Subsequently, we will compare the training effects of the WaveNet network using the transfer learning strategy and not using the transfer learning strategy. Finally, we compare the results of WaveNet with previous related work.

Table 5 outlines the hardware and software environment during the experiment.

**Table 5.** Parameters of experimental conditions.

| Hardware Equipment | Software Environment |
|---|---|
| CPU: Intel(R) Xeon(R) Gold 5218R 2.10 GHz | Rocky Linux 8 |
| RAM:32GB | CUDA 11.4 |
| GPU: NVIDIA RTX 3090 | Pytorch 1.12.1 |

### *4.1. Analysis of Generated Images Using the Modified DCGAN*

In order to analyze the data distribution generated by the modified DCGAN more intuitively, the t-distributed stochastic neighbor embedding (t-SNE) [26] dimensionality reduction method is used in this paper. t-SNE is an algorithm for visualizing high-dimensional datasets by measuring the distance between each data point and other data points to calculate their correlation [27]. We randomly select 200 images from each category of the original data and the modified DCGAN-generated data for testing (as shown in Figure 8), where different colors represent different labels.

The results of t-SNE dimensionality reduction highlight the effectiveness of the modified DCGAN in generating synthetic remote sensing images that closely resemble real-world images while introducing valuable diversity. The key takeaways from this research include:

1. Realism and Diversity: The modified DCGAN has demonstrated its capability to produce synthetic remote sensing images that exhibit realism, making them highly similar to authentic remote sensing data. Additionally, the diversity observed within the generated images is a significant advantage. The ability to generate diverse samples within each class contributes to a more comprehensive and representative dataset.

2. Matching Feature Distribution: The use of t-SNE for dimensionality reduction and visualization has substantiated that the feature distribution within the generated images closely aligns with that of real images. This alignment suggests that the modified DCGAN has successfully captured and retained essential features present in real remote sensing data, enhancing the data's quality.

3. Enhanced Data for Training: By effectively generating realistic and diverse data samples, the modified DCGAN provides a valuable resource for training neural network

models. This expanded dataset can be instrumental in improving the generalization capabilities of recognition models, as it exposes them to a wider range of scenarios and variations present in the real world.

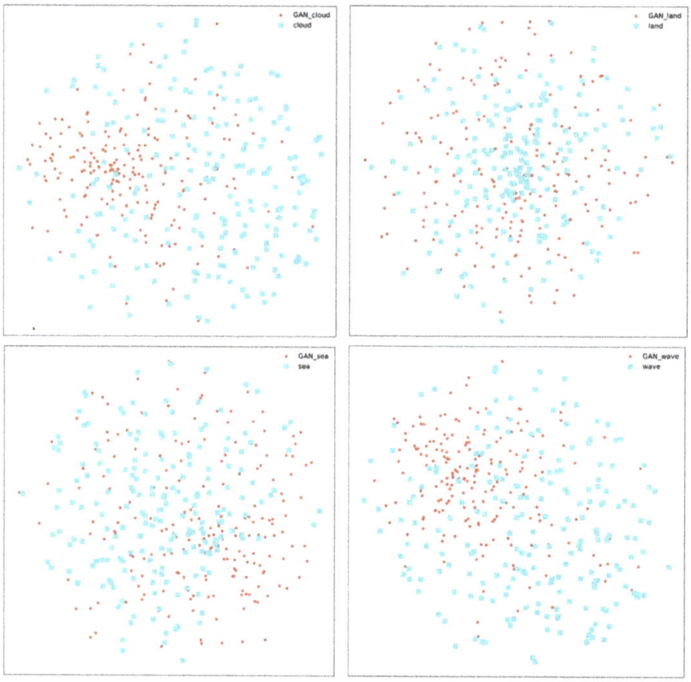


**Figure 8.** T-SNE dimensionality reduction is applied to different data augmentation datasets, where red represents the data generated by the modified DCGAN and blue represents the original data.

### 4.2. Comparison of Classification Results of the WaveNet Network Using Different Training Sets

This article evaluates the WaveNet model using the overall accuracy (OA) and average accuracy (AA) metrics for classification. The confusion matrix $S$ is a $L \times L$ matrix, $L$ that represents the number of classes. $S_{ij}$ represents the number of test samples that belong to the class i and were classified as a class $j$. The total number of test samples is $M = \sum_i^L \sum_j^L S_{i,j}$.

The overall accuracy (OA) metric provides a good description of the overall classification accuracy, where OA is calculated by dividing the number of correctly classified samples by the total number of test samples, which can be represented as follows:

$$\mathrm{OA} = \frac{\sum_i^L S_{i,j}}{M} \times 100\% \tag{10}$$

The average accuracy (AA) metric provides a good description of the classification performance differences among each class, representing the average classification accuracy for each class, which can be represented as follows:

$$\mathrm{AA} = \frac{\sum_i^L \left(\frac{S_{i,j}}{\sum_j^L S_{i,j}}\right)}{L} \times 100\% \tag{11}$$

Precision is a metric widely used to evaluate the performance of classification models. It measures the accuracy of the model in predicting positive examples. The calculation formula of accuracy is as follows:

$$Precision_i = \frac{S_{i,i}}{\sum_j^L S_{j,i}} \times 100\% \quad (12)$$

In this formula, $Precision_i$ represents the precision for class $i$. $S_{i,i}$ denotes the number of samples correctly classified as class i, and $\sum_j^L S_{j,i}$ represents the total number of samples predicted as class $i$.

Recall is also one of the indicators widely used to evaluate the performance of classification models. It measures the proportion of actual positive examples identified by the model, that is, how many of all actual positive examples were correctly predicted as positive by the model. The calculation formula of recall rate is as follows:

$$Recall_i = \frac{S_{i,i}}{\sum_j^L S_{i,j}} \times 100\% \quad (13)$$

In this formula, $Recall_i$ represents the recall for class $i$. $S_{i,i}$ denotes the number of samples correctly classified as class $i$, and $\sum_j^L S_{i,j}$ represents the total number of samples actually belonging to class $i$.

To investigate whether the images generated by the modified DCGAN network possess features similar to those of real images and whether they are more effective as a data augmentation technique than traditional methods, enhancing the network's generalization ability, we conducted a series of comparative experiments. The overall accuracy (OA) and average accuracy (AA) experimental results are shown in Table 6. The precision of each category is shown in Table 7. The recall rates of each category are shown in Table 8.

**Table 6.** Accuracy results of WaveNet under the same experimental parameters using training sets based on traditional data augmentation and DCGAN-generated data augmentation.

| | **Traditional Data Augmentation/%** | **Data Augmentation Based on the Modified DCGAN/%** |
|---|---|---|
| Overall accuracy | 93.188 | 98.625 |
| Accuracy of cloud recognition | 97.000 | 99.750 |
| Accuracy of land recognition | 90.750 | 97.250 |
| Accuracy of ocean recognition | 90.250 | 98.500 |
| Accuracy of oceanic internal waves recognition | 94.750 | 99.000 |

**Table 7.** Results of precision of WaveNet under the same experimental parameters using a training set based on traditional data augmentation and DCGAN-generated data augmentation.

| | **Traditional Data Augmentation/%** | **Data Augmentation Based on the Modified DCGAN/%** |
|---|---|---|
| Precision of cloud recognition | 92.38 | 98.28 |
| Precision of land recognition | 91.44 | 98.73 |
| Precision of ocean recognition | 91.62 | 98.25 |
| Precision of oceanic internal waves recognition | 97.43 | 99.25 |

By using the modified DCGAN data augmentation method proposed in this paper, we observed an overall improvement of 5.437% in the classification accuracy of the test set, with all four types of remote sensing images showing increased recognition rates. Specifically, the recognition accuracy of clouds, land, oceans and internal waves improved by 2.75%, 6.5%, 8.25%, and 4.25%, respectively. Thanks to the modified DCGAN data enhancement method proposed in this article, the WaveNet model has also been greatly improved on the test set in terms of precision and recall indicators of each category. For example, the recognition accuracy of land increased by 7.29%, while the internal accuracy of identifying waves increased by 1.82%; the recall rate of identifying oceans increased by 8.25%; and the recall rate of identifying internal waves also increased by 4.25%.

**Table 8.** Results of recall of WaveNet under the same experimental parameters using a training set based on traditional data augmentation and DCGAN-generated data augmentation.

| | Traditional Data Augmentation/% | Data Augmentation Based on the Modified DCGAN/% |
|---|---|---|
| Recall of cloud recognition | 97.00 | 99.75 |
| Recall of land recognition | 90.75 | 97.25 |
| Recall of ocean recognition | 90.25 | 98.50 |
| Recall of oceanic internal waves recognition | 94.75 | 99.00 |

*4.3. Discussion*

For fair comparison, all works are pre-trained on the EuroSAT dataset. And the hyperparameter settings are as shown in Table 9.This article shows the performance comparison results of each model on the test set in Table 10. Our proposed WaveNet achieves the highest accuracy, which further illustrates the effectiveness of the WaveNet model. Other models have a high number of layers and a large number of parameters, but their performance is poor. This may be because the larger the model, the larger the dataset required for learning to ensure that the model is effective.

Figure 9 shows the t-SNE 2D visualization of semantic features extracted by WaveNet on datasets enhanced by traditional data augmentation and data augmentation based on the modified DCGAN, respectively. The feature distribution of the same type of data has a large overlap, while the data of different types are far apart. The feature distribution of each type of image shows an obvious balloon-like distribution, which shows that for WaveNet, these subtle features are distinguishable.

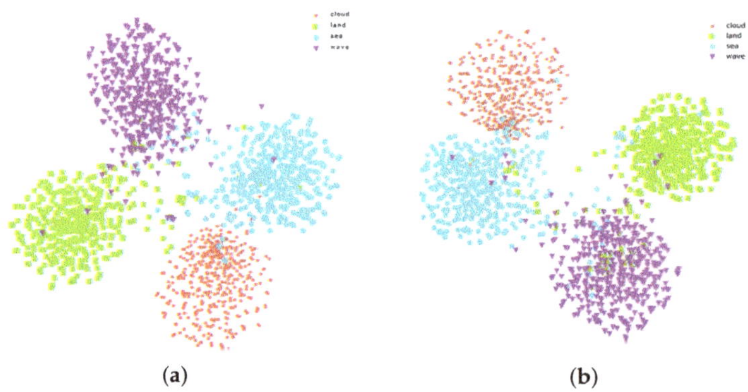


**Figure 9.** Classification results of WaveNet using different data augmentation methods. (**a**) Classification results of WaveNet using the DCGAN-augmented dataset. (**b**) Classification results of WaveNet using the traditionally augmented dataset.

**Table 9.** Hyperparameters for training.

| Batch Size | Learning Rate | Training Epochs | Optimizer |
|---|---|---|---|
| 128 | 0.001 | 500 | Adam |

**Table 10.** Performance comparison on the dataset.

| Methods | Overall Accuracy/% | Accuracy of Oceanic Internal Waves Recognition/% |
|---|---|---|
| AlexNet [28] | 95.00 | 95.25 |
| VGG11 [29] | 97.875 | 96.75 |
| VGG16 [29] | 97.50 | 96.25 |
| GoogLeNet [30] | 98.50 | 98.75 |
| Resnet18 [24] | 98.125 | 98.75 |
| WaveNet (ours) | 98.625 | 99.00 |

### 4.4. Display of Test Results

The results shown in Figure 10 demonstrate that by cropping the original remote sensing images and inputting them into the WaveNet network for classification, the ocean internal waves in the MODIS images were successfully identified with high accuracy. This is a notable achievement as accurately identifying oceanic internal waves is a challenging task in remote sensing image analysis.

By outlining the areas classified as "internal waves" in the original image, the figure provides a visual representation of the algorithm's effectiveness in detecting these features. This successful identification of oceanic internal waves has practical implications for various applications, including oceanography and environmental monitoring, where the detection and tracking of such phenomena are crucial.

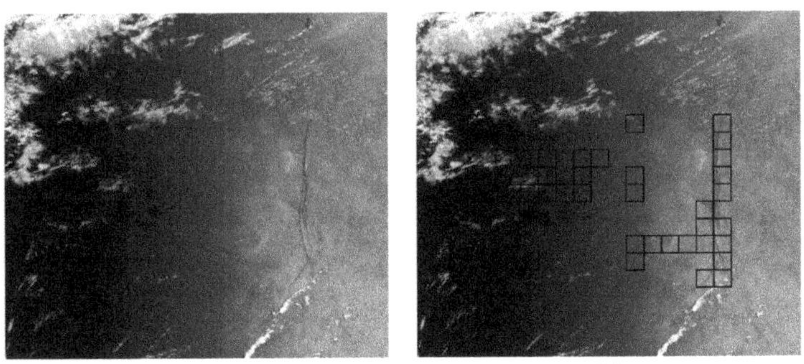

(a) (b)

**Figure 10.** Display of test results: (**a**) a whole remote sensing image; (**b**) remote sensing images after being detected.

## 5. Conclusions

This paper uses MODIS remote sensing data to produce different types of remote sensing image samples, including internal waves, processes these images, and generates a database including internal waves. In addition, this paper proposes an end-to-end method that uses deep learning technology to improve the recognition performance of MODIS remote sensing images. One of the key contributions is the use of a modified DCGAN network for data augmentation, which significantly improves the diversity of the dataset and enhances the generalization ability of the recognition model. This method has the following advantages: (1) Reduce data collection costs: collecting and labeling large-scale remote sensing datasets is expensive and time-consuming. By using the modified DCGAN to generate data, the time and resources for collecting real data can be effectively reduced; (2) Increase data diversity: The data generated by the modified DCGAN are very similar to real remote sensing images. The data it generates contain a wider range of variations and scenarios, helping network models learn and process more complex real-world data more effectively; (3) Improve generalization ability: The combination of real data and synthetic data improves the generalization ability of the model. This means that the trained model is more able to accurately classify and identify new, unknown data.

Another contribution lies in the design of the WaveNet network with excellent recognition performance. By adding a channel attention mechanism to the deep convolutional layer, WaveNet can pay different attention to each channel of the feature map, thereby more effectively learning features related to remote sensing classification. Moreover, the convolutional layers of WaveNet have a residual structure, which allows WaveNet to avoid overfitting problems caused by too deep layers in actual training.

By combining the above methods, we have improved the recognition accuracy of remote sensing image classification tasks. This has practical applications in many fields, such as environmental monitoring, early warning of marine disasters, detection of internal ocean waves, etc. However, due to the fact that the data come from optical remote sensing satellites, it may be difficult to obtain sea surface data when the weather is bad, resulting in certain limitations in actual citation.

In future research, we plan to further explore how to integrate geographical location information into remote sensing images to further improve the practicality of the research. Specifically, we plan to collect data on the geographical coordinates of where remote sensing images were taken and associate these data with the images. Through this association we can achieve the following goals.

Geographical information feature extraction: We plan to use geolocation information to extract information related to geographical features in images. For example, we can determine the distance of waves within the ocean from the shoreline in an image, as well as the shape and wavelength of the waves. This information is of great significance to fields such as ocean research.

Environmental monitoring and management: Geolocation information can also be used for environmental monitoring and management. We plan to use this information to track changes in specific areas, such as changes in land use or changes in ocean water quality. This will lead to a better understanding and management of natural resources.

Geographic Information System (GIS): We also plan to use remote sensing images in conjunction with GIS technology to create a geographic information system. This will enable users to better visualize and analyze geographic data and support a variety of applications, from urban planning to natural disaster management.

These works will help apply remote sensing images to a wider range of fields and improve the practicality and adaptability of models.

**Author Contributions:** Methodology, Z.J. and X.G.; Project administration, Z.J.; Resources, Z.J., L.S., and N.L.; Writing—original draft preparation, X.G.; Writing—review and editing, L.S. and L.Z. All authors have read and agreed to the published version of the manuscript.

**Funding:** The authors gratefully acknowledge the co-funding and support of the work by the Jiangsu Key Research and Development Plan (BE2021012-2 and BE2021012-5), the Changzhou Science and Technology Support Program (CE20225034), and the Digital Twin Technology Engineering Research Center for petrochemical process support program (DTEC202002 ).

**Institutional Review Board Statement:** Not applicable.

**Informed Consent Statement:** Not applicable.

**Data Availability Statement:** The data that support the findings of this study are available from the corresponding author on reasonable request.

**Conflicts of Interest:** The authors declare no conflict of interest.

## References


1. Chen, C.-Y.; Hsu, J.R.-C.; Cheng, M.; Chen, H.-H.; Kuo, C.-F. An investigation on internal solitary waves in a two-layer fluid: Propagation and reflection from steep slopes. *Ocean Eng.* **2007**, *34*, 171–184. [CrossRef]
2. Seas, C.; Liu, A.K.; Steve, Y.; Liang, N.K. Evolution of nonlinear internal waves in the East and South China Seas. *J. Geophys. Res. Ocean.* **1998**, *103*, 7995–8008. [CrossRef]
3. Whalen, C.B.; de Lavergne, C.; Naveira Garabato, A.C.; Klymak, J.M.; MacKinnon, J.A.; Sheen, K.L. Internal wave-driven mixing: Governing processes and consequences for climate. *Nat. Rev. Earth Environ.* **2020**, *1*, 606–621. [CrossRef]
4. van Haren, H.; Gostiaux, L. Energy Release Through Internal Wave Breaking. *Oceanography* **2012**, *25*, 124–131. [CrossRef]
5. Cai, S.-Q.; Gan, Z.-J. Progress in the study of the internal solition in the northern south china sea. *Adv. Earth Sci.* **2001**, *16*, 215–219. [CrossRef]
6. Barnes, W.L.; Xiong, X.; Salomonson, V.V. Status of Terra MODIS and Aqua MODIS. In Proceedings of the IEEE International Geoscience and Remote Sensing Symposium, Toronto, ON, Canada, 24–28 June 2002; Volume 2, pp. 970–972. [CrossRef]
7. Cortes, C.; Vapnik, V.N. Support-Vector Networks. *Mach. Learn.* **1995**, *20*, 273–297. [CrossRef]
8. Samadzadegan, F.; Hasani, H.; Schenk, T. Simultaneous feature selection and SVM parameter determination in classification of hyperspectral imagery using Ant Colony Optimization. *Can. J. Remote. Sens.* **2012**, *38*, 139–156. [CrossRef]
9. Haut, J.M.; Paoletti, M.E.; Plaza, J.; Li, J.Y.; Plaza, A.J. Active Learning With Convolutional Neural Networks for Hyperspectral Image Classification Using a New Bayesian Approach. *IEEE Trans. Geosci. Remote Sens.* **2018**, *56*, 6440–6461. [CrossRef]
10. Bao, S.; Meng, J.; Sun, L.; Liu, Y. Detection of ocean internal waves based on Faster R-CNN in SAR images. *J. Oceanol. Limnol.* **2019**, *38*, 55–63. [CrossRef]
11. Yu, D.; Xu, Q.; Guo, H.; Zhao, C.; Lin, Y.; Li, D. An Efficient and Lightweight Convolutional Neural Network for Remote Sensing Image Scene Classification. *Sensors* **2020**, *20*, 1999. [CrossRef]
12. Li, Y.; Zhang, Y.; Zhu, Z. Error-Tolerant Deep Learning for Remote Sensing Image Scene Classification. *IEEE Trans. Cybern.* **2020**, *51*, 1756–1768. [CrossRef] [PubMed]
13. Zheng, Y.-G.; Zhang, H.-S.; Qi, K.; Ding, L.-Y. Stripe segmentation of oceanic internal waves in SAR images based on SegNet. *Geocarto Int.* **2021**, *37*, 8567–8578. [CrossRef]
14. Tao, M.; Xu, C.; Guo, L.; Wang, X.; Xu, Y. An Internal Waves Data Set From Sentinel-1 Synthetic Aperture Radar Imagery and Preliminary Detection. *Earth Space Sci.* **2022**, *9*, e2022EA002528. [CrossRef]
15. Serebryany, A.; Khimchenko, E.; Zamshin, V.; Popov, O.Y. Features of the Field of Internal Waves on the Abkhazian Shelf of the Black Sea according to Remote Sensing Data and In Situ Measurements. *J. Mar. Sci. Eng.* **2022**, *10*, 1342. [CrossRef]
16. Guo, A.J.X.; Zhu, F. A CNN-Based Spatial Feature Fusion Algorithm for Hyperspectral Imagery Classification. *IEEE Trans. Geosci. Remote Sens.* **2018**, *57*, 7170–7181. [CrossRef]
17. Ramp, S.R.; Yang, Y.J.; Bahr, F.L. Characterizing the nonlinear internal wave climate in the northeastern South China Sea. *Nonlinear Process. Geophys.* **2010**, *17*, 481–498. [CrossRef]
18. Meng, J.; Sun, L.; Zhang, H.; Hu, B.; Hou, F.; Bao, S. Remote sensing survey and research on internal solitary waves in the South China Sea-Western Pacific-East Indian Ocean (SCS-WPAC-EIND). *Acta Oceanol. Sin.* **2022**, *41*, 154–170. [CrossRef]
19. Goodfellow, I.J.; Pouget-Abadie, J.; Mirza, M.; Xu, B.; Warde-Farley, D.; Ozair, S.; Courville, A.C.; Bengio, Y. Generative Adversarial Nets. In Proceedings of the 28th Conference on Neural Information Processing Systems—NIPS 2014, Montreal, QC, Canada, 8–13 December 2014. [CrossRef]
20. Zhang, H.; Xu, T.; Li, H.; Zhang, S.; Wang, X.; Huang, X.; Metaxas, D.N. StackGAN: Text to Photo-Realistic Image Synthesis with Stacked Generative Adversarial Networks. In Proceedings of the 2017 IEEE International Conference on Computer Vision (ICCV), Venice, Italy, 22–29 October 2017; pp. 5908–5916. [CrossRef]
21. Wei, G.; Luo, M.; Liu, H.; Zhang, D.; Zheng, Q. Progressive generative adversarial networks with reliable sample identification. *Pattern Recognit. Lett.* **2020**, *130*, 91–98. [CrossRef]
22. Radford, A.; Metz, L.; Chintala, S. Unsupervised Representation Learning with Deep Convolutional Generative Adversarial Networks. *arXiv* **2015**, arXiv:1511.06434.
23. Nair, V.; Hinton, G.E. Rectified Linear Units Improve Restricted Boltzmann Machines. In Proceedings of the ICML'10: Proceedings of the 27th International Conference on International Conference on Machine Learning, Haifa, Israel, 21–24 June 2010. [CrossRef]
24. He, K.; Zhang, X.; Ren, S.; Sun, J. Deep Residual Learning for Image Recognition. In Proceedings of the 2016 IEEE Conference on Computer Vision and Pattern Recognition (CVPR), Las Vegas, NV, USA, 27–30 June 2016; pp. 770–778. [CrossRef]
25. Helber, P.; Bischke, B.; Dengel, A.R.; Borth, D. EuroSAT: A Novel Dataset and Deep Learning Benchmark for Land Use and Land Cover Classification. *IEEE J. Sel. Top. Appl. Earth Obs. Remote Sens.* **2017**, *12*, 2217–2226. [CrossRef]
26. van der Maaten, L.; Hinton, G.E. Visualizing Data using t-SNE. *J. Mach. Learn. Res.* **2008**, *9*, 2579–2605.
27. van der Maaten, L. Learning a Parametric Embedding by Preserving Local Structure. In Proceedings of the International Conference on Artificial Intelligence and Statistics, Clearwater Beach, FL, USA, 16–18 April 2009.
28. Krizhevsky, A.; Sutskever, I.; Hinton, G. E. ImageNet classification with deep convolutional neural networks. *Commun. ACM* **2012**, *60*, 84–90. [CrossRef]

29. Simonyan, K.; Zisserman, A. Very Deep Convolutional Networks for Large-Scale Image Recognition. *arXiv* **2014**, arXiv:1409.1556.
30. Szegedy, C.; Liu, W.; Jia, Y.; Sermanet, P.; Reed, S.E.; Anguelov, D.; Erhan, D.; Vanhoucke, V.; Rabinovich, A. Going deeper with convolutions. In Proceedings of the 2015 IEEE Conference on Computer Vision and Pattern Recognition (CVPR), Boston, MA, USA, 7–12 June 2015; pp. 1–9. [CrossRef]

*applied sciences*

MDPI

*Article*

# Remote Sensing Multimodal Image Matching Based on Structure Feature and Learnable Matching Network

**Songlai Han [1], Xuesong Liu [1], Jing Dong [1,*] and Haiqiao Liu [2]**

[1] Research Institute of Aerospace Technology, Central South University, Changsha 410017, China
[2] School of Electrical and Information, Hunan Institute of Engineering, Xiangtan 411228, China
* Correspondence: dongjing@csu.edu.cn

**Abstract:** Matching remotely sensed multimodal images is a crucial process that poses significant challenges due to nonlinear radiometric differences and substantial image noise. To overcome these difficulties, this study presents a novel and practical template-matching algorithm specifically designed for this purpose. Unlike traditional approaches that rely on image intensity, the proposed algorithm focuses on matching multimodal images based on their geometric structure information. This approach enables the method to effectively adapt to variations in grayscale caused by radiometric differences. To enhance the matching performance, principal component analysis calculation based on the log-Gabor filter is proposed to estimate the structural feature of the image. The proposed method can estimate the structure feature accurately even under severe noise distortion. In addition, a learnable matching network is proposed for similarity measuring to adapt to the gradient reversal caused by the radiometric difference among remotely sensed multimodal images. Infrared, visible light, and synthetic aperture radar images are adopted for the evaluation, to verify the performance of the proposed algorithm. Based on the results, the proposed algorithm has a distinct advantage over other state-of-the-art template-matching algorithms.

**Keywords:** remotely sensed image; multimodal image; template matching; principal component analysis; structure feature

## 1. Introduction

Multimodal image matching is the process of overlaying two or more images of the same scene captured by different sensors [1]. Since the imaging methods are based on different physical effects, remotely sensed multimodal images acquired by different sensors can capture different object characteristics, which provide complementary information. Multimodal image matching can integrate the complementary information by registering different multimodal images into one identical map, which is an important step for many remote sensing image processing tasks, such as image fusion [2], vision-based satellite attitude determination [3], and image-to-map rectification [4].

Automatic high-performance matching for remotely sensed multimodal images remains a problematic task because of the severe radiometric deformation produced by different types of sensors.

Traditional image-matching methods can be classified into two categories: feature-based [5–8] and area-based methods [9–11]. The scale-invariant feature transform (SIFT), which is invariant to scale and rotation changes, is the most representative feature-based method [12]. The SIFT-based method has been widely employed in the registration of remotely sensed multimodal images [6–8,13,14]. However, these SIFT-based descriptors were developed to handle the geometric affine variation of images with linear intensity changes. Therefore, these methods cannot resolve complicated nonlinear intensity changes caused by radiometric variations among remotely sensed multimodal images [14,15]. Ye et al. proposed a descriptor based on the local histogram of phase congruency to adapt to nonlinear intensity variation [16]. This method can accurately register remotely sensed multimodal images if the overlapping areas in the images are sufficiently large.

Feature correspondence-based methods usually fail to match remotely sensed multimodal images because repeatable and reliable feature detection is considerably difficult for these images. To address this problem, locality-preserving matching [17] and matching based on local linear transformation [18] are proposed to achieve reliable feature correspondence.

The area-based method also called the correlation-like or template-matching method [1], compares the template to each candidate window on the base image; the corresponding window with maximum similarity is then selected as the matching result. Compared with the feature-based matching method, the area-based method exhibits better performance in matching images with few features or noise distortion [11].

There are two problems in using the area-based method for matching remotely sensed multimodal images. One is that some remotely sensed images contain intense noise. For example, synthetic aperture radar (SAR) images have randomly distributed speckle noise introduced by the interference of ground objects or surfaces to the backward reflection of electromagnetic waves.

Another problem is the significant nonlinear radiometric difference among remotely sensed multimodal images. This difference introduces nonlinear intensity variation, indicating that the same part of an object may be represented by different intensities in the images captured by different modalities [19]. It is impossible to calculate (even roughly) the intensity variation among multimodal images through a single mapping function [20]. Therefore, it is not ideal to match two remotely sensed multimodal images directly based on grayscale via commonly used similarity measurements techniques, such as the sum of squared differences (SSD), the sum of absolute differences (SAD), normalized cross-correlation (NCC), and matching by tone mapping (MTM) [21].

In addition, radiometric differences can cause gradient reversal, i.e., the gradients of corresponding parts of images change their orientation in the opposite direction [19], as shown in Figure 1. In this case, if the image feature is represented by a descriptor based on the texture gradient or structure orientation, then the two images of the same object display opposite geometric information. However, gradient reversal does not always occur, and the location may be unknown, making this problem more intractable.

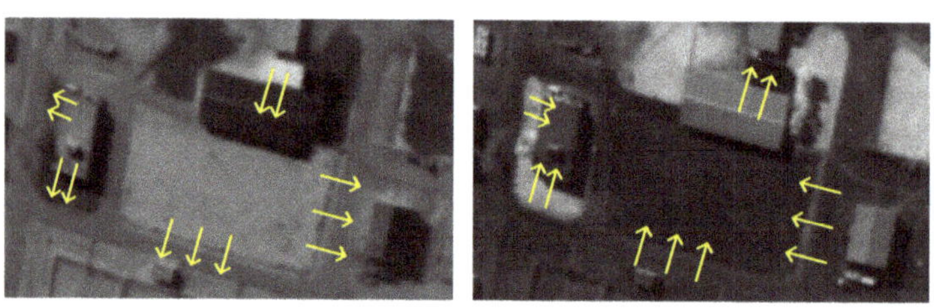

(a) an infrared image (b) a visible light image

**Figure 1.** Gradient reversal between infrared and visible light images. Arrows show gradient orientation from brighter areas to darker areas.

In recent years, deep learning-based methods are proposed for addressing the challenge problems of image matching [22–26]. Their testing results show that the deep learning based matching methods can achieve significant improvement compared with the traditional matching methods.

The main advantage of deep learn based methods is that they employ convolutional neural networks to learn powerful feature descriptors which are more robust to appearance changes than the classical descriptors. However, the feature learning networks are usually pre-trained in large datasets such as ImageNet [27] which consists of visible light images with rich and clear features. Therefore, the performance of these deep learning-based methods could drop significantly for matching SAR or infrared images, and the retraining or the fine tuning is also not idea ways if the application dataset is small.

In order to address the aforementioned challenges, this study introduces a template-matching algorithm that aims to achieve accurate matching of remotely sensed multimodal images. The algorithm proposed in this research enhances the estimation of structural features by incorporating principal component analysis (PCA) and employs a learnable matching network (LMN) to measure the similarity between two images. The primary contributions of this study can be summarized as follows:

1. Novel descriptor based on PCA-enhanced structure feature. As introduced, nonlinear intensity variation can significantly decrease the gray-level correlation among images. Instead of directly matching images based on the image grayscale, a descriptor based on the structure feature for capturing the geometric information of the image is introduced. Since the structure feature may be distorted by noise affection, a PCA-enhancing method is proposed to reduce the noise component in signals and estimate the local dominant orientation. The structure feature can be accurately calculated by the proposed descriptor even in images with severe noise distortion.
2. Improved similarity measurement based on LMN. Severe miscalculations can result if SAD or SSD is used to measure the similarity of the structure feature because the complicated radiometric variation causes gradient reversal. To solve this problem, a similarity measurement based on the LMN is proposed. The correlation layer and the regression network of the LMN can handle the gradient reversal and significantly improve the matching of remotely sensed multimodal images, as described in the experimental section.
3. A Novel combined matching method for application with a small dataset. It is very hard to train a deep convolutional network to extract robust cross-modal features with a small dataset. Therefore, the PCA-enhanced structure feature is adopted, which is a handcraft stable cross-modal feature. For addressing the complicated gradient reversal and radiometric variation between multi-modal images, we developed a light learnable matching network to learn the similarity measurement and regress the transformation parameters.

The remainder of this paper is organized as follows. Section 2 introduces related works. The proposed PCA–LMN template-matching algorithm is described in Section 3. In Section 4, the performance of the proposed algorithm is evaluated. Conclusions are presented in Section 5.

## 2. Related Work

Complex grayscale variation is a major problem for area-based multimodal image matching. Some template-matching algorithms have attempted to solve this problem by improving the similarity measurement [21,28,29]. These algorithms usually assume that the gray distortion caused by different imaging conditions, or the spectral sensitivity of sensors can satisfy a mapping model [1]. Therefore, gray distortion can be resolved by developing a similarity measurement that ignores the grayscale variation, conforming to the mapping model. The NCC, which is invariant to linear gray changes, is the most commonly used similarity measurement approach for adapting gray distortion among images [29]. Even under conditions with monotonic nonlinear gray variation, the NCC usually performs well, because these variations can be typically assumed as locally linear. However, the NCC cannot handle complex gray distortions, such as non-monotonic nonlinear gray differences, or situations where the gray mapping between two images is not function mapping [21]. Visual examples of gray mapping between remote sensing multimodal images can be found in [30].

Hel-Or proposed a fast-matching measurement called MTM [21], which is invariant to nonlinear gray variation. It can be regarded as a generalization of the NCC for non-linear mappings and reduces to the NCC when the mappings are linear. Although the computational time of MTM is the same as that of the NCC, it exhibits better matching performance. However, the MTM also assumes that the grayscale mapping between two images is function mapping.

The mutual information (MI) technique is a similarity measurement approach commonly used in multimodal image matching [28]. This technique measures the gray statistical dependency between two images, without requiring their grayscale mapping to be function mapping. Moreover, compared with the NCC and MTM, MI affords advantages in adapting the nonlinear gray variation among multimodal images [10]. However, it requires the construction of a local histogram for each candidate window during the search process, thereby leading to high computational costs. Additionally, the MI technique is sensitive to the size of histogram bins for joint density estimation [21].

Measurement improvement is not the only approach for resolving multimodal image matching. Some area-based methods match remotely sensed multimodal images with dense feature descriptors based on structural information.

The histogram of oriented gradient (HOG) is a commonly used descriptor that employs the orientation and amplitude of gradients to capture the structural features of an image [31]. This descriptor was successfully applied to many image-matching methods. Sibiryakov proposed a template-matching algorithm based on the projected and quantizing histograms of oriented gradients (PQ-HOG). It transforms the images into dense binary codes to improve their computational efficiency [32]. The HOG is considerably resistant to illumination change or contrast variation; however, it cannot adapt to the complex nonlinear grayscale distortion among remotely sensed multimodal images. In addition, the gradient-based descriptor is usually sensitive to image noise.

Schechtman and Irani proposed the local self-similarity (LSS) descriptor [33], which had been previously applied to various template-matching methods [14,34]. However, the LSS cannot effectively capture informative features for multimodal matching in textureless areas [35], and its discriminative power is considerably limited [36].

The phase congruency model proposed by Kovesi [37] can capture the structure magnitude of the image, which is invariant to the complex nonlinear grayscale distortion among multimodal images. However, this model cannot capture the structure orientation of an image which is crucial for multimodal image matching. To solve this problem, Ye et al. extended the phase congruency model to build a dense descriptor called histogram of oriented phase congruency (HOPC) [9]. They used the log-Gabor odd-symmetric wavelets to calculate the orientation of phase congruency and construct a descriptor using the orientation and amplitude of phase congruency.

Compared with the HOG, the HOPC is more robust for matching remotely sensed multimodal images. Ye et al. demonstrated that the performance of the template-matching algorithm based on the HOPC is superior to those based on the NCC, MTM, or MI for remotely sensed multimodal images [9]. Recently, a novel template-matching method based on the channel features of oriented gradients (CFOGs) was proposed. This novel feature is an extension of the pixel-wise HOG descriptor [35]. Compared with the HOPC, the CFOG is more robust and efficient in matching multimodal images [35]. However, both the HOPC and CFOG handle the gradient reversal in a problematic manner, as described in part 2 of Section 3. Furthermore, they are sensitive to noise distortion, as discussed in Section 4.

In recent years, deep learning-based methods are proposed for matching multimodal images or aerial images. X. Han et al. [23] proposed a unified approach for feature and metric learning, dubbed Match-Net. They developed a deep convolutional network to extract features from images and a network of three fully connected layers to measure the similarity. Match-Net can achieve better performance compared with the state-of-the-art handcraft methods according to their testing results. I. Rocco et al. [24] proposed a trainable end-to-end matching network, which is not just for learning the feature and the similarity, but also estimating the transformation parameters with a regression network. This method is further developed for aerial image matching in [25,26], and the testing results confirmed that deep learning-based matching methods can achieve significant improvement compared with the traditional matching methods.

## 3. Template Matching Based on PCA–LMN

The proposed matching algorithm is a full-search template-matching method that compares the template with a candidate window of the same size on the base image to identify the position of the target window (Figure 2). Since the method only searches in translation, a preliminary correction must be performed before the matching, so the direction and scale of the template and the direction and scale of the base image are approximately the same. Usually, the preliminary correction can be automatically performed according to the altitude and attitude information provided by the onboard navigation system [38].

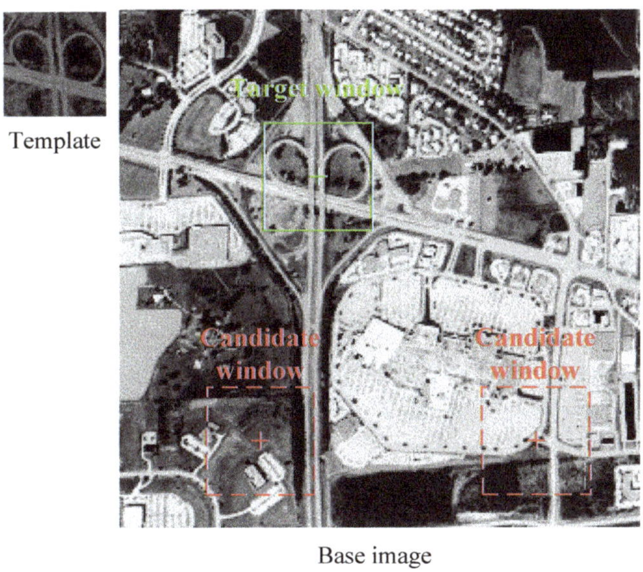


**Figure 2.** Example of template matching with the multimodal image.

The template-matching algorithm based on the PCA–LMN consists of three main steps: log-Gabor orientation estimation, orientation enhancement using the PCA, and similarity measurement and translation estimation based on the LMN; the algorithm flowchart is shown in Figure 3.

### *3.1. Orientation Estimation and Enhancing*

A significant challenge in matching remotely sensed multimodal images lies in the severe distortion of grayscale relationships among the images. As shown in Figure 4, the grayscale change between infrared image (a) and visible light image (b) is considerably significant; however, the structural orientation, which is present in images (c) and (d), is considerably more stable than the gray part of the image. Accordingly, the structure orientation based on log-Gabor is employed by some methods [7,9] for matching multimodal images. These methods exhibit a significant advantage in terms of matching performance over template-matching algorithms directly based on the image grayscale.

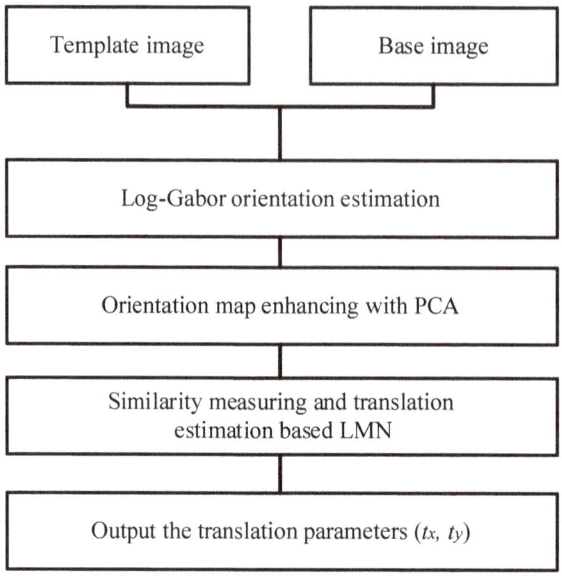


**Figure 3.** Flowchart of PCA–LMN template-matching algorithm.

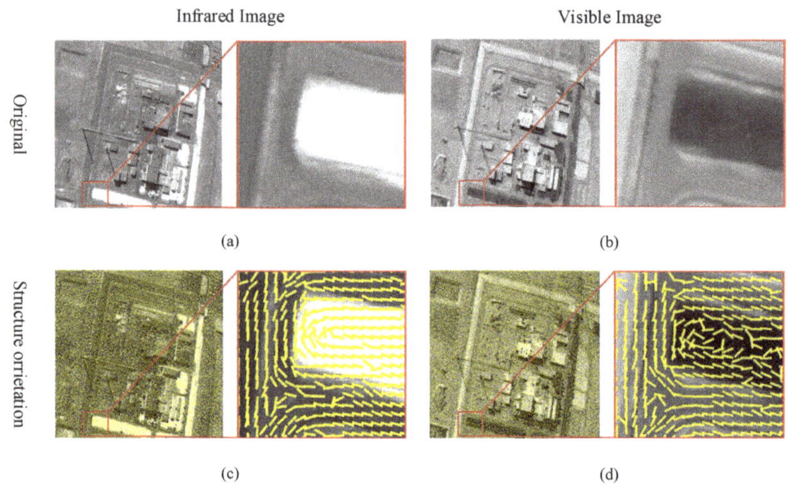


**Figure 4.** Multimodal images and their orientation maps: (**a**) infrared image; (**b**) visible light image; (**c**) orientation map of infrared image; and (**d**) orientation map of visible light image. Short yellow lines indicate the structure orientation direction, which is also the image edge direction.

However, the structure orientation estimated with the log-Gabor filter is sensitive to noise distortion. The orientation map is estimated via log-Gabor filters. Note that the orientation map is disturbed by noise distortion.

To improve the noise adaptiveness, the PCA is employed to enhance the structure orientation estimated using log-Gabor. The PCA is typically used to calculate the dominant vectors of a given dataset that can reduce the noise component in signals and estimate the local dominant orientation.

For each pixel, the PCA can be applied to the local gradient vectors to obtain their local dominant direction. In general, the PCA can be implemented in two ways: eigenvalue decomposition (EVD) of the data covariance matrix and singular value decomposition

(SVD) of the data matrix. In this work, because of the superiorities in flexibility and robustness [39], SVD is employed to calculate the PCA.

Given the original image (I) with its horizontal derivative image ($a$) and vertical derivative image ($b$), an $N \times 2$ local gradient matrix is constructed for each pixel:

$$G = \begin{bmatrix} \mathrm{A}^T & \mathrm{B}^T \end{bmatrix} \quad (1)$$

where $N$ is determined by the size of the local window of the PCA calculation. For example, if the size of the window for each pixel is $3 \times 3$, then the value of $N$ is 9. The vectors of the local derivatives, i.e., A and B, can be calculated using the following expression.

$$\mathrm{A} = \{a_1 a_2, \ldots, a_N\},\ \mathrm{B} = \{b_1 b_2, \ldots, b_N\} \quad (2)$$

where $a_{1,2,\ldots,N}$ can be calculated according to Equation (1) and $b_{1,2,\ldots,N}$ can be calculated according to Equation (2).

The dominant orientation can be estimated by determining a unit vector, $u$, perpendicular to the local gradient vectors (Figure 5). This can be formulated as the following minimization problem.

$$\left\| u^T G \right\| = \sum_{i=1}^{N} [a_i, b_i]^T u \quad (3)$$

**Figure 5.** Dominant structure orientation and local gradient vectors.

This can be solved by applying SVD to the local gradient matrix, $G$.

$$G = USV^T \quad (4)$$

where $U$ is an $N \times N$ orthogonal matrix; $S$ is an $N \times N$ matrix; $V = \begin{bmatrix} v & u \end{bmatrix}$ is a $2 \times 2$ orthogonal matrix; and $v$ indicates the local dominant orientation of the local gradient vectors. The PCA-enhanced structure orientation can be calculated according to the following equation.

$$\varphi = \frac{1}{2} \angle(u_1, u_2) \quad (5)$$

where $\varphi \in [0°, 180°)$, $u_1$ is the element in the first row of $u$ and $u_1$ is the element in the second row of $u$. Note that the structure orientation is orthogonal to the orientation of the gradient vector.

The structure orientation enhanced by PCA is more robust against noise than the structure orientation directly estimated with log-Gabor filters. The enhanced structure orientation, which can be represented by $(u_1, u_2)$, is estimated for each pixel. Suppose the size of an image is $w \times h$, and the size of its feature map is $w \times h \times 2$.

### *3.2. Learnable Matching Network*

The gradient reversal among remotely sensed multimodal images is considerably common. This causes orientation reversal (Figure 1), which is a critical problem for the similarity measurement of structure orientation.

To solve this problem, some methods remap the structure orientation to range [0–180°] by adding 180° to the negative value [9,10,35]. However, miscalculations can result if the orientation is not appropriately reversed. For example, suppose that a structure orientation vector in one image is 5°. The orientation of the corresponding structure in the other image should be changed to −5° because of the noise effect. According to the remapping rule, the orientation difference between them is 170°, which is evidently unreasonable. The SAR and infrared images sometimes contain intense noise; hence, these methods may encounter problems in matching the images.

Instead of changing the orientation mapping, a learnable matching network (LMN) is proposed, which consists of a correlation layer and a regression network.

### 3.2.1. Correlation Layer

Suppose the feature map of a base image is $f_B \in R^{w_b \times h_b \times 2}$, and the feature map of a template is $f_T \in R^{w_t \times h_t \times 2}$, the correlation layer between them is shown in Figure 6.

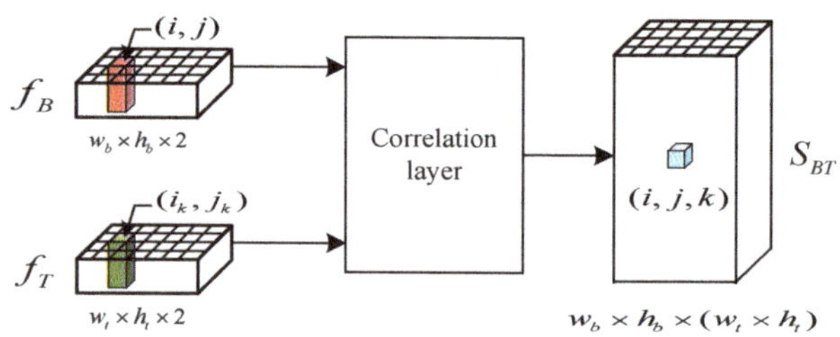


**Figure 6.** Simulation map computation by correlation layer.

The similarity between the feature map of the base image and the feature map of the template is calculated with the following equation,

$$S_{BT}(i,j,k) = \mathrm{f_B}(i,j)^T \mathrm{f_T}(i_k, j_k) \quad (6)$$

where $(i,j)$ indicates the individual feature position in the feature map of the base image, and $(i_k, j_k)$ indicates the individual feature position in the feature map of the image. The correlation layer output $\mathrm{S}_{BT}$ contains all pairs of similarities between individual features of $\mathrm{f_B} \in f_B$ and $\mathrm{f_T} \in f_T$.

### 3.2.2. Regression Network

The similarity map is passed through a regression network for translation estimation, which can be represented by the function F:

$$F : R^{w_b \times h_b \times w_t \times h_t} \rightarrow R^{\mathrm{n}} \quad (7)$$

where $n$ is the number of parameters to regress, $n$ = 2 for translation.

As shown in Figure 7, The regression network consists of two blocks, each block contains a convolutional layer, followed by batch normalization and ReLU. The last layer is a fully connected layer that regresses to the parameters of the transformation.

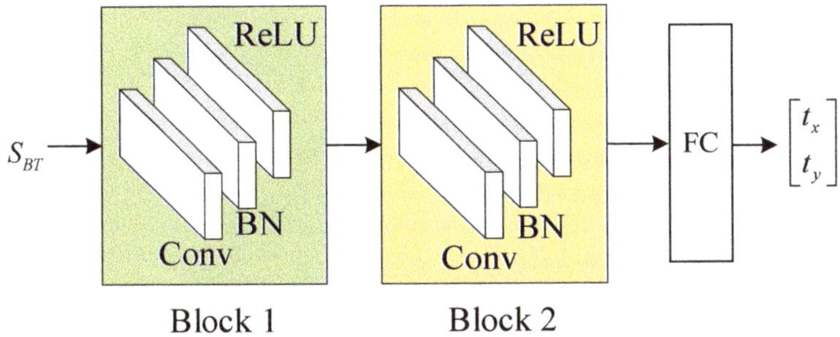


**Figure 7.** Transformation parameters estimation based on regression network.

3.2.3. Loss Function

Each training pair includes a template and a base image. Suppose the ground truth transformation between them is $H_{gt}$, and the transformation estimated between the sub-areas of this training pair is $\hat{H}_k$, the loss function is calculated by the following:

$$loss = \sum_{k=1}^{4} \sum_{i=1}^{N} ||H_{gt}(x_i, y_{i'}) - (\hat{H}_k(x_i - t_x, y_i - t_y)) + (t_x, t_y)||^2 \tag{8}$$

where $N$ is the number of grid points, and $(t_x, t_y)$ is the translation between the sub-area and the template, as shown in Figure 7. The loss function is based on the transformed grid loss [24], which minimizes the discrepancy between the estimating transformation and the ground truth transformation.

The partition approach increases the resolution of the input image and the resolution of the feature extraction, which helps improve the matching precision. In the partition approach, the four subarea pairs share the regression network, which means the number of the parameters of the network is not increased. This facilitates the retraining processing and the deployment of the network, which will eventually enhance the matching performance with a small training dataset.

The other way of enhancing the precision is inputting image with high resolution, but that may introduce an enormous increase of parameters to the network, which makes the retraining process (for cross-modal images) difficult and eventually lower the performance of the inference.

## 4. Experiment

The evaluation and comparison of the matching performance of PCA–LMN with the MTM [21], MI [28], HOPC [9], CFOG [35], and deep homography estimation (DHE) [26] are presented in this section. MTM, MI, HOPC, and CFOG are traditional image-matching methods. MTM and MI match images directly based on image gray, while HOPC and CFOG match images based on handcraft features. PCA–LMN is also based on the handcrafted feature, but it employs a learnable matching network to measure the similarity and regress the transformation parameters. DHE is a deep learning-based end to end-trainable method.

For the ablation study, PCA–LMN is also compared with the PCA–SAD, which uses the similarity measurement based on the SAD, and LGO–LMN, which directly estimates the structure orientation based on the log-Gabor filters.

### *4.1. Dataset and Training*

To test the proposed algorithms, 200 remotely sensed multimodal image pairs were used, which were taken from areas such as urban airports, plantations, harbors, and hilly terrain. Some examples of multimodal image pairs are shown in Figure 8. Significant radiometric differences among these remotely sensed multimodal images are observed. In addition, the SAR image contains severe noise distortions and lacks details.

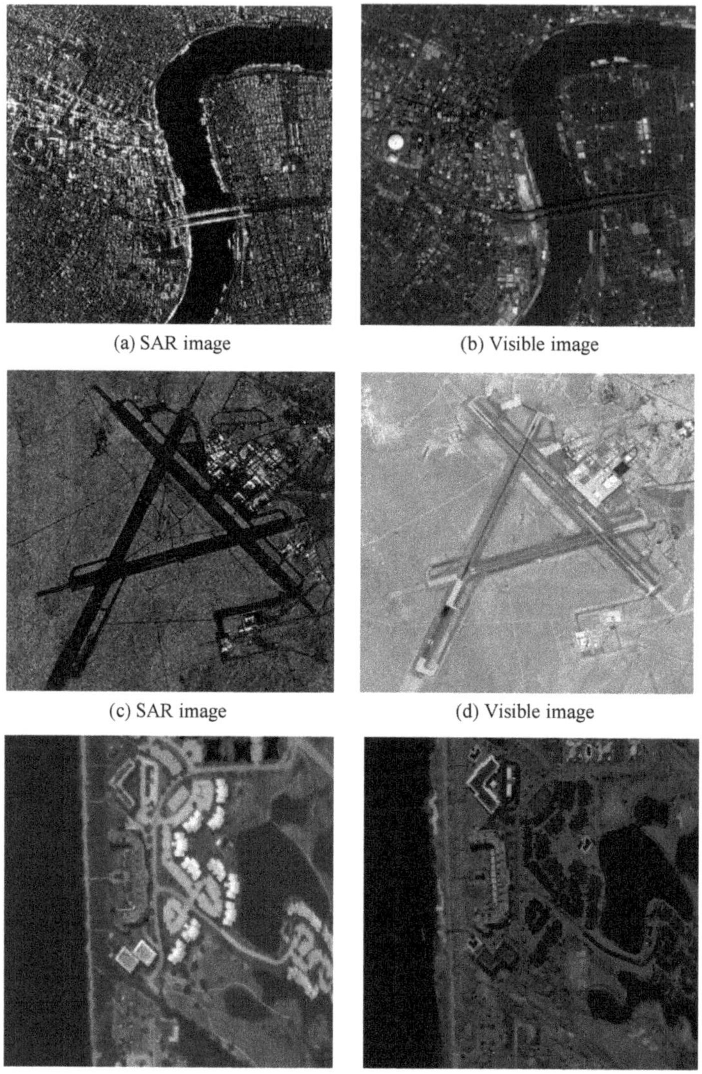

(a) SAR image (b) Visible image

(c) SAR image (d) Visible image

(e) Infrared image (f) Visible image

**Figure 8.** Examples of multimodal image pairs.

For each remotely sensed image, 100 templates of different sizes were randomly selected and matched to the base image. Since the dataset was small, 60% of the samples were employed for training and 20% of the samples were employed for validation and 20% were employed for testing. The final result was generated with the testing set. Data augmentation techniques such as grayscale variation, noise injection, and random erasing were adapted to the training set. Since our dataset was small, DHE used the pre-trained model provided by [25] and fine-tuned it with our dataset. The regression network of the LMN was totally trained with our dataset.

To evaluate the algorithms, 16 tests were performed. Table 1 summarizes the test information. Before the testing, the sensed image was manually corrected to the same coordinates as the base image. After the correction, the true position of the template in the based image and the position of the template selected from the sensed image were the same. In addition, Gaussian noise with different variances was added to test the noise adaptiveness of the algorithms.

**Table 1.** Testing information.

| Test | Variance of Gaussian Noise | Size of Base Image | Size of Template |
|---|---|---|---|
| $T_1$ | Without noise | 512 × 512 | 64 × 64 |
| $T_2$ | Without noise | 512 × 512 | 96 × 96 |
| $T_3$ | Without noise | 512 × 512 | 128 × 128 |
| $T_4$ | Without noise | 512 × 512 | 160 × 160 |
| $T_5$ | 0.01 | 512 × 512 | 64 × 64 |
| $T_6$ | 0.01 | 512 × 512 | 96 × 96 |
| $T_7$ | 0.01 | 512 × 512 | 128 × 128 |
| $T_8$ | 0.01 | 512 × 512 | 160 × 160 |
| $T_9$ | 0.03 | 512 × 512 | 64 × 64 |
| $T_{10}$ | 0.03 | 512 × 512 | 96 × 96 |
| $T_{11}$ | 0.03 | 512 × 512 | 128 × 128 |
| $T_{12}$ | 0.03 | 512 × 512 | 160 × 160 |
| $T_{13}$ | 0.05 | 512 × 512 | 64 × 64 |
| $T_{14}$ | 0.05 | 512 × 512 | 96 × 96 |
| $T_{15}$ | 0.05 | 512 × 512 | 128 × 128 |
| $T_{16}$ | 0.05 | 512 × 512 | 160 × 160 |

### 4.2. Evaluation Criteria

The correct matching rate (CMR) was selected as the evaluation criterion; CMR = CM/M; where M is the number of total matched point pairs, and CM is the number of correctly matched results. A matching result is correct if the overlapping area ratio (OAR) between the matching and true positions reaches 90%. The OAR is calculated according to the following equation:

$$OAR = \frac{f(TW - \triangle x)f(TW - \triangle y)}{TW^2} \tag{9}$$

where $TW$ is the template length; $\Delta x$ and $\Delta y$ denote the errors between the matching and true positions, respectively; and $f(\cdot)$ is a truncate function, which is defined as follows:

$$f(x) = \begin{cases} x, & x > 0 \\ 0, & x \leq 0 \end{cases} \tag{10}$$

### 4.3. Results and Analysis

The matching results of the tests (i.e., $T_{1-4}$) are shown in Figure 9. Notably, the CMRs of the investigated algorithms increase with the size of the templates. This is because a larger template aids in avoiding repetitive patterns in the base images.

Overall, the PCA–LMN achieved the best matching performance in the tests. The average CMR of PCA–LMN was 91.69%, which is 10.63% higher than that of the CFOG (the best traditional method in this test). It is assumed that this is because the PCA–LMN benefits from the PCA orientation enhancement and LMN. The PCA-enhanced orientation can accurately capture the structural features, even in the presence of significant noise distortions. This makes PCA–LMN more reliable in matching multimodal images with noise distortion, such as the optical SAR matching pairs in Figure 10c. In addition, LMN can be trained to adapt to the gradient reversal caused by the radiometric difference among remotely sensed multimodal images. The measurement function trained from LMN can be more sophisticated and accurate than the remapping function employed by the CFOG and HOPC.

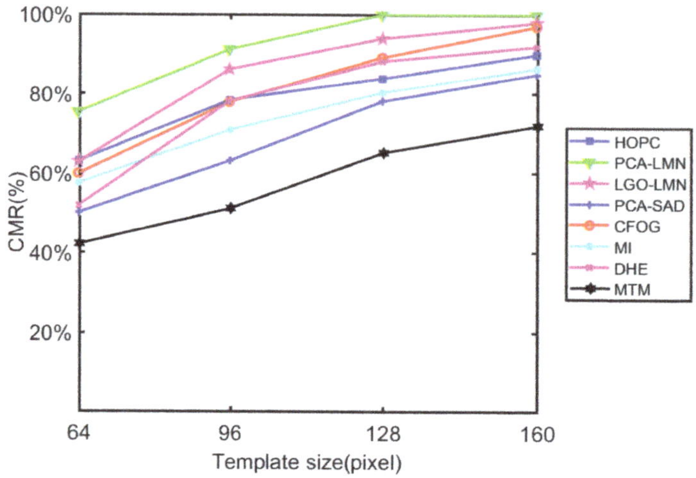


**Figure 9.** Matching results of test T1–T4.

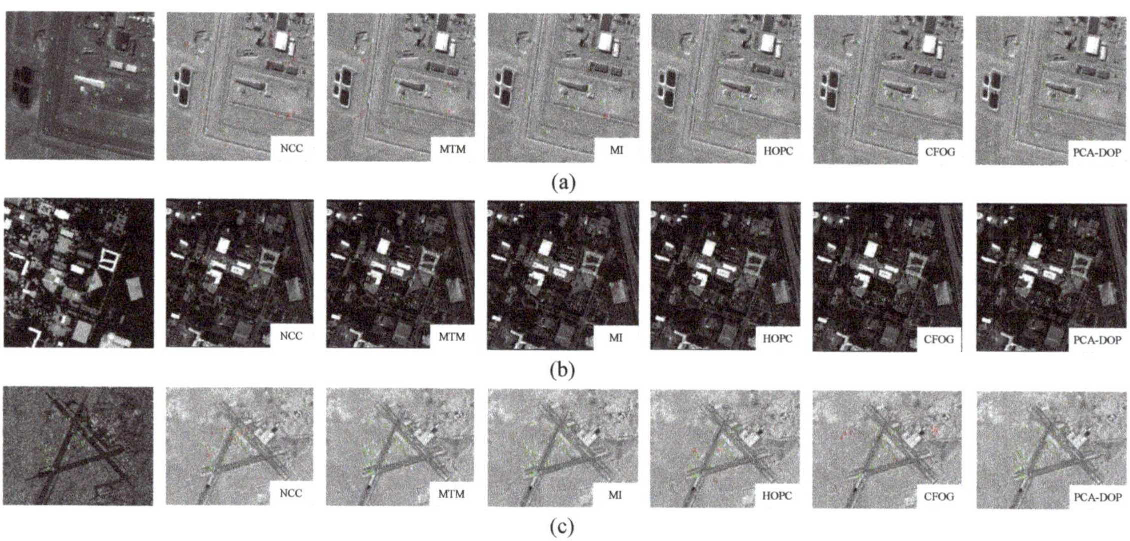


**Figure 10.** (**a–c**) Matching results of investigated algorithms on multimodal images. The first column shows true positions on the base image; columns 2–7 show matching positions of investigated algorithms on the sensed image; red and green numbers indicate erroneous and correct matching results, respectively.

The performance of DHE is not ideal in the tests, the average CMR of DHE is 3.44% lower than that of the CFOG and 14.07% lower than PCA–LMN. We believe the reason for these results is that the testing dataset is too small to train the deep feature extraction network of DHM, while the pre-trained dataset is not including multi-modal image pairs, which is very different from the testing dataset.

A clear contribution of the PCA can be observed by comparing the PCA–LMN and LGO–LMN. The average CMR of PCA–LMN is 6.38% higher than the average CMR of LGO–LMN in $T_{1\text{–}4}$, as shown in Figure 9. This is because the proposed PCA-enhanced method is more stable and can accurately capture the structure direction of remotely sensed multimodal images when compared with the log-Gabor orientation method, as described in parts 1 and 2 of Section 3. A clear contribution of the LMN can be observed by comparing the PCA–LMN and PCA–SAD. The average CMR of PCA–LMN is 22.50% higher than that of the PCA–SAD in $T_{1\text{–}4}$, as shown in Figure 9. This indicates that gradient reversal is a significant problem in matching remotely sensed images, as emphasized by numerous other works in this field of research [9,10,35]. Clearly, the matching performance considerably can be improved by resolving the gradient reversal problem with the proposed LMN.

In tests $T_{5\text{–}16}$, Gaussian noise is added to the test images to evaluate the noise adaptability of the investigated algorithms; Figure 11 shows the matching results. Notably, the CMRs of the algorithms decrease as the noise level increases. The CMRs of the structure feature-based algorithms (i.e., HOPC, CFOG, and PCA–LMN) decrease faster than the three algorithms directly based on the image grayscale (MTM, and MI). This indicates that the structure feature-based methods are more sensitive to noise distortion than the methods directly based on the image grayscale. This trend is assumed to occur because structural features can be easily distorted by image noise. For example, the log-Gabor orientation used in the HOPC, and the gradient orientation employed by the CFOG, are considerably sensitive to noise distortion. However, the structure orientation of the proposed method is enhanced by the PCA, which has considerably better noise adaptiveness than the log-Gabor and gradient orientations. Therefore, compared with the CFOG and HOPC, the PCA–LMN shows a significant advantage in tests $T_{5-16}$. In these tests, the average CMR of PCA–LMN is 80.27%, which is 6.54% and 11.29% higher than that of the CFOG and HOPC, respectively.

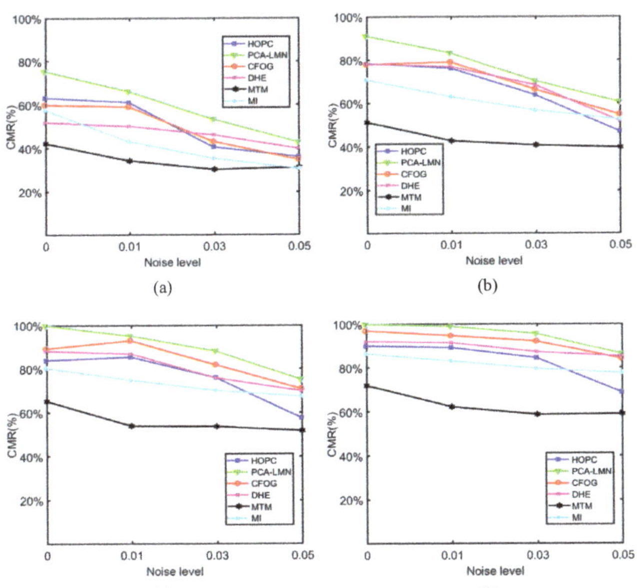


**Figure 11.** Matching results of T5–T16 template sizes: (**a**) 64 × 64; (**b**) 96 × 96; (**c**) 128 × 128; (**d**) 160 × 160.

## 5. Conclusions

In this paper, we propose a novel approach that combines PCA-based noise adaptiveness with a learnable matching network to address the challenge of matching remotely sensed multimodal images. Our method focuses on enhancing the noise adaptiveness of structural features and providing a robust trainable measurement of similarity while regressing transform parameters.

By integrating the learnable matching network with PCA-enhanced structure features, our proposed method effectively handles the complex radiometric variations that exist among remotely sensed multimodal images. This adaptability is crucial in achieving robust image matching. As demonstrated in the experiments, the PCA–LMN approach achieves the best matching performance among all the methods evaluated. The average CMR achieved by PCA–LMN is 91.69%, which surpasses the CMR of CFOG, the best traditional method, by 10.63%.

The ablation study in Section 4.3 further revealed that the improved performance of PCA–LMN can be attributed to two main factors. Firstly, PCA–LMN benefits from the PCA orientation enhancement, which enables accurate capture of structural features even in the presence of significant noise distortions. This enhancement plays a vital role in matching multimodal images with challenging noise distortions. Secondly, LMN is trained to adapt to gradient reversal caused by radiometric differences among remotely sensed multimodal images. This adaptability allows for more precise measurements and better handling of radiometric variations compared to traditional methods.

Furthermore, our method offers the advantage of not requiring the training of a deep convolutional neural network for feature extraction. This makes it easy to train and deploy, and it can achieve stable performance even with small training datasets. The testing results in Section 4.3 show that PCA–LMN does have a significant advantage over DHE (which employs deep convolutional neural network for feature extraction) when the training dataset is small.

While our proposed method demonstrates superior performance, it is important to note that it lacks the ability to handle rotation and scale variations between images. As a prerequisite for accurate matching, it is necessary to correct the template's direction and scale to align them with the base image approximately. Failure to do so can result in changes to directional features due to rotation and alterations to the scale of extracted features caused by image scaling.

**Author Contributions:** Conceptualization, J.D. and S.H.; methodology, S.H. and J.D.; software, J.D., S.H. and H.L.; validation, S.H., J.D. and H.L.; formal analysis, X.L.; investigation, X.L.; resources, J.D.; data curation, X.L.; writing—original draft preparation, J.D., S.H. and X.L.; writing—review and editing, J.D. and X.L.; visualization, H.L.; supervision, J.D. and S.H.; project administration, H.L.; funding acquisition, S.H. and H.L. All authors have read and agreed to the published version of the manuscript.

**Funding:** This work was supported in part by the National Natural Science Foundation of China under Grant 62203163; and in part by the Scientific Research Foundation of Hunan Provincial Department of Education under Grant 21B0661.

**Institutional Review Board Statement:** Not applicable.

**Informed Consent Statement:** Not applicable.

**Data Availability Statement:** There are no new data is created.

**Conflicts of Interest:** The authors declare no conflict of interest.

## References

1. Zitová, B.; Flusser, J. Image registration methods: A survey. *Image Vis. Comput.* **2003**, *21*, 977–1000. [CrossRef]
2. Simone, G.; Farina, A.; Morabito, F. Image fusion techniques for remote sensing applications. *Inf. Fusion* **2002**, *3*, 3–15. [CrossRef]
3. Kouyama, T.; Kanemura, A.; Kato, S.; Imamoglu, N.; Fukuhara, T.; Nakamura, R. Satellite Attitude Determination and Map Projection Based on Robust Image Matching. *Remote Sens.* **2017**, *9*, 90. [CrossRef]
4. Dave, C.P.; Joshi, R.; Srivastava, S.S. A Survey on Geometric Correction of Satellite Imagery. *Int. J. Comput. Appl.* **2015**, *116*, 24–27.
5. Fan, J.; Wu, Y.; Li, M.; Liang, W.; Zhang, Q. SAR image registration using multiscale image patch features with sparse representation. *IEEE J. Sel. Top. Appl. Earth Obs. Remote Sens.* **2016**, *10*, 1483–1493. [CrossRef]
6. Yi, Z.; Cao, Z.; Yang, X. Multi-spectral remote image registration based on SIFT. *Electron. Lett.* **2008**, *44*, 107–108. [CrossRef]
7. Hossain, M.; Lv, G.; Teng, S.; Lu, G.; Lackmann, M. Improved symmetric-sift for multi-modal image registration. In Proceedings of the International Conference on Digital Image Computing Techniques and Applications (DICTA), Noosa, QLD, Australia, 6–8 December 2011; pp. 197–202.
8. Wang, L.; Niu, Z.; Wu, C. A robust multisource image automatic registration system based on the SIFT descriptor. *Int. J. Remote Sens.* **2012**, *33*, 3850–3869. [CrossRef]

9. Ye, Y.; Shen, L. HOPC: A novel similarity metric based on geometric structural properties for multi-modal remote sensing image matching. In Proceedings of the 23rd ISPRS Congress, Prague, Czech Republic, 16–19 July 2016; pp. 9–16.
10. Ye, Y.; Shan, I.; Bruzzone, L.; Shen, L. Robust registration of multimodal remote sensing images based on structural similarity. *IEEE Trans. Geosci. Remote Sens.* **2017**, *55*, 2941–2958. [CrossRef]
11. Korman, S.; Reichman, D.; Tsur, G.; Avidan, S. Fast-match: Fast affine template matching. *Int. J. Comput. Vis.* **2017**, *121*, 2331–2338. [CrossRef]
12. Lowe, D.G. Distinctive image features from scale-invariant keypoints. *Int. J. Comput. Vis.* **2004**, *60*, 91–110. [CrossRef]
13. Sedaghat, A.; Ebadi, H. Distinctive order based self-similarity descriptor for multi-sensor remote sensing image matching. *ISPRS J. Photogramm. Remote Sens.* **2015**, *108*, 62–71. [CrossRef]
14. Ye, Y.; Shan, J. A local descriptor based registration method for multispectral remote sensing images with non-linear intensity differences. *ISPRS J. Photogramm. Remote Sens.* **2014**, *90*, 83–95. [CrossRef]
15. Kelman, A.; Sofka, M.; Stewart, C. Keypoint descriptors for matching across multiple image modalities and non-linear intensity variations. In Proceedings of the IEEE Conference on Computer Vision and Pattern Recognition, Minneapolis, MN, USA, 17–22 June 2007; pp. 3257–3264.
16. Ye, Y.; Shan, J.; Hao, S.; Bruzzone, L.; Qin, Y. A local phase based invariant feature for remote sensing image matching. *ISPRS J. Photogramm. Remote Sens.* **2018**, *142*, 205–221. [CrossRef]
17. Ma, J.; Zhao, J.; Jiang, J.; Zhou, H.; Guo, X. Locality preserving matching. *Int. J. Comput. Vis.* **2019**, *127*, 12–31. [CrossRef]
18. Ma, J.; Zhou, H.; Zhao, J.; Gao, Y.; Jiang, J.; Tian, J. Robust feature matching for remote sensing image registration via locally linear transforming. *IEEE Trans. Geosci. Remote Sens.* **2015**, *53*, 6469–6481. [CrossRef]
19. Pluim, J.P.; Maintz, J.A.; Viergever, M.A. Image registration by maximization of combined mutual information and gradient information. *IEEE Trans. Med. Imaging* **2000**, *19*, 809–814. [CrossRef]
20. Kagarlitsky, S.; Moses, Y.; Helor, Y. Piecewise-consistent color mappings of images acquired under various conditions. In Proceedings of the IEEE International Conference on Computer Vision, Kyoto, Japan, 29 September–2 October 2009; pp. 2311–2318.
21. Hel-Or, Y.; Hel-Or, H.; David, E. Matching by tone mapping: Photometric invariant template matching. *IEEE Trans. Pattern Anal. Mach. Intell.* **2013**, *36*, 317–330. [CrossRef]
22. Ma, J.; Jiang, X.; Fan, A.; Jiang, J.; Yan, J. Image matching from handcrafted to deep features: A survey. *Int. J. Comput. Vis.* **2021**, *12*, 23–79. [CrossRef]
23. Han, X.; Leung, T.; Jia, Y.; Sukthankar, R.; Berg, A.C. Matchnet: Unifying feature and metric learning for patch-based matching. In Proceedings of the IEEE Conference on Computer Vision and Pattern Recognition, Boston, MA, USA, 7–12 June 2015.
24. Rocco, I.; Arandjelovic, R.; Sivic, J. Convolutional neural network architecture for geometric matching. In Proceedings of the IEEE Conference on Computer Vision and Pattern Recognition, Honolulu, HI, USA, 21–26 July 2017.
25. Park, J.H.; Nam, W.J.; Lee, S.W. A two-stream symmetric network with bidirectional ensemble for aerial image matching. *Remote Sens.* **2020**, *12*, 465. [CrossRef]
26. Oh, M.S.; Lee, Y.J.; Lee, S.W. Precise Aerial Image Matching based on Deep Homography Estimation. *arXiv* **2021**, arXiv:2107.08768.
27. Russakovsky, O.; Deng, J.; Su, H.; Krause, J.; Satheesh, S.; Ma, S.; Huang, Z.; Karpathy, A.; Khosla, A.; Bernstein, M.; et al. ImageNet Large Scale Visual Recognition Challenge. *Int. J. Comput. Vis. (IJCV)* **2015**, *115*, 211–252. [CrossRef]
28. Viola, P.; Wells, W. Alignment by maximization of mutual information. In Proceedings of the IEEE International Conference on Computer Vision, Cambridge, MA, USA, 20–23 June 1995; pp. 16–23.
29. Buntinga, P.; Labrosseb, F.; Lucasa, R. A multi-resolution area-based technique for automatic multi-modal image registration. *Image Vis. Comput.* **2010**, *28*, 1203–1219. [CrossRef]
30. Inglada, J.; Giros, A. On the possibility of automatic multisensor image registration. *IEEE Trans. Geosci. Remote Sens.* **2004**, *42*, 2104–2120. [CrossRef]
31. Dalal, N.; Triggs, B. Histograms of oriented gradients for human detection. In Proceedings of the IEEE Conference on Computer Vision and Pattern Recognition 2005, San Diego, CA, USA, 20–25 June 2005; pp. 886–893.
32. Sibiryakov, A. Fast and high-performance template matching method. In Proceedings of the IEEE Conference on Computer Vision and Pattern Recognition, Colorado Springs, CO, USA, 20–25 June 2011; pp. 1417–1424.
33. Shechtman, E.; Irani, M. Matching local self-similarities across images and videos. In Proceedings of the IEEE Conference on Computer Vision and Pattern Recognition, Minneapolis, MN, USA, 17–22 June 2007; pp. 1744–1752.
34. Kim, S.; Min, D.; Ham, B.; Ryu, S.; Do, M.N.; Sohn, K. DASC: Dense adaptive self-correlation descriptor for multi-modal and multi-spectral correspondence. In Proceedings of the IEEE Conference on Computer Vision and Pattern Recognition, Boston, MA, USA, 7–12 June 2015.
35. Ye, Y.; Shan, I.; Bruzzone, L. Fast and robust matching for multimodal remote sensing image registration. *IEEE Trans. Geosci. Remote Sens.* **2019**, *57*, 9059–9070. [CrossRef]
36. Kim, S.; Min, D.; Lin, S.; Sohn, K. Deep self-correlation descriptor for dense cross-modal correspondence. In *Computer Vision–ECCV 2016: 14th European Conference, Amsterdam, The Netherlands, 11–14 October 2016, Proceedings, Part VIII 14*; Springer: Berlin/Heidelberg, Germany, 2016; pp. 679–695.
37. Kovesi, P. Image features from phase congruency. *J. Comput. Vis. Res.* **1999**, *1*, 1–26.

38. Geometric Corrections in Remote Sensed Image [OL]. Available online: https://geolearn.in/geometric-corrections-in-remote-sensing-images/ (accessed on 25 May 2023).
39. Alexandris, N.; Gupta, S.; Koutsias, N. Remote sensing of burned areas via PCA, Part 1; centering, scaling and EVD vs. SVD. *Open Geospat. Data Softw. Stand.* **2017**, *2*, 17. [CrossRef]

applied sciences

MDPI

Article

# Multi-Scale Dynamic Analysis of the Russian–Ukrainian Conflict from the Perspective of Night-Time Lights

**Le-Lin Li [1,*], Peng Liang [2], San Jiang [3] and Ze-Qiang Chen [4]**

1 Hunan Provincial Key Laboratory of Geo-Information Engineering in Surveying, Mapping and Remote Sensing, Hunan University of Science and Technology, Xiangtan 411201, China
2 School of Earth Science and Space Information Engineering, Hunan University of Science and Technology, Xiangtan 411201, China
3 School of Computer Science, China University of Geosciences, Wuhan 430074, China
4 Collaborative Innovation Centre of Geospatial Technology, Wuhan 430074, China
* Correspondence: lilelin@hnust.edu.cn

**Abstract:** Under the influence of various forces, the conflict between Russia and Ukraine is violent and changeable. The obtaining of battlefield data by conventional means is difficult but necessary in order to ensure security, reliability, and comprehensiveness. The use of remote sensing technology can make up for the deficiencies of conventional methods. By using night-time light data, the total number of night-time lights in the built-up areas of Ukrainian cities within 36 days of the outbreak of the Russian–Ukrainian conflict is compiled in this paper. Furthermore, the dynamic changes in night-time light at the national, regional, and urban scales are analyzed by using the night-time light ratio index and the dynamic degree model combined with the time-series night-time light data. The results show that (1) after the outbreak of the war, more than 60% of the night-time lights in Ukrainian cities were lost. In terms of the night-time light recovery speed, the night-time lights in the pro-Russian areas recovered significantly faster, followed by Russian-controlled areas, and the recovery speed in areas of conflict was the lowest. (2) Decision-making by belligerents affects non-combatant activities and thus corresponds to light at night. The loss of night-time light will be reduced if military operations are reduced and mitigated if humanitarian operations are increased. (3) The changes in night-time light reflect the changes in the conflict situation well. When the conflict between Russia and Ukraine intensifies, the overall change of night-time light shows a downward trend. In this context, night-time light data can be used as an effective source to deduce and predict battlefield situations.

**Keywords:** night-time lights; NPP VIIRS; Russian–Ukrainian conflict; multiscale analysis; built-up area

## 1. Introduction

Since the official outbreak of the Russian–Ukrainian conflict on 24 February 2022, local armed conflicts and riots have had an irreversible impact on the economic development, infrastructure, and population movements in the war zone in the short term. There is a strong link between war and refugees, and wars between nations can cause internal displacement as refugees flee combat zones [1]. On March 11, the United Nations refugee agency stated that 2.5 million people had fled Ukraine due to the ongoing conflict, which represents the fastest increase in refugees over a period of time since World War II. Meanwhile, the U.N. refugee agency predicts that as the war continues, the number of Ukrainian refugees may exceed 4 million.

Given the reality of war, there are significant limitations concerning the acquisition of spatio-temporal information in conflict areas, such as the personal safety of journalists, traffic congestion, and information transparency. Due to the difficulty of conducting a comprehensive field survey in a combat zone, the acquisition of information is completely dependent on official announcements and witness reports, which can result in insufficient data richness and doubtful authenticity. In contrast to this, the acquisition of ground data through remote sensing satellites stands out for its low cost and high-efficiency characteristics, and remote sensing technology has become an indispensable and important part of many decision-making processes [2,3].

Night-time light data can reflect human activities very well, and in good weather conditions, the dazzling light emitted in human-inhabited areas can be observed from space [4]. It is well known that scale is an important factor in remote sensing image analysis [5]. Due to the differences in scale of geographic information, even from the same type of data, the research results obtained by using single-scale data in a specific area cannot be effectively generalized to other study areas. To address this issue, in this research, the 36-day VIIRS daily data from 24 February 2022, to 31 March 2022 is used as the data for the conflict period, and the monthly VIIRS data in December 2021 is used as the stable pre-war data. In terms of dimensions, three scales are combined, namely at the national level, state level, and city level, in order to analyze the time-series changes in night-time lights.

## 2. Literature Reveiew

The remote sensing technology of night-time light which is an indirect manifestation of human activities is widely used for the study of population dynamic distribution [6–9], urban scale evolution [10–19], socioeconomic research [20,21], energy consumption estimation [22–24], and various other fields.

Today, there are two main types of night light data used worldwide, namely, the DMSP/OLS Night-time light data and the NPP-VIIRS Night-time light data. Night-time light data has been widely used in studies reflecting human activities and can be used as an ideal data source for obtaining spatio-temporal information in conflict areas. The level of violent conflict and quality of life in Baghdad, Iraq, was studied by using DMSP/OLS data [25]. The DMSP/OLS data has been used to estimate Sri Lanka's GDP and electricity consumption, as affected by the civil war [26]. The impact of war in Russia and the Caucasus, e.g., in countries such as Georgia, has also been studied and monitored with DMSP/OLS data, and it has been verified that the use of DMSP/OLS data can reflect long-term burning of fires and large-scale population movements [27]. The application value of night-time light data, such as DMSP/OLS images in the research of humanitarian crises, for instance, in Syria, has been verified, and the causes of large losses of urban lights can be speculated upon [28].

Compared with DMSP/OLS data, the NPP/VIIRS data has more advantages in terms of spatial resolution, and studies have shown that the observation accuracy of human activities based on NPP/VIIRS data is higher than the former [29,30]. The feasibility of using NPP-VIIRS data in three areas of human activity that have a large impact on society: light pollution, gas flaring, and armed conflict is verified by analyzing the VIIRS data for a long time [31]. Estimates of power shortages and their affected populations with VNP46A2 data during the Ukrainian–Russian conflict have been made [32]. Like other satellites data, NPP-VIIRS has the problem of cloud overcover [33]. The application performance of NPP-VIIRS can be improved by combining it with other multi-source remote sensing data. VIIRS was combined with multi-source data to estimate the total area of fires caused by war in the first month after the outbreak of the conflict between Russia and Ukraine [34].

Differences in spatial and radiative properties between DMSP-OLS and NPP-VIIRS make it difficult to perform time-consistent analyses using both datasets. A cross sensor calibration model of DMSP-OLS and NPP VIIRS noctilucent images has been established [35]. The NPP-VIIRS data was used to simulate DMSP-OLS data to study the urbanization process in Southeast Asia [36]. Simulated DMSP/OLS images with NPP/VIIRS images were also used to assess the dynamics of urban lights in Syria during the civil war [37]. Furthermore, by correcting NPP-VIIRS night-time light using DMSP-OLS, it has been shown that war has darkened Yemen, providing support for international humanitarian aid organizations [38].

In this study, stable night-time light data from the built-up areas of Ukrainian cities are extracted through statistical methods, and data from each city is counted as a sub-area in order to obtain the sum of the night-time light intensity of each city. The night-time light ratio index is obtained by comparing the sum of the night-time light intensity during the war with the sum of the stable night-time lights before the war, so as to reflect the spatial and temporal changes in the relative night-time light intensity affected by the conflict [39]. The NLRI has different meanings at different scales. When the scale is large, NLRI can reflect the trend of human activities in an entire country. As the scale shrinks, the changes to the NLRI are more sensitive, and regional events show continuous directional changes. Such directional changes can be used to reflect the trends of a conflict. This study mainly considers the night-time light changes in areas with a significant human presence, as well as the light emissions from road lamps and traffic flows. It does not focus on rural lighting and vegetation fire but instead conducts a multi-scale analysis of night-time light in areas of major human activity, aiming to explore the value of daily night-time light data in the Russia–Ukraine conflict.

## 3. Study Area and Data

### 3.1. Study Area

Ukraine is located in the east of Europe, bordered by Russia to the east; Belarus to the north; Slovakia, Poland, and Hungary to the west; and Moldova and Romania to the south, where the Sea of Azov and the Black Sea are also located. The geographical location of Ukraine is shown in Figure 1. The geographical location of Ukraine is very important, as it lies at a geopolitical intersection between the European Union and the CIS, and especially given its proximity and historical relationship with Russia. Ukraine is extremely rich in natural resources, with its borders encompassing two-fifths of the world's black soil area and more than 70 types of mineral resources [40].

**Figure 1.** Map of Ukraine's geographic location.

According to the civilian casualty figures updated by the United Nations Office of the High Commissioner for Human Rights on 26 March, since the full-scale outbreak of the conflict between Russia and Ukraine, there have been 2788 civilian casualties in Ukraine, of which 1081 persons have been killed and 1707 persons have been injured. Under the combined effect of the Russia–Ukraine conflict and sanctions against Russia, the global food supply, microchip manufacturing materials, and energy prices have all been affected to varying degrees.

### 3.2. Data Sets and Preprocessing

The Suomi National Polar Partnership satellite SNPP/VIIRS global night-time light DNB data is used in this study as the daily data, and the images were obtained from the official website of the National Oceanic and Atmospheric Administration (https://ngdc.noaa.gov/(accessed on 2 April 2022)). The imaging period of the daily night-time light images was selected from the conflict eruption between 24 February to 1 March, a total of 36 days. The monthly night-time light imaging period was selected as December 2021. The night-time light data for 2021, as released by the Payne Institute for Public Policy (https://www.mines.edu/), was used to calculate the built-up area. The night-time light imaging area is 75N/060W, using the coordinate system Select WGS_1984_UTM_Zone_50N. The administrative boundary data were downloaded from the Global Administrative Division Database (https://gadm.org/). The road network data was obtained from the OpenStreetMap (https://www.openstreetmap.org/ (accessed on 27 November 2022)).

Daily NPP VIIRS DNB data has been applied to assess the impact of natural or anthropogenic phenomena on human social activities [41]. In the process of constructing multi-time-series VIIRS images, it is necessary to ensure the continuity of image time and space [42]. Within the time-frame studied in this paper, on some days the NPP VIIRS data was partially missing, such as 19 March and 31 March. According to the idea of averaging, the missing night-time light data on a given day is obtained by interpolating the night-time lights of the nearest two days, the formula for which is as follows:

$$2\,\boldsymbol{D_i^t} = \boldsymbol{D_i^{t+1}} + \boldsymbol{D_i^{t-1}} \tag{1}$$

In Formula (1), $\boldsymbol{D_i^t}$, $\boldsymbol{D_i^{t+1}}$ and $\boldsymbol{D_i^{t-1}}$ represent the radiance value of the $\boldsymbol{i}$-th pixel on days $\boldsymbol{t}$, $\boldsymbol{(t+1)}$, and $\boldsymbol{(t-1)}$, respectively.

To infer the changing trends of night-time light in a region affected by war, the minimum night-time light value of the research region in the study period should be extracted and compared to the minimum value of night-time light on a day prior to the outbreak of war. If the minimum value during a conflict period is smaller than the minimum value before the conflict period, this means that the area is affected by the conflict and has suffered from the loss of night-time lights.

## 4. Methods

### 4.1. Multiscale Analysis Frame of the Night-Time Light Data

The range of urban built-up areas is extracted using statistical methods, and the thresholds of light intensity at night are continuously iterated in order to compare the areas of extracted light with the actual statistical areas under each threshold, until the two areas are the closest, and then the threshold is determined.

Because NPP/VIIRS data are sensitive to night-time light brightness, night-time light not only includes artificial light but also other types of night-time lights, such as forest fires, gas flares, and volcanic eruptions, as well as background noise. There are negative and extremely positive values in the image data, and outliers need to be handled in the image preprocessing stage. In this paper, the built-up area mask is used to remove abnormal lighting outside cities, and thus the night-time light image within the built-up area is de-noised and the data pre-processing is realized using an efficient method. Pixels with negative values are replaced with zero, and pixels with large positive values are replaced with the maximum value of night-time lights in major cities in Ukraine. In the following order, the corrected night-time light images, built-up area masks, and level 1 administrative boundary data are used to calculate the sum of night-time lights in built-up areas. At various scales, i.e., national scale, region scale, city scale, and road scale, the total level of night-time lights is extracted and a multi-scale dynamic analysis of the situation in the Russia–Ukraine conflict is carried out. The technical method is shown in Figure 2.

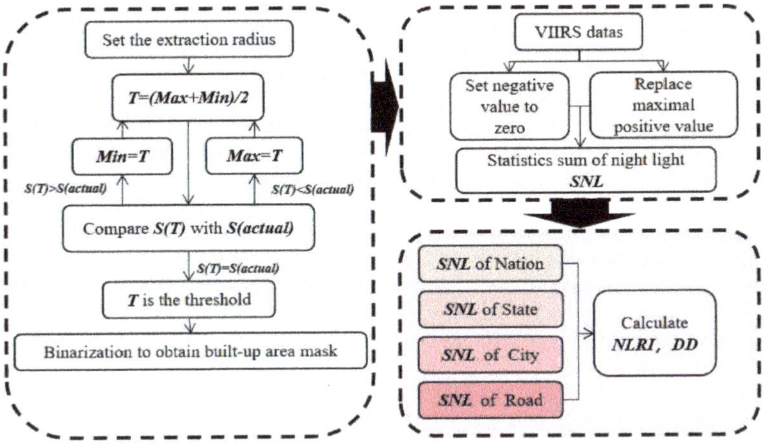


**Figure 2.** Conceptual diagram of the proposed multi-scale analysis frame.

### 4.2. City Mask Extraction

By extracting urban built-up areas from night-time light data, the lack of spatio-temporal characteristics caused by using statistical data alone in urban evolution studies can be solved [43]. In this paper, the method of obtaining a built-up area mask is divided into the following three phases:

(1) Taking the city of Kiev as an example, the center of the city's urban area was first found in Google Earth, and then a circle around the city center that contained the radial built-up area of the city was drawn, thus establishing Kiev's built-up area.

(2) By changing the maximum or minimum value, the night-time light data threshold is constantly constrained, and the area of each night-time light concentration is counted and compared with the statistical areas of urban land. When the areas of the night-time light concentrations are larger than the statistical areas of urban land, it indicates that the threshold is too small. In this case, the threshold is replaced by the original minimum value, so that the estimated area is close to the real area; alternatively, the maximum value is replaced for iteration.

(3) When the threshold is determined, the threshold is used to binarize the stable night-time light data, assigning a value of 1 for the built-up areas and 0 for the non-built-up areas, and thus, a binarized urban mask grid is obtained.

### 4.3. Night-Time Light Ratio Index

Generally speaking, the intensity of human activity in a certain area can be reflected by using the total night-time light index of the area in question. In this paper, the time-series night-time light data are used to analyze the light dynamics in Ukrainian cities. For the built-up areas of the Ukrainian regions, the total amount of night-time light, SNL, is calculated using the following formula:

$$SNL = \sum_{i=1}^{n} x_i d_i \quad (2)$$

In Formula (2), *SNL* represents the total light index of the city at night; n is the total number of pixels in the city region; $x_i$ represents the value of the i-th pixel in the built-up area and non-built-up area—in the built-up area, $x_i = 1$, and outside the built-up area (that is, in the non-built-up area) $x_i = 0$; and $d_i$ represents the brightness value of the ***i-th*** pixel.

As Li [44] noted, the night-time light ratio index can indicate the night-time light dynamics of a city, and this research also utilizes this index to represent the relative change of night-time light intensity. The transformation formula is:

$$NLRI_i = \frac{S_i}{S_k} \tag{3}$$

In Formula (3), $NLRI_i$ represents the night-time light ratio index on the ***i-th*** day; $S_i$ represents the total urban lighting index on the ***i-th*** day during the war; $S_k$ represents the total urban lighting data from before the war.

*4.4. Night-Time Light Dynamics*

The dynamic model is used to reflect the changing range and speed of specific ground object information in time units in the study area [45,46], which can effectively display the changing pattern of night-time lights over time. In this paper, the dynamic degree model is used to analyze the night-time light change rate.

$$DD = \frac{d_{ij} - d_{ii}}{d_{ii}} \times \frac{1}{T} \times 100\% \tag{4}$$

In Formula (4), $DD$ represents the change rate of night-time lights in a specific period, $d_{ii}$ and $d_{ij}$ represent the sum of the intensity of night-time light changes in a certain area at the beginning and end of the research period, respectively, and $T$ is the length of the time interval.

## 5. Results

*5.1. Dynamic Analysis of Night-Time Light Changes at the National Scale*

The urban built-up area mask is extracted using the statistical method. When the threshold is equal to 10.02, the built-up area of Kiev is 393.89 square kilometers, which is close to the statistical area of 394 square kilometers. This threshold is set as the overall threshold in order to extract data for the whole country. The built-up area is shown in Figure 3.

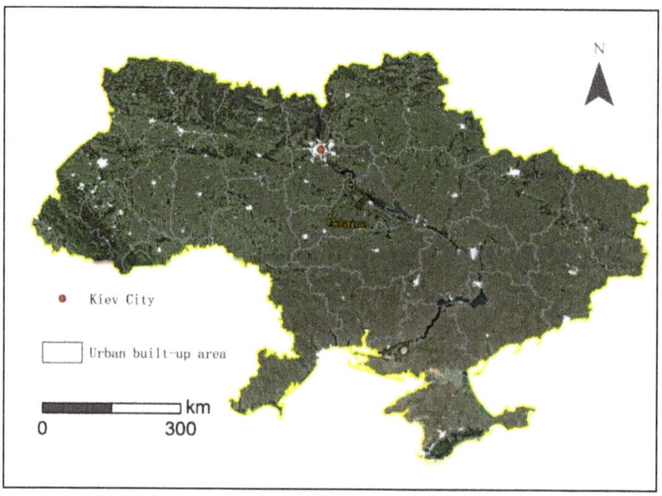


**Figure 3.** Ukrainian urban built-up areas.

The built-up area data were binarized, where the city value was set to 1 and the non-city value was set to 0. The data were first multiplied with the corrected night-time light data to retain the night-time light intensity of the built-up area, then the unstable night-time light was removed and the city night-time light images were obtained. Following this, the administrative boundary division was then used to analyze the above image and obtain the total amount of night-time light in Ukraine.

More than 60 percent of the country's night-time light was lost after the conflict broke out. We analyzed the dynamic change of night-time light by combining the military clashes and strategies between Russia and Ukraine. By screening the data published by online media, key information affecting the development of the conflict, such as urban control, humanitarian aid, and other key data, can be obtained, as shown in Table 1. Several cities are taken as examples to determine the ways in which different events of the conflict have affected night-time light.

**Table 1.** Ukrainian cities in military conflict situations. There are three types of cities analyzed here, namely cities in the RC (Russian-controlled area), cities in the UC (Ukrainian-controlled area), and cities in the B (belligerence/conflict area). The plus sign indicates whether a specific event has promoted night-time light restoration, and a minus sign indicates no light restoration prompted by an event.

| State | Type | Outbreak of Conflict | The First Negotiation | The Second Negotiation | Humanitarian Corridor | The Third Negotiation |
|---|---|---|---|---|---|---|
| | Date | 2.24 — | 0228 | 0303 | 3.5 — | 3.7 |
| | Impact | - | + | + | + | + |
| Cherkasy | UC | – | – | – | – | – |
| Chernivtsi | B | IN | IN | IN | IN | IN |
| Dnipropetrovs'k | B | IN | IN | IN | IN | IN |
| Donets'k | RC | IN | IN | – | IN | – |
| Kharkiv | B | IN | IN | IN | IN | IN |
| Kiev City | B | IN | IN | IN | IN | IN |
| Zaporizhzhya | RC | IN | IN | IN | IN | – |
| **State** | **Type** | **Multinational Intervention** | **The Fourth Negotiation** | **Conflict Escalation** | **The Fifth Negotiation** | **Russian Troops Withdraw** |
| | Date | 3.14 — | 3.15 | 3.20 — | 3.29 | |
| | Impact | - | + | - | + | + |
| Cherkasy | UC | IN | – | – | – | – |
| Chernivtsi | B | IN | IN | IN | IN | – |
| Dnipropetrovs'k | B | – | IN | IN | IN | – |
| Donets'k | RC | – | – | – | – | – |
| Kharkiv | B | – | IN | IN | IN | – |
| Kiev City | B | IN | IN | IN | IN | IN |
| Zaporizhzhya | RC | – | – | – | – | – |

Figure 4 shows the time-dependent changes in the Ukrainian national night-time light ratio index. Different columns in the figure represent the five negotiations that took place within the research period of this paper, namely on 28 February, 3 March, 7 March, 15 March, and 29 March. Peace negotiations are closely related to military operations. During the talks, military operations often stagnate. The negotiation results often affect subsequent military operations, which in turn affect human activity. The intensity of night-time light should reflect these changes.

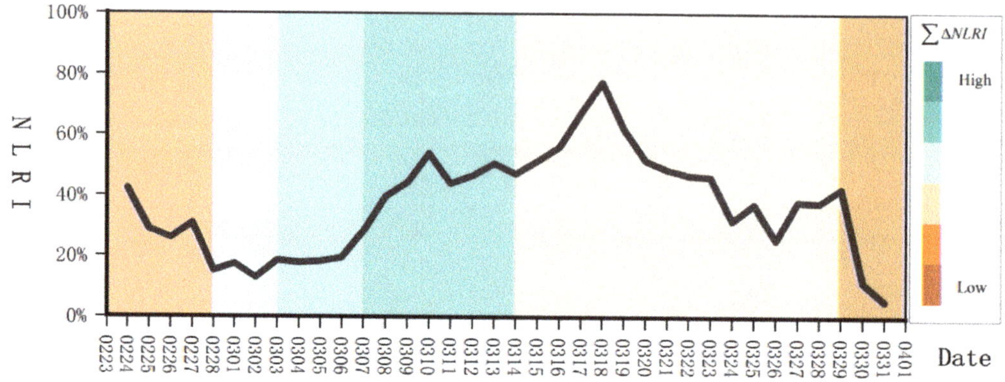


**Figure 4.** Ukraine's night-time light ratio index from 23 February to 1 April.

After the Russian–Ukrainian conflict broke out on 24 February, the intensity of night-time lights in Ukraine dropped sharply. On 26 February, the Russian side proposed negotiations with the Ukrainian side and halted relevant military operations. On 27 February, the Ukrainian side agreed to the talks. The index image shows a brief rise at this point. The first Russia–Ukraine negotiation did not yield results, but the two sides reached a consensus for further negotiations. Before the second Russia–Ukraine negotiations took place on 3 March, night-time lights did not change significantly.

The second Russia–Ukraine negotiations mainly focused on humanitarian relief and achieved obvious results. The two sides reached a consensus on the formation of a humanitarian corridor for civilians and announced a temporary ceasefire to ensure the evacuation of civilians. It can be seen from Figure 4 that after the second meeting, the night-time lights remained stable and showed an upward trend.

In the third meeting, Russia and Ukraine again reached a consensus on humanitarian issues, but the negotiations were not successful, and Russian military operations continued. During this period, night-time lights showed a "rising-falling-rising" trend.

The fourth negotiation was also unsuccessful, and after four rounds of negotiations, the military operations of Russia and Ukraine slowed down and briefly reached a stalemate, with the conflict situation initially showing signs of easing. At this point, the intensity of the night-time lights was greatly restored; however, with the re-escalation of the situation on 20 March, the Russian side again carried out a large number of military operations, and night-time lights continued to fall.

During the fifth negotiation, the two sides reached considerably more consensus regarding the security of Ukraine. The Russian side agreed to withdraw its troops from Kiev. The phenomenon of night-time light loss that occurred after the fifth negotiation is presumed to have been caused by the large-scale movement of the army and subsequent panic among civilians.

In conclusion, NTRL at the national level can depict how a war situation affects the entire nation. NTRL often increases during Russia and Ukraine's peace negotiations; but, when the war between the two countries worsened, the change was undone.

*5.2. Dynamic Analysis of Night-time Light Changes at the Regional Scale*

Using the ratio of the total amount of night-time light during the conflict and the total level of night-time light before the conflict, in turn, the daily night-time light ratio index of the Ukrainian regions is attained, with the date as the horizontal axis and the night-time light ratio index as the vertical axis on the presented line graphs; the vertical axis scale is adapted to each region. The night-time light ratio index is shown in Figure 5. When NLRI = 1, it means a return to the pre-war night-time light level.

Overall, almost all of the region's night-time light ratio indexes show a decreasing trend in the early stages of the war, then an increase in the middle period, and then decrease again in the later period. This research divides the period into three segments. The first lasted from 28 February to 18 March, and the average night-time light ratio index of each region is shown in Table 2. After the outbreak of the Russian–Ukrainian conflict, most regions lost more than 50% of their night-time light. Within five days of the outbreak of the conflict, from 24 February to 28 February, the states that responded quickly to the war showed a short "L-shaped" or sinking segment in the line chart, and most of these were in Eastern and Northern Ukraine, where Luhans'k Oblast lost 90% of its night-time light, Kharkiv lost 87% of its night-time light, Chernihiv Oblast lost 88% of its night-time light, and Zaporizhzhya Oblast lost 87% of its night-time light. At this time, these regions were in the conflict zone between Russia and Ukraine. The fighting during this period was very fierce, and the night-time lights reduced rapidly in these areas. On the other hand, the night-time light response speed of the regions located in central Ukraine was relatively slow overall, but due to the rapid advance of the Russian army, the night-time light loss rate of the central cities was still very fast. A transition period of night-time light loss in these areas can be seen, which is shown in the line chart to have a long "L-shaped" opening section. Of these regions, Kiev Oblast lost 74% of its night-time light, Kirovohrad Oblast lost 64% of its night-time light, and Vinnytsya Oblast lost 60% of its night-time light, as they were affected by the different geographical locations of the major cities in each region in regard to the outbreak of the conflict. The change in night-time lights is also varied, and the closer a region is to the combat zone, the more dramatic the reaction. Most of the regions with the "N-shaped" beginning are located on the western side of Ukraine. These regions, such as Volyn and L'viv, are relatively far away from the combat zone and have seen no large-scale advance of the Russian army.

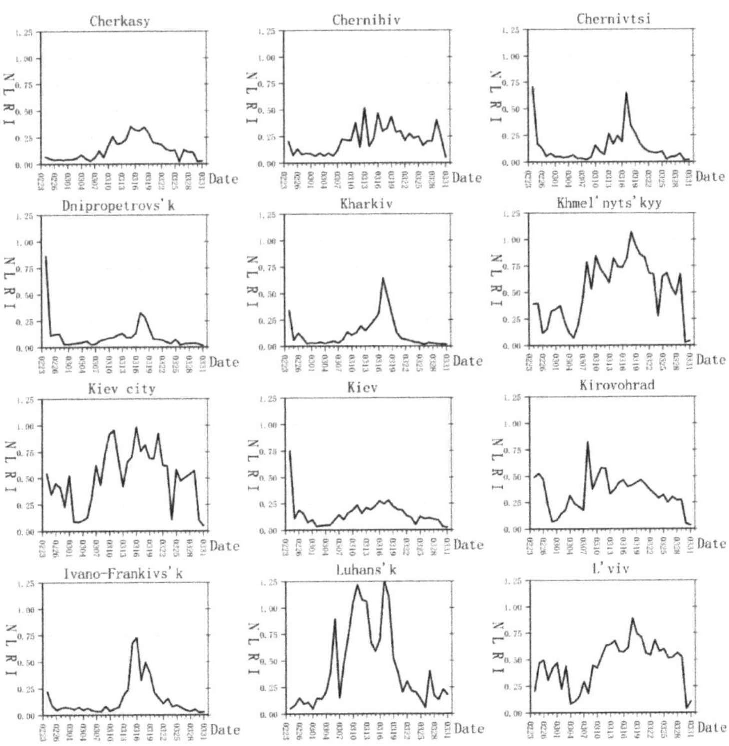


**Figure 5.** *Cont.*

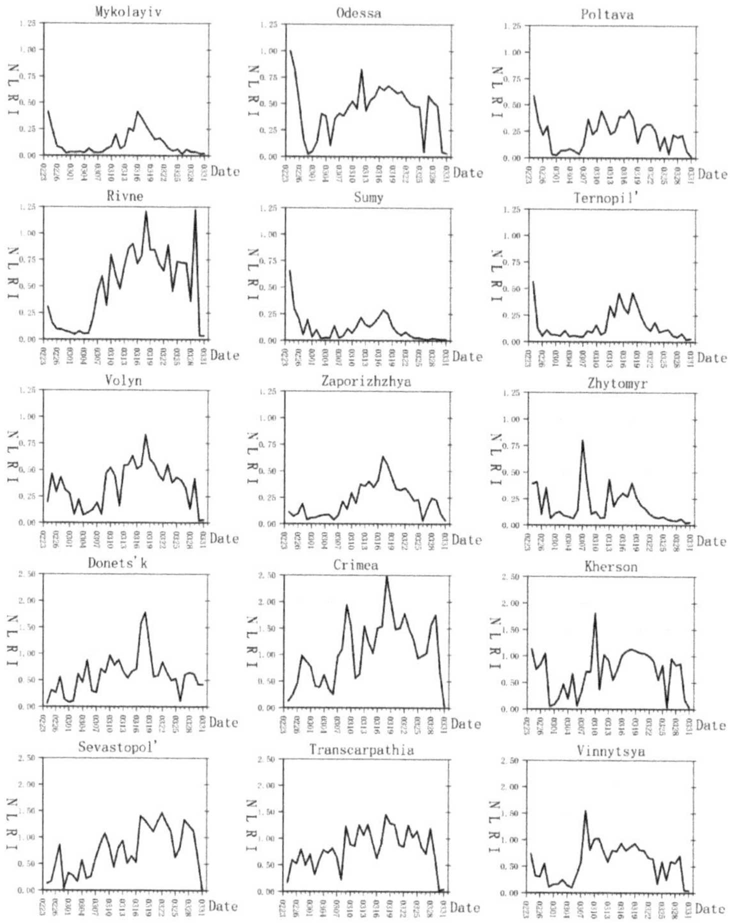


**Figure 5.** Schemes follow the same formatting.

From 1 March to 18 March, almost all the regions showed an upward trend in terms of the total amount of night-time light. The slowing of the battlefield situation, the repeated joint talks between Russia and Ukraine, and the active silence of the Russian side have somewhat stabilized civilian life and thus brought about the restoration of night-time lights. The states with pro-Russian power concentrations are highlighted by the rapid restoration of night-time light to pre-war night-time light levels. For instance, the Donets'k region returned to 87% of its pre-war night-time light levels on 5 March and the Luhans'k region achieved the same on 6 March, when it returned to 89% of its night-time light before the war. These states have seen fewer overt military–civilian conflicts, and civilians have made fewer attempts at fleeing. The loss of night-time lights was therefore quickly remedied. In regard to the Russian-controlled areas, except for Donets'k and Luhans'k, the Crimean Autonomous Region, Sevastopol' Port, and the Kherson Region have all returned to pre-war levels of light. The night-time light levels on the Crimean Peninsula have increased to more than twice the level of night-time lights before the war, which is presumed to be the influence of the military lights caused by the mass accumulation of troops by the Russian army in Crimea.

**Table 2.** The average luminous ratio of each region in the three time periods (2.24–2.28, 3.1–3.18, 3.19–3.31).

| Region | Average Night-Time Light Ratio Index | | | Region | Average Night-Time Light Ratio Index | | | Region | Average Night-Time Light Ratio Index | | |
|---|---|---|---|---|---|---|---|---|---|---|---|
| Cherkasy | 0.05 | 0.17 | 0.11 | Khmel'nyts'kyy | 0.28 | 0.59 | 0.50 | Rivne | 0.15 | 0.53 | 0.59 |
| Chernihiv | 0.12 | 0.22 | 0.23 | Kiev | 0.26 | 0.16 | 0.10 | Sevastopol' | 0.34 | 0.71 | 0.99 |
| Chernivtsi | 0.23 | 0.15 | 0.07 | Kiev City | 0.40 | 0.56 | 0.46 | Sumy | 0.28 | 0.12 | 0.03 |
| Crimea | 0.54 | 1.12 | 1.19 | Kirovohrad | 0.36 | 0.38 | 0.27 | Ternopil' | 0.18 | 0.18 | 0.09 |
| Dnipropetrovs'k | 0.25 | 0.10 | 0.04 | L'viv | 0.48 | 0.57 | 0.48 | Transcarpathia | 0.51 | 0.89 | 0.78 |
| Donets'k | 0.27 | 0.72 | 0.54 | Luhans'k | 0.10 | 0.64 | 0.20 | Vinnytsya | 0.40 | 0.67 | 0.46 |
| Ivano-Frankivs'k | 0.10 | 0.20 | 0.08 | Mykolayiv | 0.17 | 0.13 | 0.06 | Volyn | 0.44 | 0.47 | 0.32 |
| Kharkiv | 0.13 | 0.17 | 0.04 | Odessa | 0.50 | 0.46 | 0.39 | Zaporizhzhya | 0.10 | 0.26 | 0.20 |
| Kherson | 0.76 | 0.71 | 0.65 | Poltava | 0.29 | 0.23 | 0.17 | Zhytomyr | 0.26 | 0.22 | 0.07 |

The sporadic night-time light ratio in the Zaporizhzhya region showed a relatively stable rise but never returned to its pre-war level. There are many factors affecting this. Strategically, Zaporizhzhya, which connects the Russian-controlled Crimea and Donets'k, is presumably a key target for Russia. The humanitarian corridor from Mariupol to Zaporizhzhya via Berdyans'k restricts the military operations of Russia and Ukraine to a certain extent, making it difficult to carry out large-scale conflicts. Under the bidirectional effect, the change in the night-time light ratio in Zaporizhzhya Oblast is most likely to be reflective of the overall trend of night-time light changes in Ukraine.

During this period, the night-time light growth in other Ukrainian cities can be divided into two types one is a slow growth type, mainly concentrated in the regions within conflict zones. For example, in the Sumy, Kiev, Mykolayiv, and Kharkiv regions, the maximum light intensity at night has only been restored to 30–60% of the pre-war level. One possible reason for this is that the overall conflict situation briefly slowed, but smaller-scale battles still continued. The other is the abrupt growth type, occurring mainly in central and western Ukraine, far from the war zones, such as in the Khmel'nyts'kyy, Transcarpathia, and Vinnytsya regions, where changes to the night-time light ratio index are characterized by a return to pre-war levels and multiple peaks, with the continuous influx of refugees from the war zones, which has led to a continuous increase in night-time lights. From 19 March to 1 April, due to the re-escalation of the situation between Russia and Ukraine, the intensity of night-time light in Ukraine fell once again.

In conclusion, the level of a region's sensitivity to conflict can be indicated by its region scale NTRL. The already-conflicted territory and the one next door are more susceptible to war; in terms of civilian sentiments, the region with a higher proportion of pro-Russian inhabitants is less sensitive to conflict.

### *5.3. Dynamic Analysis of Night-Time Light Changes at the Urban Scale*

Using the five peace negotiation events, the time series of this study was divided into four periods, and urban built-up areas of more than two square kilometers were selected as the objects of this study of night-time light dynamics. A total of 176 small and larger cities were chosen. Taking the night-time light intensity after the first peace negotiation as the starting benchmark, the change in night-time light is visualized using a stacked graph, as shown in Figure 6. The orange segment represents the amount of night-time light lost in this period compared to the previous period, the yellow segment is the amount of night-time light in this period, the yellow segment plus the orange segment represents the night-time light in the previous period, and the orange segment is the level of reduction of night-time light. When night-time light recovers, the cyan segment represents the level of night-time light increased in this period compared to the previous period, the yellow segment depicts the night-time light of the previous period, the cyan segment plus the yellow segment represents the night-time light in the current period, and the ratio between the cyan segment and the yellow segment is the increasing rate of night-time lights. In the base map, at the time of writing, the colored area is controlled by Russia and the striped area is the conflict zone between Russia and Ukraine.

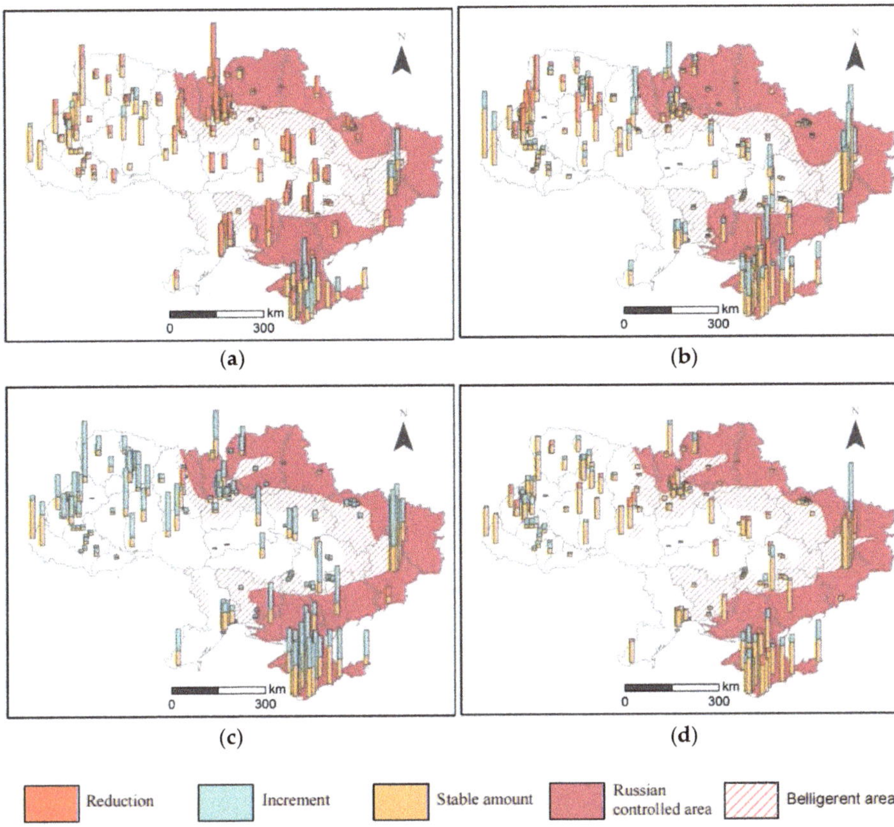


**Figure 6.** Dynamics of night-time light in Ukraine during the five negotiation events: (**a**) dynamics of night-time lighting during the first and second negotiations; (**b**) dynamics of night-time lighting during the second and third negotiation; (**c**) dynamics of night-time lighting during the third and fourth negotiations; and (**d**) dynamics of night-time lighting during the fourth and fifth negotiations.

In Figure 6a, when compared with the first period, it can be seen that the night-time light intensity of the southeastern region, such as the Crimean Peninsula and Donets'k, increased rapidly, with a growth rate of over one. Secondly, with the spread of the war to the interior, there was a large area of abrupt night-time light loss in cities in central Ukraine, and the loss of night-time light is generally greater than 60%. The Russian side advanced faster in the early stage of the conflict, and prior to the second peace negotiation, it controlled many cities in Donets'k, Luhans'k, and Crimea, where night-time light recovered quickly. As the Russian army entered the Kiev Oblast and surrounded the city of Kiev, the night-time light around the city of Kiev was greatly reduced. The rapid spread of war has led to the loss of night-time light in cities in central and western Ukraine.

In Figure 6b, due to the consensus reached by Russia and Ukraine on humanitarian issues, humanitarian passages to Mariupol and Vornovaha were opened and night-time lights along Zaporizhzhya and Donets'k correspondingly appeared. The results of the negotiations were not successful and Russian military operations continued, but the speed of their advancement was significantly slowed. The light levels of the central cities did not change much in terms of night-time light intensity, while the western Ukrainian cities experienced a significant night-time light intensity decline, such as in L'viv, during this period. The light loss exceeded 70%. Foreign players entered the war using the western side of Ukraine as an entrance.

Panic caused by civilians also affected the loss of night-time lights. In Figure 6c, most of the cities depicted show an increasing night-time light trend. While Russia and Ukraine have reached a certain degree of consensus on humanitarian issues, Russia has repeatedly announced that it has entered a state of silence and has stopped negotiating on military operations, thus making the overall war situation worse. There has been some recovery, and some public activities have resumed.

In Figure 6d, the direction of night-time light loss has been reversed, the speed of night-time light loss in the war zone has decreased, and the stable night-time light in the original Russian-controlled area has been lost. There may be two reasons. First, the long-term Russian offensive led to the military becoming fatigued, and the Ukrainian side tried to launch a counter-offensive in some areas, which caused a reverse loss of night-time light. Second, the Russian war strategy has been adjusted, and the large-scale army transfer has caused panic among the people. Based on this, one might also simply say that there may have been a slight change in the subsequent development of the situation in Russia and Ukraine. Analysis of the temporal and spatial changes of night-time light based on the dynamic degree can intuitively reflect the changes of night-time light, which is confirmed in this paper.

In summary, the NTRL change at the urban scale possesses directivity. In the control zone, the night-time light loss was less severe than it was in the combat zone. Humanitarian aid can stabilize the local combat situation, causing a modest increase in night-time light output.

### 5.4. Dynamic Analysis of Night-Time Light Changes at the Road Scale

Twelve regions affected by the earliest phases of the conflict were selected as examples in this study, including Chernihiv, Dnipropetrovs'k, Donets'k, Kharkiv, Kherson, Kiev City, Kiev, Luhans'k, Odesa, Sumy, and Volyn. Ukrainian road network data were obtained from OpenStreetMap to calculate the density of the road network by region. Specifically, Kiev has the highest road network density, with the majority of road types being highways. The road types seen in Dnipropetrovs'k and Chernihiv are the second most common, also known as trunk roads. Table 3 shows the dynamic changes in road lighting in 12 regions by using the five negotiations between Russia and Ukraine as time points.

In Figure 7, during the period from February 28 to March 3, the nighttime road lights in most states increased, with the largest increase of 185% seen in Donets'k, as compared to February 28, and the largest decreases in Kiev City, Kiev, Sumy and Chernihiv. After the first negotiation, the conflict between Russia and Ukraine eased to some extent. Donets'k, as the first prefecture controlled by Russia in the conflict, has a higher number of pro-Russia citizens than elsewhere in the country and, as the main channel connecting Russia and Ukraine for both civilians and the military, road lights were quickly restored. Between March 3 and March 7, the night-time lights on roads in all regions except Kiev City showed a downward trend; however, between March 7 and 15, the night-time lights on roads in all states except Kiev City showed an upward trend. Kiev was heavily affected by the early outbreak of the conflict on 24 February and lost a significant number of night-time lights, so it was easy to recover those in later stages. During the second half of March, as the war dragged on, the night-time lights on the roads declined. Between 15 March and 29 March, there was little change in Donets'k and Luhans'k, and the night-time lights on roads in other states showed a downward trend, except in the city of Kiev.

**Table 3.** The night-time road light dynamics of each region in the five times negotiation periods (28.02., 03.03., 03.07., 03.15. and 03.29.).

| Region | Density of Road Network (km/km²) | The Total of Road Lights at Night (nW/cm²/sr) | | | | | Road Night-Time Light Dynamics (%) | | | |
|---|---|---|---|---|---|---|---|---|---|---|
| | | 2.28 | 3.3 | 3.7 | 3.15 | 3.29 | 2.28–3.3 | 3.3–3.7 | 3.7–3.15 | 3.15–3.29 |
| Chernihiv | 0.105 | 47,242 | 31,410 | 20,109 | 38,625 | 27,851 | −33.51% | −35.98% | 92.08% | −27.89% |
| Dnipropetrovs'k | 0.110 | 26,337 | 32,984 | 21,979 | 31,017 | 14,347 | 25.24% | −33.37% | 41.12% | −53.74% |
| Donets'k | 0.099 | 24,064 | 68,628 | 41,640 | 44,522 | 46,742 | 185.19% | −39.32% | 6.92% | 4.99% |
| Kharkiv | 0.084 | 24,226 | 26,644 | 18,410 | 25,941 | 11,820 | 9.98% | −30.90% | 40.91% | −54.44% |
| Kherson | 0.058 | 10,703 | 14,014 | 10,985 | 16,622 | 11,615 | 30.93% | −21.61% | 51.31% | −30.13% |
| Kiev City | 0.559 | 15,016 | 12,711 | 21,671 | 12,809 | 28,899 | −15.35% | 70.49% | −40.89% | 78.69% |
| Kiev | 0.100 | 41,577 | 28,148 | 24,758 | 37,665 | 34,302 | −32.30% | −12.04% | 52.13% | −8.93% |
| Luhans'k | 0.073 | 20,656 | 31,233 | 25,029 | 31,625 | 31,346 | 51.20% | −19.86% | 26.36% | −0.88% |
| Odessa | 0.094 | 23,144 | 40,629 | 31,860 | 46,573 | 31,550 | 75.55% | −21.58% | 46.18% | −32.26% |
| Sumy | 0.096 | 32,106 | 18,703 | 16,827 | 22,837 | 9245 | −41.75% | −10.03% | 35.71% | −59.52% |
| Volyn | 0.094 | 17,099 | 20,682 | 12,947 | 26,851 | 16,801 | 20.95% | −37.40% | 107.40% | −37.43% |
| Zaporizhzhya | 0.078 | 14,477 | 19,163 | 14,620 | 21,866 | 16,973 | 32.37% | −23.71% | 49.56% | −22.37% |

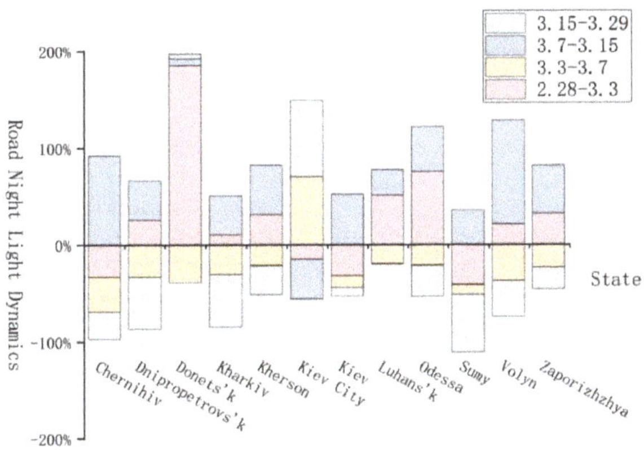


**Figure 7.** The night-time road light dynamics of twelve regions.

In conclusion, region and road lights at night are strongly associated. The production of night-time light from the roadways encircling the city increases as the region's night-time light increases.

## 6. Discussion and Conclusions

After the outbreak of the conflict between Russia and Ukraine, the night-time light in the whole country decreased sharply, and most provinces lost more than 60% of their night-time lights due to the war. As a good representation of human activities, the night light data reflects socio-economic development, population migration, and energy consumption. Possible reasons for the reduction of night lights in the war zone of the Russia-Ukraine conflict are as follows: (1) city light and power systems were damaged during the conflict; (2) the exodus of people caused a loss of urban population and thus a reduction of night-time light; and (3) the government's curfew policy affected the emission of light at night. The possible reasons for the large increase in night-time light intensity in some areas after the outbreak of the conflict are as follows: (1) civilian life in the regions with a larger number of pro-Russian individuals had an easier return to normal after being controlled by the Russian side; (2) military lighting caused by the deployment of troops and abnormal ignition points caused by the conflict; and (3) the refugee corridor for humanitarian aid inhibits surrounding military combat to a certain extent, which may lead to enhanced local night-time light.

In this paper, the night-time light ratio index and dynamics on different scales are calculated from the 36-day corrected Ukrainian NPP VIIRS data, and the quantitative analysis of the night-time light damage caused by the conflict outbreak is carried out over the time and space of the conflict. The following three conclusions are drawn:

(1) The unique perspective of night-time light data can clearly and intuitively monitor the impact of war on residents' social activities and can reflect various types of information, including the spread of conflict and refugee migration.

(2) Combined with time sequence analysis of the restoration and the loss of night-time light, dynamic laws can often echo the changes in actual conflict situations. Using night-time light data to analyze and predict the trends of the conflict can enable researchers to view the battlefield situation at a macro level.

(3) The observation of the battlefield situation from the perspective of night-time light can reduce or eliminate the casualty risks faced by collectors of field survey and ground survey data and compensates for the lack of statistics obtained through credible, reliable, and efficient means.

In this paper, the use of daily night-time light data to examine the conflict situation has a lag of at least one day, and it is impossible to monitor conflict changes caused by emergencies in real-time. In addition, due to the limitations of the resolution of night-time light data, it is difficult to identify areas with relatively small economies such as villages and towns. The total level of night-time light in the built-up areas is extremely small and, as the correlation with the evolution of the conflict situation is insufficient, it should be combined with other high-resolution remote sensing data. The application of remote sensing earth observation in emergencies not only breaks the restrictions of national boundaries and geographical conditions but also breaks the time boundary between the past and the present. The combination of remote sensing and emergency investigation has many advantages, such as breaking through the limitation of visual space-time and frequency spectrum, revealing the law of multi-scale, and maintaining the objectivity of news. A function model between the light intensity attenuation index and the severity of a war or the degree of economic contraction from the perspective of night-time light intensity response to the regional economy may be established through further research.

**Author Contributions:** Conceptualization, L.-L.L. and P.L.; methodology, L.-L.L. and P.L.; software, P.L.; validation, P.L., S.J. and Z.-Q.C.; formal analysis, S.J. and Z.-Q.C.; resources, L.-L.L. and P.L.; data curation, L.-L.L. and P.L.; writing—original draft preparation, L.-L.L. and P.L.; writing—review and editing, L.-L.L. and Z.-Q.C.; visualization, P.L.; supervision, L.-L.L. All authors have read and agreed to the published version of the manuscript.

**Funding:** This research was funded by the Key Research and Development Projects of Hunan Science and Technology Plan (Grant NO.2015GK3027).

**Institutional Review Board Statement:** Not applicable.

**Informed Consent Statement:** Not applicable.

**Data Availability Statement:** The SNPP/VIIRS global night-time light DNB data were obtained from the official website of the National Oceanic and Atmospheric Administration (https://ngdc.noaa.gov/ (accessed on 2 April 2022)).

**Acknowledgments:** The authors would like to thank the editor and the anonymous reviewers for their constructive comments on an earlier version of this paper.

**Conflicts of Interest:** The authors declare no conflict of interest.

## References

1. AbuZayd, K.; Sullivan, D.J.; Akram, S.M.; Roy, S. The Syrian Humanitarian Crisis: What Is to Be Done? *Middle East Policy* **2015**, *22*, 1–29. [CrossRef]
2. Li, X.C.; Zhou, Y.Y.; Zhao, M.; Zhao, X. A harmonized global nighttime light dataset 1992–2018. *Sci. Data* **2020**, *7*, 168. [CrossRef] [PubMed]
3. Yu, B.; Shi, K.; Hu, Y.; Huang, C.; Chen, Z.; Wu, J. Poverty Evaluation Using NPP-VIIRS Nighttime Light Composite Data at the County Level in China. *IEEE J. Sel. Top. Appl. Earth Obs. Remote Sens.* **2015**, *8*, 1217–1229. [CrossRef]

4. Li, X.; Chen, F.; Chen, X. Satellite-Observed Nighttime Light Variation as Evidence for Global Armed Conflicts. *IEEE J. Sel. Top. Appl. Earth Obs. Remote Sens.* **2013**, *6*, 2302–2315. [CrossRef]
5. Collin, G.H.; Cameron, L.A.; Debra, K.; Spencer, J.S. Multi-scale remote sensing sagebrush characterization with regression trees over Wyoming, USA: Laying a foundation for monitoring. *Int. J. Appl. Earth Obs. Geoinf.* **2012**, *14*, 233–244. [CrossRef]
6. Sutton, P.; Roberts, D.; Elvidge, C.; Baugh, K. Census from Heaven: An estimate of the global human population using night-time satellite imagery. *Int. J. Remote Sens.* **2001**, *22*, 3061–3076. [CrossRef]
7. Bharti, N.; Tatem, A.J.; Ferrari, M.J.; Grais, R.F.; Djibo, A.; Grenfell, B.T. Explaining Seasonal Fluctuations of Measles in Niger Using Nighttime Lights Imagery. *Science* **2011**, *334*, 1424–1427. [CrossRef]
8. Levin, N.; Zhang, Q.L. A global analysis of factors controlling VIIRS nighttime light levels from densely populated areas. *Remote Sens. Environ.* **2017**, *190*, 366–382. [CrossRef]
9. Bagan, H.; Yamagata, Y. Analysis of urban growth and estimating population density using satellite images of nighttime lights and land-use and population data. *GISci. Remote Sens.* **2015**, *52*, 765–780. [CrossRef]
10. Liu, Z.F.; He, C.Y.; Zhang, Q.F.; Huang, Q.; Yang, Y. Extracting the dynamics of urban expansion in China using DMSP-OLS nighttime light data from 1992 to 2008. *Landsc. Urban Plan.* **2012**, *106*, 62–72. [CrossRef]
11. Yi, K.; Tani, H.; Li, Q.; Zhang, J.; Guo, M.; Bao, Y.; Wang, X.; Li, J. Mapping and Evaluating the Urbanization Process in Northeast China Using DMSP/OLS Nighttime Light Data. *Sensors* **2014**, *14*, 3207–3226. [CrossRef] [PubMed]
12. Ma, T.; Zhou, Y.; Zhou, C.; Haynie, S.; Pei, T.; Xu, T. Night-time light derived estimation of spatio-temporal characteristics of urbanization dynamics using DMSP/OLS satellite data. *Remote Sens. Environ.* **2015**, *158*, 453–464. [CrossRef]
13. Ma, X.; Tong, X.; Liu, S.; Luo, X.; Xie, H.; Li, C. Optimized Sample Selection in SVM Classification by Combining with DMSP-OLS, Landsat NDVI and GlobeLand30 Products for Extracting Urban Built-Up Areas. *Remote Sens. Multidiscip. Digit. Publ. Inst.* **2017**, *9*, 236. [CrossRef]
14. Li, G.; Zhang, H.; Chen, S.; Qiu, J.; Wang, X. Assessing the impact of urban development on net primary productivity during 2000–2010 in Taihu Basin. *Environ. Earth Sci.* **2016**, *75*, 1266. [CrossRef]
15. Zhang, Q.; Seto, K.C. Mapping urbanization dynamics at regional and global scales using multi-temporal DMSP/OLS nighttime light data. *Remote Sens. Environ.* **2011**, *115*, 2320–2329. [CrossRef]
16. Li, X.; Zhao, L.; Li, D.; Xu, H. Mapping Urban Extent Using Luojia 1-01 Nighttime Light Imagery. *Sensors* **2018**, *18*, 3665. [CrossRef]
17. Liu, L.; Li, Z.; Fu, X.; Liu, X.; Li, Z.; Zheng, W. Impact of Power on Uneven Development: Evaluating Built-Up Area Changes in Chengdu Based on NPP-VIIRS Images (2015–2019). *Land* **2022**, *11*, 489. [CrossRef]
18. Levin, N.; Kark, S.; Crandall, D. Where have all the people gone? Enhancing global conservation using night lights and social media. *Ecol. Appl.* **2015**, *25*, 2153–2167. [CrossRef]
19. Jiang, S.; Wei, G.; Zhang, Z.; Wang, Y.; Xu, M.; Wang, Q.; Das, P.; Liu, B. Detecting the Dynamics of Urban Growth in Africa Using DMSP/OLS Nighttime Light Data. *Land* **2021**, *10*, 13. [CrossRef]
20. Elvidge, C.D.; Sutton, P.C.; Ghosh, T.; Tuttle, B.T.; Baugh, K.E.; Bhaduri, B.; Bright, E. A global poverty map derived from satellite data. *Comput. Geosci.* **2009**, *35*, 1652–1660. [CrossRef]
21. Shi, K.; Yu, B.; Huang, Y.; Hu, Y.; Yin, B.; Chen, Z.; Chen, L.; Wu, J. Evaluating the Ability of NPP-VIIRS Nighttime Light Data to Estimate the Gross Domestic Product and the Electric Power Consumption of China at Multiple Scales: A Comparison with DMSP-OLS Data. *Remote Sens.* **2014**, *6*, 1705–1724. [CrossRef]
22. Letu, H.; Hara, M.; Yagi, H.; Naoki, K.; Tana, G.; Nishio, F.; Shuhei, O. Estimating energy consumption from night-time DMPS/OLS imagery after correcting for saturation effects. *Int. J. Remote Sens.* **2010**, *31*, 4443–4458. [CrossRef]
23. Zhong, Y.; Lin, A.; Xiao, C.; Zhou, Z. Research on the Spatio-Temporal Dynamic Evolution Characteristics and Influencing Factors of Electrical Power Consumption in Three Urban Agglomerations of Yangtze River Economic Belt, China Based on DMSP/OLS Night Light Data. *Remote Sens.* **2021**, *13*, 1150. [CrossRef]
24. Elvidge, C.D.; Ziskin, D.; Baugh, K.E.; Tuttle, B.T.; Ghosh, T.; Pack, D.W.; Erwin, E.H.; Zhizhin, M. A Fifteen Year Record of Global Natural Gas Flaring Derived from Satellite Data. *Energies* **2009**, *2*, 595–622. [CrossRef]
25. Agnew, J.; Gillespie, T.; Gonzalez, J.; Min, B. Baghdad Nights: Evaluating the US Military 'Surge' Using Nighttime Light Signatures. *Environ. Plan. A* **2008**, *40*, 2285–2295. [CrossRef]
26. Pathmasiri, E.; Kim, M. Influence of intra annual calibration methods in changing the preciseness of the obtainable information from DMSP-OLS NLT images. *Int. J. Res. Publ.* **2018**, *4*, 1–15.
27. Witmer, F.D.W.; O'Loughlin, J. Detecting the Effects of Wars in the Caucasus Regions of Russia and Georgia Using Radiometrically Normalized DMSP-OLS Nighttime Lights Imagery. *GISci. Remote Sens.* **2011**, *48*, 478–500. [CrossRef]
28. Li, X.; Li, D. Can night-time light images play a role in evaluating the Syrian Crisis? *Int. J. Remote Sens.* **2014**, *35*, 6648–6661. [CrossRef]
29. Elvidge, C.D.; Baugh, K.E.; Zhizhin, M.; Hsu, F.C. Why VIIRS data are superior to DMSP for mapping nighttime lights. *Proc. Asia Pac. Adv. Netw.* **2013**, *35*, 62. [CrossRef]
30. Dou, Y.; Liu, Z.; He, C.; Yue, H. Urban Land Extraction Using VIIRS Nighttime Light Data: An Evaluation of Three Popular Methods. *Remote Sens.* **2017**, *9*, 175. [CrossRef]
31. Ajmar, A.; Arco, E.; Eusebio, A. The VIIRS Nighttime Lights average annual global dataset: Exploratory and brisk trend analysis on three different domains. In Proceedings of the IEEE 21st Mediterranean Electrotechnical Conference (MELECON), Palermo, Italy, 14–16 June 2022; pp. 454–459. [CrossRef]

32. Zheng, Z.; Wu, Z.; Cao, Z.; Zhang, Q.; Chen, Y.; Guo, G.; Yang, Z.; Guo, C.; Wang, X.; Marinello, F. Estimates of Power Shortages and Affected Populations during the Initial Period of the Ukrainian-Russian Conflict. *Remote Sens.* **2022**, *14*, 4793. [CrossRef]
33. Elvidge, C.D.; Zhizhin, M.; Ghosh, T.; Hsu, F.-C.; Taneja, J. Annual Time Series of Global VIIRS Nighttime Lights Derived from Monthly Averages: 2012 to 2019. *Remote Sens.* **2021**, *13*, 922. [CrossRef]
34. Nicolae, R.A. Using NASA's Fire Information for Resource Management System (FIRMS) to evaluate the impact of war in Ukraine on environment during the first month of conflict. In Proceedings of the 17th Present Environment and Sustainable Development, Iași, Romania, 3 June 2022; p. 62.
35. Zheng, Q.M.; Weng, Q.H.; Wang, K. Developing a new cross-sensor calibration model for DMSP-OLS and Suomi-NPP VIIRS night-light imageries. *ISPRS J. Photogramm. Remote Sens.* **2019**, *153*, 36–47. [CrossRef]
36. Hao, M.; Zhou, Y.Y.; Li, X.C.; Zhou, C.H.; Cheng, W.M.; Li, M.C.; Huang, K. Building a Series of Consistent Night-Time Light Data (1992–2018) in Southeast Asia by Integrating DMSP-OLS and NPP-VIIRS. *IEEE Trans. Geosci. Remote Sens.* **2019**, *58*, 1–14. [CrossRef]
37. Li, X.; Li, D.; Xu, H.; Wu, C. Intercalibration between DMSP/OLS and VIIRS night-time light images to evaluate city light dynamics of Syria's major human settlement during Syrian Civil War. *Int. J. Remote Sens.* **2017**, *38*, 1–18. [CrossRef]
38. Jiang, W.; He, G.; Long, T.; Liu, H. Ongoing Conflict Makes Yemen Dark: From the Perspective of Nighttime Light. *Remote Sens.* **2017**, *9*, 798. [CrossRef]
39. Yang, W.T.; Deng, M.; Tang, J.B.; Luo, L. Geographically weighted regression with the integration of machine learning for spatial prediction. *J. Geogr. Syst.* **2022**. [CrossRef]
40. Baumann, M.; Kuemmerle, T.; Elbakidze, M.; Ozdogan, M.; Radeloff, V.C.; Keuler, N.S.; Prishchepov, A.V.; Kruhlov, I.; Hostert, P. Patterns and drivers of post-socialist farmland abandonment in Western Ukraine. *Land Use Policy* **2011**, *28*, 552–562. [CrossRef]
41. Zhao, X.; Yu, B.; Liu, Y.; Yao, S.; Lian, T.; Chen, L.; Yang, C.; Chen, Z.; Wu, J. NPP-VIIRS DNB Daily Data in Natural Disaster Assessment: Evidence from Selected Case Studies. *Remote Sens.* **2018**, *10*, 1526. [CrossRef]
42. Li, X.; Liu, S.; Jendryke, M.; Li, D.; Wu, C. Night-Time Light Dynamics during the Iraqi Civil War. *Remote Sens.* **2018**, *10*, 858. [CrossRef]
43. Zhou, Y.; Smith, J.S.; Elvidge, D.C.; Zhao, K.; Thomson, A.; Imhoff, M. A cluster-based method to map urban area from DMSP/OLS nightlights. *Remote Sens. Environ.* **2014**, *147*, 173–185. [CrossRef]
44. Li, X.; Zhang, R.; Huang, C.; Li, D. Detecting 2014 Northern Iraq Insurgency using night-time light imagery. *Int. J. Remote Sens.* **2015**, *36*, 3446–3458. [CrossRef]
45. Li, Q.; Lu, L.; Weng, Q.; Xie, Y.; Guo, H. Monitoring Urban Dynamics in the Southeast U.S.A. Using Time-Series DMSP/OLS Nightlight Imagery. *Remote Sens.* **2016**, *8*, 578. [CrossRef]
46. Zhao, Z.; Cheng, G.; Wang, C.; Wang, S.; Wang, H. City Grade Classification Based on Connectivity Analysis by Luojia I Night-Time Light Images in Henan Province, China. *Remote Sens.* **2020**, *12*, 1705. [CrossRef]

applied sciences

MDPI

Article

# Using HJ-1 CCD and MODIS Fusion Data to Invert HJ-1 NBAR for Time Series Analysis, a Case Study in the Mountain Valley of North China

**Huaiyuan Li [1,2], Zhiyuan Han [1,2,*] and Heng Wang [1,2]**

[1] National Engineering Research Center of Port Hydraulic Construction Technology, Tianjin Research Institute for Water Transport Engineering, M.O.T., Tianjin 300456, China
[2] Key Laboratory of Engineering Sediment of Ministry of Transport, Tianjin Research Institute for Water Transport Engineering, M.O.T., Tianjin 300456, China
* Correspondence: tkshzy@foxmail.com; Tel.: +86-22-59812345-6416

**Abstract:** HJ-1 charge-coupled device (CCD) data with high temporal and medium spatial resolution are widely used in environmental and disaster monitoring in China. However, due to bad weather, it is difficult to obtain sufficient time-continuous HJ-1 CCD data for environmental monitoring. In this study, the mountain valley with farmland and forestland in North China is selected as the experimental area, and HJ-1 CCD and moderate resolution imaging spectroradiometer (MODIS) data are used in the case study. An improved method of fusing data and inversing surface reflectivity is presented to obtain the HJ-1 inversion network-based application resolution (NBAR) data using linear matching of the Ross Thick-Li Sparse Reciprocal (RTLSR) model, and then predicted reflectivity using the seasonal autoregressive integrated moving average (SARIMA) model. The fusion data have advantages of high spatial and temporal resolution, as well as meeting the requirements of high quality and quantity of small-scale regional data. This case study provides a feasibility method for the HJ-1 satellites to produce the secondary products for small-scale remote sensing ground surface research. It also provides a reference for dynamic information acquisition and application of small satellite data, contributing to the improvement in RS estimation of surface environment variables.

**Keywords:** spatial–temporal fusion; HJ-1 CCD; MODIS; NBAR time series

## 1. Introduction

HJ-1 satellites, with high spatial and moderate temporal and spectral resolution data, are the first small satellite constellations dedicated to environmental and disaster monitoring and forecasting in China [1,2]. HJ-1 charge-coupled device (CCD) data are widely used in land use classification [3], and also used in wetland classification based on object-oriented classification methods, resulting in a low-cost and effective technical means to obtain high-precision wetland distribution information [4]. In other applications, the improved Carnegie–Ames–Stanford Approach (CASA) model is used to estimate the net primary productivity of the vegetation [5,6], and the soil organic matter content in the experimental area was previously monitored by HJ-1 data [7–9].

However, due to bad weather, the limited quantity restricts the quality of remote sensing (RS) extraction, the quality of HJ-1 two-level products is poor, geometric calibration is not accurate, and available data are reduced. HJ-1 CCD data are often used in combination with other RS data [10]. Fusing with multi-sensor observations is an effective way to improve the observation frequency [11,12]. The leaf area index (LAI) is usually calculated using the multi-sensor observation data set, which is constructed by sensors of HJ-1 CCD and Landsat 8 OLI data, and results have shown that the fusion data set can produce LAI products with reliable precision and continuous time resolution [13,14].

The temporal characteristics of surface reflectivity is a key factor for land surface process research, and has a great significance for the prediction of future surface reflectivity, such as hydrological forecasting, crop pest and disease disaster prediction, environmental pollution control, and ecological balance [15]. Compared to traditional surface reflectivity, the nadir bidirectional reflectance distribution function (BRDF)-adjusted reflectivity (NBAR) has advantages of being able to provide comparable dynamic data, thus avoiding the angular impact, and reflecting the dynamic variation patterns of the vegetated target area [16,17]. Li et al. provided an effective method to simulate the temporal changes in the MODIS NBAR time series of typical farmland surfaces by means of the season-trend statistics [18], and they also used the improved NBAR data as matching values for the LAI neural network estimation, making the predicted LAI time series more continuous than the MODIS LAI [14,19].

In this study, HJ-1 CCD and MODIS data were selected as data sources to invert the HJ-1 NBAR with 30 m resolution based on the MODIS NBAR production algorithm. The inversion data have advantages of high spatial and temporal resolution, as well as meeting the requirements of high quality and quantity of small-scale regional data. This provides a feasibility method for the HJ-1 satellites to produce the secondary products for small-scale RS ground surface research, and provides a reference for dynamic information acquisition and application of small satellite data, contributing to the improvement in RS estimation of surface environment variables. Moreover, the NBAR time series is a special sequence of surface observations, which has a pronounced seasonal change. It contains an important information about the four seasons and solar radiation; its predictions can contribute to early warning of pests and phonological changes, and it can be useful in other studies.

## 2. Materials and Methods

### *2.1. Study Areas and Materials*

#### 2.1.1. Study Areas

The study area is located in Huailai County in northern Hebei Province of North China, adjacent to Beijing in the East (See Figure 1a for location). The central position is about 115° east longitude and 40° north latitude, with a total area of 1801 km$^2$. Huailai County is located in the north of the Yanshan Mountains and the upper reaches of Yongding River. It is located in the semi-arid area of the middle temperate zone and belongs to the temperate continental monsoon climate.

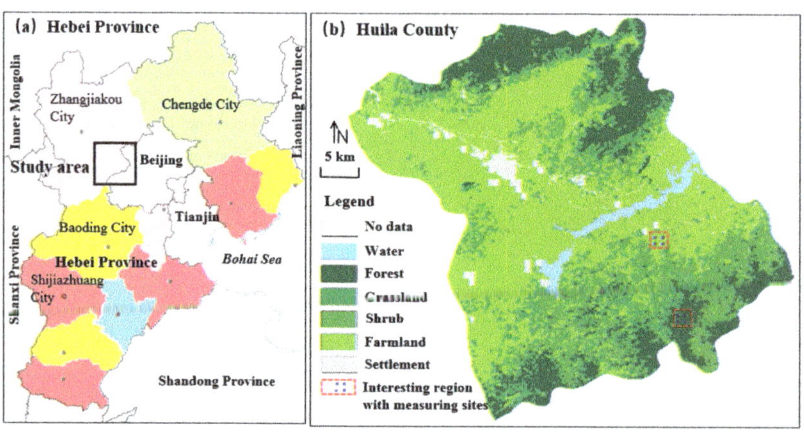


**Figure 1.** Location of the study area, Huailai County in northern Hebei Province of North China (**a**), and two interesting regions containing the measuring sites (**b**) in Huailai County.

Two interesting regions (2 × 2 km) with eight measuring sites were selected as the research areas (see Figure 1b for locations). The north region (40.30°–40.32° N and 115.74°–115.76° E) is covered by farmlands, and corn is planted over 90% of the total area, and the south region

(40.19°–40.21° N and 115.78°–115.80° E) is covered by forest; there are four measuring sites in each region.

2.1.2. Materials

The satellite data used in this study are listed in Table 1:

(1) HJ-1A/HJ-1B data: the two-level system geometric correction products of HJ-1A/HJ-1B satellites, which have spatial resolution of 30 m, a revisit period of 4 days, viewing swath width of 700 km, and include a total of four bands (blue, green, red, and near-infrared).
(2) MODIS NBAR products: the MODIS NBAR products (MOD43A4), which have spatial resolution of 500 m, three-level terrestrial standard, and 16 days of synthetic data products.
(3) MODIS daily products: the MODIS global daily reflectivity products (MOD09GA), for which the spatial resolution is 500 m.
(4) The surface measured data: the measured reflectance data collected at the same time as the test sites, measured by ASD spectrometer.

**Table 1.** Experimental data and band comparison.

| Products | Sensor | Spatial Resolution | Time (Year) | Spectral Region (nm) | | | | Data Sources |
|---|---|---|---|---|---|---|---|---|
| | | | | Blue | Green | Red | NIR | |
| HJ-1A/HJ-1B | HJ-1 CCD | 30 m | 2011–2014 | 430–520 | 520–600 | 630–690 | 760–900 | http://www.cresda.com (accessed on 2 April 2019) |
| MOD43A4/ MOD09GA | MODIS | 500 m | 2011–2014 | 459–479 | 545–565 | 620–670 | 841–846 | https://glovis.usgs.gov/ (accessed on 2 April 2019) |

The red (R) and near-infrared (NIR) bands of the NBAR data were chosen for the time series analysis, as they were sensitive to the vegetation growth and were typically used in most quantifications of vegetation inversion, such as LAI and NDVI, among others.

## 2.2. Data Preprocessing

### 2.2.1. HJ-1 Data Preprocessing

To obtain more accurate data, we used the earth resources data analysis system (ERDAS) software, based on the Landsat 8 data of the same region, to georeference the HJ-1 data again. Based on the size of the target object and filed survey, 20 control points for geometric correction were selected through a visual recognition method. By selecting the ground object points with obvious visual features on the image, such as road intersections and waterway junction lines, we ensured that the control points were evenly distributed on the two images.

The results showed that the geometric correction only caused shift correction, and there were few changes before and after image correction. The accuracy of Chinese satellite data was gradually increased.

Due to the influence of the atmosphere, product data with geometric corrections were distorted when analyzing the true surface reflectivity. To obtain the real surface information from satellite images, the next step was to accurately remove the atmospheric effect, or atmospheric correction, which eliminates the influence of atmospheric aerosol on clutter reflections and obtains the real surface reflectivity.

To improve the efficiency of atmospheric correction, we calculated the lookup table offline based on the 6S model [20]. The table contained different atmospheric aerosol optical thickness, solar zenith angle, observed zenith angle, and relative azimuth angle. It provided the atmospheric correction coefficient to achieve atmospheric correction of HJ-1 A/B data. To omit steps, we used the absolute atmospheric correction methods, while removing the effect of atmosphere in the image, to convert digital number (DN) value into surface reflectivity.

In the 6S model, the atmospheric correction is based on the following equation:

$$\begin{cases} y = x_a * L_i - x_b \\ \rho = \frac{y}{x_c * y + 1} \end{cases} \quad (1)$$

where $x_a$, $x_b$, and $x_c$ represent the three atmospheric correction coefficients; $L_i$ represents the radiance value measured for the $i$-th band of the sensor; $\rho$ represents the surface reflectance corrected by atmosphere.

When entering a parameter, we selected the predefined standard mode in 6S as the input. The two selected atmospheric models were mid-latitude summer and mid-latitude winter. The aerosol mode selected was the continental type, and assumed that the surface has uniform Lambertian reflection characteristics.

The aerosol optical thickness at 550 nm was set to 0.0, 0.1, 0.2, 0.3, 0.4, 0.5, 0.8, 1.0, 1.5, and 2.0. Six sensor zeniths were set up within the viewing angle of the CCD camera. Nineteen relative azimuth angles were set up within the range of 0–85°. Eighteen solar zenith angles were set up within the range of 0–180° [21]. After atmospheric correction, we obtained more realistic surface information, as indicated in Figure 2a.

After removing the atmospheric effect, radiometric calibration of the data was carried out to convert the DN value into the surface reflectivity, based on the spectral response function of the HJ and MODIS data. The four-band spectral response functions of HJ-1A CCD1 and MODIS are shown in Figure 2b.

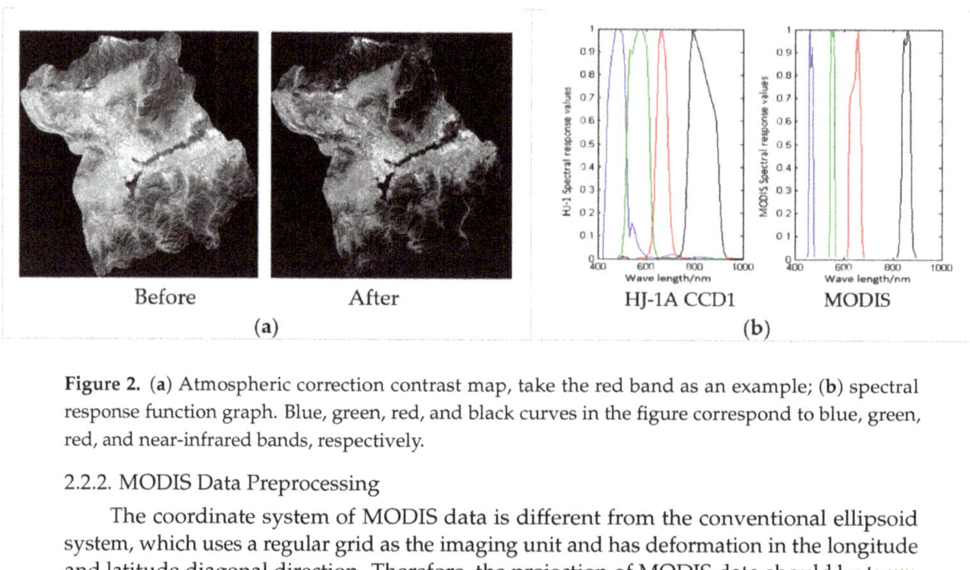


**Figure 2.** (**a**) Atmospheric correction contrast map, take the red band as an example; (**b**) spectral response function graph. Blue, green, red, and black curves in the figure correspond to blue, green, red, and near-infrared bands, respectively.

### 2.2.2. MODIS Data Preprocessing

The coordinate system of MODIS data is different from the conventional ellipsoid system, which uses a regular grid as the imaging unit and has deformation in the longitude and latitude diagonal direction. Therefore, the projection of MODIS data should be transformed into the HJ-1 projection coordinate system of UTM_zone_50N, and then the second geometric correction adjustment on the MODIS data is conducted based on Landsat 8 data, causing the MODIS data and HJ-1 data to overlap accurately.

### 2.2.3. Measured Data Preprocessing

To measure the surface reflectivity, we used the ASD spectrometer. The ASD spectrometer has a wavelength range of 350–2500 nm, and has reference reflectivity plates of 40%, 50%, and 99%. We set the ASD spectrometer to vertical. To eliminate the effects of random noise, we used the mean surface measured data for 20 min before and after the HJ-1 transit, and collected the solar zenith angles at the same time.

Through the RTLSR model, we calculated the NBAR values, and the average NBAR values at these eight sample points in two interest areas, to evaluate the overall performance.

## 2.3. Method

An overview of the methods is shown in Figure 3, and each step is described in detail in this section.

We focused on inversing HJ-1 NBAR, based on the MOD43 product inversion method, and analysis of time series characteristics of two types of vegetation land cover: forest and farmland. The change in the vegetation within such areas can be well depicted by the time series using HJ-1 NBAR data, which has less mixed pixels than MODIS.

The red and NIR bands of HJ-1 reflectivity data and MOD09GA data were used as input data in the process. We created the HJ-1 fusion data of 16 days as a group, and 8 days as a cycle. In each cycle, it was assumed that the change in the object matter was negligible and the HJ-1 NBAR was reversed.

Using the inversion of the NBAR and the MODIS NBAR data, we conducted a time series feature analysis in both the forest and the farmland areas of interest. Through the prediction and comparison with surface measured values, we proved the practicality of the data.

However, referring to the MODIS flags file, which on behalf of the image point were cloud, cloud shadow, ice, and water, and 15 other types of coverage, we excluded unavailable data from HJ-1 data, resulting in insufficient data in the cycle. Therefore, we used the spatial and temporal adaptive reflectance fusion model (STARFM) algorithm to fuse the data, supplemented the missing data, and obtained complete 2-day HJ-1 data for inversion.

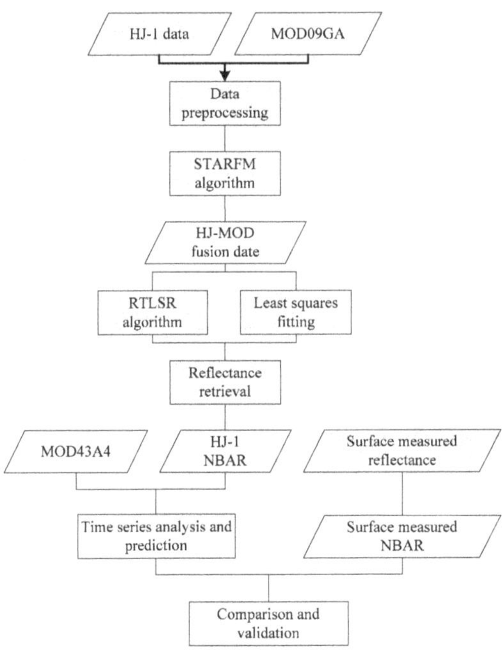


**Figure 3.** Research technical route.

### 2.3.1. STARFM Model

STARFM is a model based on reflectivity data [22]. The main purpose is to use high spatial resolution, low temporal resolution reflectivity data, and low spatial resolution, high temporal resolution reflectivity data to obtain high spatial resolution, high temporal resolution reflectivity data. This algorithm was successfully applied in the seasonal variation map of vegetation, and good results were obtained [23–25]. The algorithm has the following constructs:

$$L(x_{w/2}, y_{w/2}, T_0) = \sum_{i=1}^{w} \sum_{j=1}^{w} \sum_{k=1}^{n} W_{ijk} \times (M(x_i, y_j, T_k) + L(x_i, y_j, T_k) - M(x_i, y_j, T_0)) \quad (2)$$

where $L(x_{w/2}, y_{w/2}, T_0)$ is the HJ-1 predicted reflectivity value of time position; $W$ is the size of the moving window, we used only pixels with spectral similarity to the central pixel for prediction in the window; $M(x_i, y_j, T_k)$ and $L(x_i, y_j, T_k)$ represent MODIS and HJ-1 reflectivity values of $T_k$ time in $(x_i, y_j)$ window position, respectively; $n$ represents the number of input pairs of images.

In the STARFM algorithm, the HJ-1 pixel size has been set as the basic unit, with the moving window size at 50, which is the least common multiple of the HJ-1 space resolution (30 m) and MODIS space resolution (500 m), in order to ensure that the pixels in the window are complete. The distance weight constant has been set at the half of the window size, which is a relatively balanced value. The uncertainty of HJ-1 and MODIS data has been set at 0.2% and 0.5%, which reflects the unreliable pixel ratio and is obtained from error statistics of pixel classification. The MODIS and HJ-1 data are used as the input images, and the weight values of neighboring relative center pixels are obtained by calculating the spectral, time and spatial distance weights; thereafter, the fusion image pixel values can be calculated.

A weighted average strategy is used in the STARFM algorithm. When the difference between the two sets of input images is too large, it may cause the "time smoothing" problem. The smaller the time interval, the lesser the difference between the input MODIS images, the more similar the optional spectral pixels, and the more accurate the forecast [26]. Therefore, in the prediction, we selected the image pairs that were as close to the predicted image time as the prediction base images.

#### 2.3.2. RTLSR Model

BRDF describes the direction of the surface reflection characteristics. In the bi-directional reflectivity models, a practical method called the kernel-driven model, which has a certain physical meaning of the linear combination of the core, is used to match the surface bidirectional reflection. RTLSR is the main algorithm for MODIS diversion and albedo products are currently providing albedo and bi-directional reflectivity of global scale to users.

The RTLSR algorithm is the weighted sum of two kernels, which are called "RossThick" and "LiSparse-Reciprocal". The RTLSR algorithm has the following constructs [27]:

$$R(\theta_i, \theta_r, \varphi, \lambda) = f_{iso}(\lambda) + f_{vol}(\lambda) K_{vol}(\theta_i, \theta_r, \varphi) + f_{geo}(\lambda) K_{geo}(\theta_i, \theta_r, \varphi) \quad (3)$$

where $K_{geo}$ and $K_{vol}$ are geometrical optics kernel and bulk scattering kernel, which are functions of observing zenith angle $\theta_i$, solar zenith angle $\theta_r$ and relative azimuth angle $\varphi$, respectively. $f_{iso}$, $f_{geo}$, and $f_{vol}$ are constant coefficients, which represent the proportion of isotropic scattering, geometric optical scattering, and bulk scattering. $R(\theta_i, \theta_r, \varphi, \lambda)$ is the BRDF of the band.

The RTLSR algorithm uses the linear regression to invert the best coefficients of the matching observation data, $f_{iso}$, $f_{geo}$, and $f_{vol}$, then obtains the bi-directional reflectivity of random incident angle and observation angle by the extrapolation or interpolation of the kernel [28].

The bulk scattering kernel "RossThick" is suitable for describing dense homogeneous vegetation and leaf dip spherical distribution, proposed by Roujean [29]:

$$K_{vol}(\theta_i, \theta_r, \varphi) = \frac{(0.5\pi - g)\cos g + \sin g}{\cos\theta_i + \sin\theta_r} - \frac{\pi}{4} \tag{4}$$

where $g$ is the phase angle defined by the following formula:

$$\cos g = \cos\theta_i \cos\theta_r + \sin\theta_i \sin\theta_r \cos\varphi \tag{5}$$

The geometric optical kernel "LiSparseR" is suitable for describing a sparsely distributed canopy or other opaque geometry. It is a method based on the reciprocity principle, and it can partially improve the problems when extrapolating to the larger zenith angle without affecting the data matching ability. The expression is:

$$K_{geo}(\theta_i, \theta_r, \varphi) = A(\theta_i, \theta_r, \varphi)\sec\theta_i - B(\theta_i, \theta_r, \varphi) \tag{6}$$

where,

$$\begin{aligned}
&B(\theta_i, \theta_r, \varphi) = sec\theta_i\prime + \sec\theta_r\prime - O(\theta_i, \theta_r, \varphi) \\
&A(\theta_i, \theta_r, \varphi) = \tfrac{1}{2}(1 + cos\alpha)sec\theta_r \\
&O = \tfrac{1}{\pi}(t - \sin t\cos t)(sec\theta_i\prime + \sec\theta_r\prime) \\
&\cos t = \tfrac{h}{b}\tfrac{\sqrt{D^2 + (tan\theta_i\prime tan\theta_r\prime sin\varphi)^2}}{sec\theta_i\prime + \sec\theta_r\prime} \\
&D = \sqrt{\tan^2\theta_i\prime + \tan^2\theta_r\prime - 2\tan\theta_i\prime\tan\theta_i\prime\cos\varphi} \\
&\cos\alpha = \cos\theta_i\prime\cos\theta_r\prime + \sin\theta_i\prime\sin\theta_r\prime\cos\varphi \\
&\theta_i\prime = \arctan(\tfrac{b}{r}tan\theta_i),\ \theta_r\prime = \arctan(\tfrac{b}{r}tan\theta_r)
\end{aligned} \tag{7}$$

where $b$ is the vertical radius of the sphere; $h$ is the horizontal radius of the sphere, and $r$ is the height of the sphere. In the MODIS algorithm, $\frac{h}{b} = 2$, $\frac{b}{r} = 1$ [29].

We used this algorithm, setting the HJ-1 fusion data to 16 days as a group and 8 days as a cycle, to match a set of kernel coefficients: $f_{iso}$, $f_{geo}$, and $f_{vol}$. The least squares method was applied in matching. We set the solar zenith angle in the model to 45°, which was consistent with the solar zenith angle of MOD43 NBAR product data, and the observation zenith angle to the zenith direction. According to the red and NIR kernel coefficients of each pixel in each cycle, and the relative azimuth $\varphi$, we obtained the azimuthal direction reflectivity HJ-1 NBAR, which was adjusted by the BRDF model for each pixel in the red and NIR bands.

### 2.3.3. SARIMA Model

Generally, historical data will have a strong relationship at the potential cycle time points, especially in economics. The SARIMA model is mainly used to identify the predictions of dependent variables, influenced by seasonal fluctuations and external events. The SARIMA model contains trend and seasonal variation, which has been widely used in different fields, such as economics, statistics, and RS, forecasting a certain parameter in time series with seasonality [30–33].

According to the definition, the SARIMA model is generally referred to SARIMA $(p, d, q) \times (P, D, Q)_s$. Where $d$ and $D$ are the order of the stepwise difference and the seasonal difference, respectively; $p$ and $q$ are the order of autoregressive and moving average, respectively; $P$ and $Q$ are the order of seasonal autoregressive and seasonal moving average, respectively; $s$ is the seasonal period. The model can be expressed as follows:

$$\phi_p(B)\Phi_P(B^S)(1 - B)^d(1 - B^S)^D y_t = \theta_q(B)\Theta_Q(B^S)a_t \tag{8}$$

where,

$$\begin{aligned}
&\phi_p(B) = 1 - \phi_1 B - \phi_2 B^2 - \cdots - \phi_p B^p \\
&\theta_q(B) = 1 + \theta_1 B + \theta_2 B^2 + \cdots + \theta_q B^q \\
&\Phi_P(B^S) = 1 - \Phi_1 B^S - \Phi_2 B^{2S} - \cdots - \Phi_P B^{PS} \\
&\Theta_Q(B^S) = 1 + \Theta_1 B^S + \Theta_2 B^{2S} + \cdots + \Theta_Q B^{QS}
\end{aligned} \tag{9}$$

$\alpha_t$ is the estimated residual at time $t$, a random process of Gaussian white noise.

The general matching process includes three steps: identification, estimation, and diagnostic checking. The crucial step in the SARIMA model is the choice of orders. The main methods used are autocorrelation functions (ACF) and partial autocorrelation functions (PACF), Akaike's information criterion (AIC), and so on. First, we used the ACF and PACF methods to select several possible orders. After modeling, the AIC criterion was used to select the best order model. The $p$ value in the model was taken as 1, and the D value was taken as 1.

After the model was established, it needed to be tested. The method used here is the residual ACF test. In model testing, we are mainly concerned about whether the residual of the model is relevant, and normally distributed with a mean of zero. If the residuals of the SARIMA model are correlated and not normally distributed with a mean of zero, then the model can be further improved. On the contrary, the model fitting effect is good, and it can be considered that the model fully extracts the information of the sequence.

For the test of residual white noise, LB test was used to check whether the residual of the model is a white noise sequence. The test results are shown in Table 2, which indicates that when the residual sequence was delayed 1–12 orders, the P value of Q statistics is greater than 0.05. Therefore, at the significance level of 0.05, the original hypothesis is not rejected, meaning that the residual sequence is a white noise sequence, indicating that the fitted model had fully extracted the information in the time series.

For the residual distribution problem, we generated the model diagnosis report and analyzed the abnormal behavior of the research model, as shown in Figure 4. Figure 4a shows that the red KDE line follows the N (0, 1) line, which indicates that the residuals are normally distributed. Figure 4b shows that the residual error does not have autocorrelation, which indicates the model can help us understand the original time series data and predict the future values.

Using the HJ-1 NBAR data and MODIS NBAR data from interest areas 1 and 2 in 2011–2013 as the raw data, the model was used to predict the reflectivity of the red and NIR bands in 2014, and the results were compared with the surface measured data to discuss the prediction accuracy of the model.

**Table 2.** White noise test results of fitting model residuals.

| Lag | AC | Q | Prob (>Q) |
|---|---|---|---|
| 1.0 | −0.06261 | 2.505646 | 0.112883 |
| 2.0 | 0.016932 | 2.712496 | 0.341931 |
| 3.0 | 0.049566 | 3.637803 | 0.31302 |
| 4.0 | 0.022228 | 3.498892 | 0.424928 |
| 5.0 | −0.01017 | 3.756948 | 0.515086 |
| 6.0 | −0.02245 | 3.967206 | 0.647077 |
| 7.0 | −0.0451 | 4.997104 | 0.622196 |
| 8.0 | −0.02974 | 5.506453 | 0.680345 |
| 9.0 | 0.007786 | 6.324489 | 0.772662 |
| 10.0 | 0.0685 | 8.858177 | 0.582798 |
| 11.0 | 0.030202 | 9.274546 | 0.583861 |
| 12.0 | −0.0574 | 10.68552 | 0.511118 |

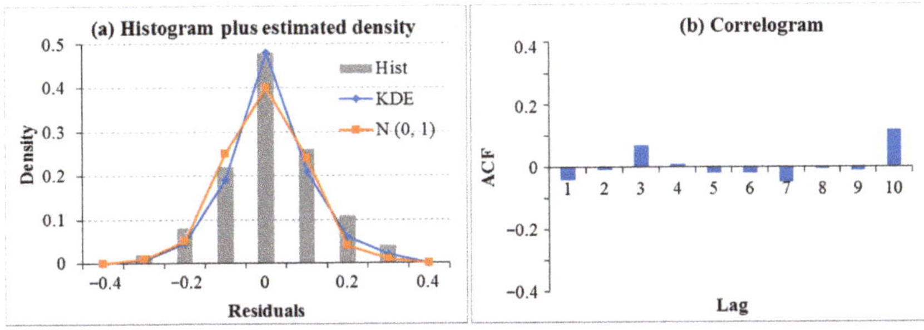


**Figure 4.** (**a**) Probability distribution diagram of model residual error, where KDE stands for kernel density estimation, N (0, 1) represents normal distribution, and Hist represents the model residual histogram; (**b**) autocorrelation function graph of residuals.

## 3. Results and Analysis

### 3.1. Data Fusion Assessments

Before the data fusion process, we selected three images for the STARFM algorithm to illustrate the accuracy of the results and thus determine availability. The experimental data were HJ-1 data from areas of interest 1 and 2, and the corresponding time MOD09GA data.

Figure 5 shows three pairs of HJ-1 and MODIS images acquired on 26 April 2012, 4 May 2012, and 8 May 2012. The first row in Figure 5 shows three MODIS reflectivity images using band 2-1-4 as the red–green–blue composite, and the second row shows three HJ-1 reflectivity images using band 4-3-2 as the red–green–blue composite. The pairs of MODIS and HJ-1 images acquired on 26 April 2012 and 8 May 2012 were selected as base images, and the STARFM algorithm was used to predict the 4 May 2012 HJ-1 fusion image.

After STARFM algorithm calculation, we compared the predictions with the 4 May 2012 observed HJ-1 reflectivity image. Figure 6 shows the observed and predicted images produced by the STARFM algorithm in regions of interest 1 and 2. The boundary information of road, water, and vegetation can be expressed clearly in the fusion image, and there is high similarity between the predicted and the measured image.

There are many commonly used error evaluation indicators for model evaluation; we used several of these indicators to refine the results of the fusion evaluation, including root mean square error (RMSE), mean absolute percent error (MAPE), average absolute difference (AAD), and coefficient of determination ($R^2$).

RMSE was used to measure the deviation between the result and the reference value; the smaller the RMSE, the closer the result is to the reference value.

MAPE indicates the accuracy between the result and the reference value; if this value is small, it indicates high accuracy, according to the model predictive ability evaluation table proposed by Xia et al. [34]. Generally, when MAPE $\leq$ 20, the forecast is valid.

Average absolute deviation (AAD) is the average of the absolute values of the difference between the result and the reference value.

R is a direct reflection of correlation between the result and the reference value. If R is positive, the correlation coefficient is positive, on the contrary, if R is negative, the two are negatively correlated. The larger the absolute value of R, the higher the degree of correlation, the greater the determination coefficient, the closer the relationship between the two sets of data.

Figure 7 shows scatter plots of the predicted and actual observations. Table 3 shows the analytical values for the predicted and actual observations. In the four scatter plots, the points are distributed on both sides of the y = x line. The distribution of the red and NIR band points of forest are closer to the straight line and more uniform, whereas the distributions of farmland are concentrated above the straight line. This means that the fusion values are slightly higher than the actual values. According to Table 3, the MAPE values of farmland are within 20%, and the $R$ values are high, which indicates that the experimental results are effective. Compared to the reference image, the mean absolute difference is small, and the predicted value is highly correlated with the actual observed value.

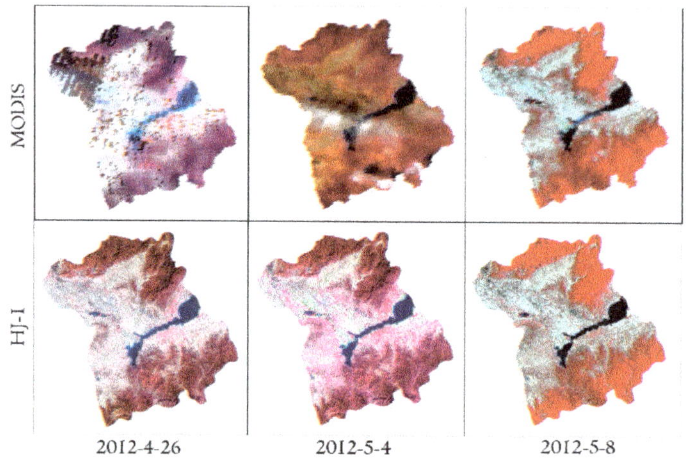


**Figure 5.** MODIS reflectivity images and HJ-1 reflectivity images for 26 April, 4 May, and 8 May 2012.

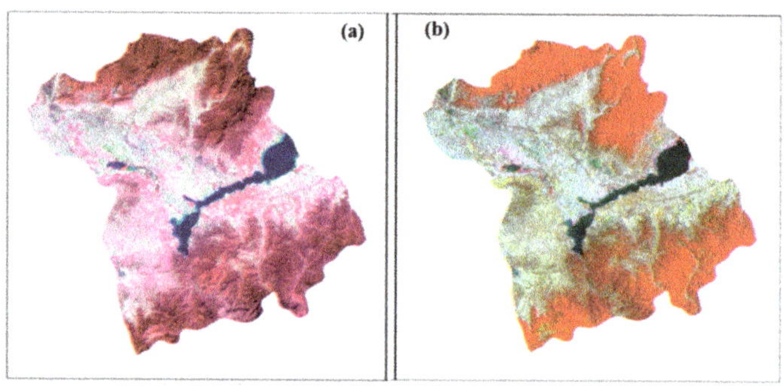


**Figure 6.** Observed image for 4 May ASD spectrometer 2012 and the predicted image: (**a**) observed image; (**b**) STARFM image.

**Table 3.** Accuracy test of four data fusion results.

| Band | RMSE | MAPE | AAD | $R^2$ |
|---|---|---|---|---|
| R-forest | 0.003 | 16.5% | 0.002 | 1 |
| NIR-forest | 0.010 | 14.3% | 0.009 | 0.998 |
| R-farmland | 0.014 | 19.6% | 0.012 | 1 |
| NIR-farmland | 0.014 | 16.8% | 0.012 | 0.998 |

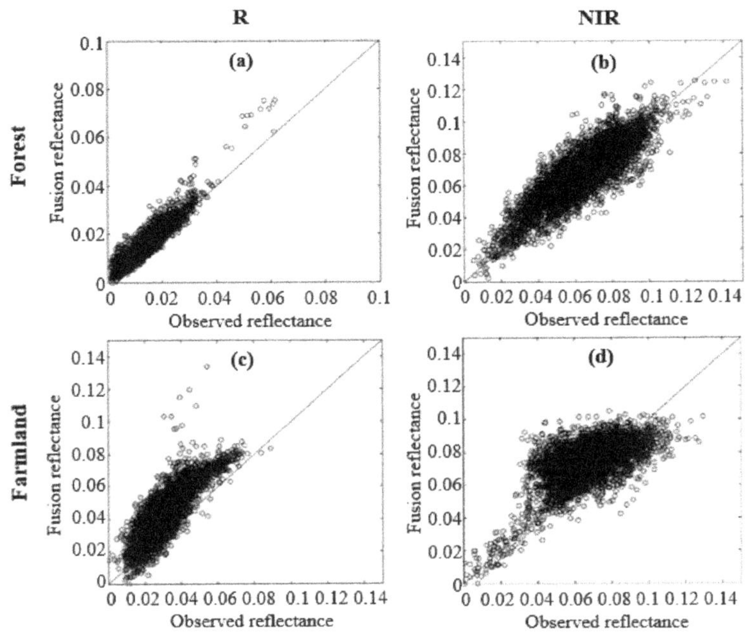


**Figure 7.** Four scatter plots of fusion reflectivity and observed reflectivity (x, observed; y, fusion): (**a**) the distribution of the red band points of forest; (**b**) the distribution of the NIR band points of forest; (**c**) the distribution of the red band points of farmland; (**d**) the distribution of the NIR band points of farmland.

The results show that the fusion images have high spatial resolution, meanwhile, preserving periodic vegetation changes. The fusion images are close to the real observation images, and can be used to study the temporal variation of vegetation. Therefore, image fusion is performed to obtain complete 2-day HJ-1 reflectivity data.

### 3.2. Inversion Results

In the process of matching the kernel coefficient, the matching precision is higher when the data sampling direction is more uniform in the spatial distribution of the hemisphere. Figure 8a shows the red band solar zenith angle, observation zenith angle, and relative azimuthal distribution for a cycle (8 May to 22 May 2012) of HJ-1 data, selecting a 3 × 3 pixel window. The sampling direction of the data in the cycle is relatively uniform.

To further determine the accuracy of the kernel coefficient matching, the kernel coefficient and the MOD43 BRDF kernel coefficient are used to create the direction reflectivity distribution. The solar zenith angle is set to 45° and the relative azimuth angle to the HJ-1 pixel relative azimuth angle. Figure 8b shows the red band inversion direction reflectivity for a cycle (8 May to 22 May 2012) of HJ-1 data and MOD43 BRDF data on 8 May 2012 in the farmland area of interest. The kernel coefficients of the matched nuclei and the MOD43 BRDF data of red band are 0.18559, 0, 0.0070382, and 0.209, 0.032, 0.049, respectively. The reflectivity of the inverted zenith direction is 0.1880 and 0.1832, respectively, whereas the MOD43 NBAR is 0.1828. Although the coefficients are different, the inversion of the reflectivity values is close. The inversion reflectivity of HJ-1 in the zenith direction is similar to that of MOD43 NBAR; it is proven that the inversion reflectivity result shows a small difference with the actual measurement result of MOD43. The results are within the error range.

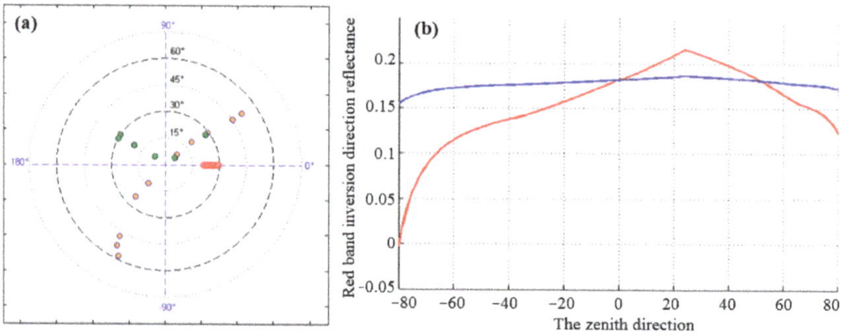


**Figure 8.** (**a**) Sun observation angle distribution, where the red circle, the green circle, and the pink circle are corresponding to the solar zenith angle, the original HJ-1 observation zenith angle, and the fusion HJ-1 data observation zenith angle, respectively; (**b**) directional reflectivity profile, where the red and blue lines are corresponding to the MOD43 and HJ-1 reflectivity, respectively.

For the inversion results, i.e., the HJ-1 NBAR data, the MODIS NBAR product is used to compare with them, analyze their similarity and variance, and analyze the accuracy of the inversion results.

In the forest and farmland areas, the average reflectivity data of HJ-1 NBAR and MOD43 NBAR zenith direction were obtained. Figure 9 shows the reflectivity of HJ-1 NBAR and MODIS NBAR time curves. It can be observed that the average reflectivity of HJ-1 NBAR is similar to that of MODIS average reflectivity, and the fluctuation range of reflectivity is almost the same. However, it can also be seen that the average reflectivity of HJ-1 NBAR fluctuates more violently. In the forest area of interest, the reflectivity of the NIR band significantly increased during the growing period, but it does not maintain a high value smoothly. There is a small fluctuation during the period, and the reflectivity curve of the red band fluctuates vigorously, and there are obvious anomalies. In the farmland area of interest, the trend of reflectivity is similar to that of MODIS, but there are smaller fluctuations before the growth period. The wet and dry conditions of the bare soil may be the reason. This result indicates that the average reflectivity of HJ-1 NBAR is more suitable to reflect the subtle changes in the surface.

Similarly, we used the quantitative indicators to calculate the difference between the HJ-1 NBAR data and the measured values, and to analyze the accuracy of the inversion data. Figure 10 shows the scatter plots of the HJ-1 NBAR and MOD43 NBAR data. Table 4 shows the accuracy test result for the HJ-1 inversion data and MODIS inversion data. In the four scatter plots, the points are distributed on both sides of the $y = x$ line. The MAPE values are within 20%, compared to the MOD43 NBAR, the mean absolute difference is small, and the predicted value is highly correlated with the actual observed value.

By comparison, it can be seen that the HJ-1 30 m resolution data are more sensitive than the MOD43 500 m resolution data. In the case of the same fluctuation trend, the HJ-1 reflectivity exhibits smaller fluctuations. Because of the underlying surface, there will be different surface reflectivities in the same vegetation growth conditions. Although there are more inflection points and a greater chance of outliers, the HJ-1 data can more realistically reflect the change in the surface during the test period.

As the results indicate, the combination of MODIS data and HJ-1 data is expected to be a method of improving the data quality of HJ-1 30 m NBAR time series. Certainly, compared with MODIS NBAR products, there is still a shortage of large-scale data usage in surface reflectivity inversion, when using the HJ-1 and MODIS joint data. However, with the use of small-scale data, the HJ-1 fusion data can highlight more detailed real surface conditions.

**Table 4.** Inversion result accuracy test.

| Band | RMSE | MAPE | AAD | $R$ | $R^2$ |
|---|---|---|---|---|---|
| R-forest | 0.003 | 16.5% | 0.002 | 1 | 1 |
| NIR-forest | 0.010 | 14.3% | 0.009 | 0.999 | 0.998 |
| R-farmland | 0.014 | 19.6% | 0.012 | 1 | 1 |
| NIR-farmland | 0.014 | 16.8% | 0.012 | 0.999 | 0.998 |

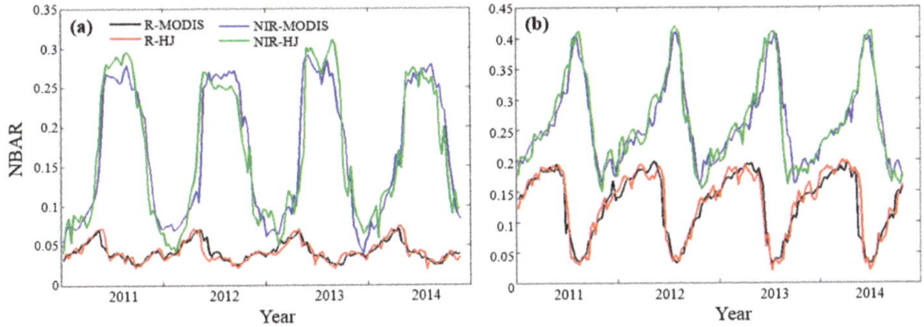


**Figure 9.** Comparison between HJ-1 and MODIS NBAR time series: (**a**) forest; (**b**) farmland. Black, red, blue, and green curves in the subfigures correspond to MODIS red band, HJ-1 red band, MODIS NIR band, and HJ-1 NIR band NBAR, respectively.

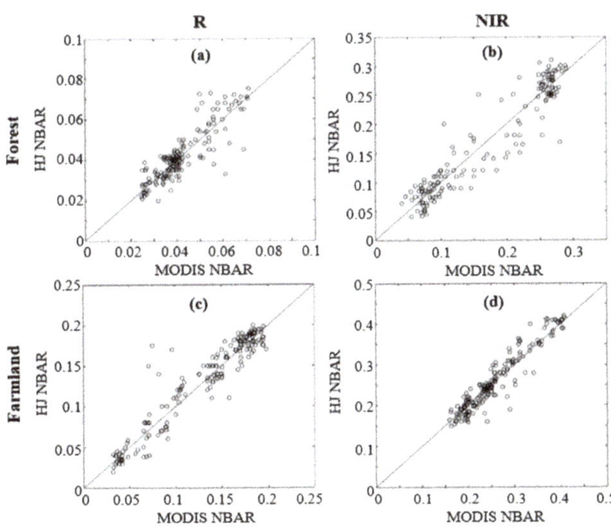


**Figure 10.** Four scatter plots of inversion reflectivity and observed reflectivity (x, observed MODIS NBAR; y, predicted HJ-1 NBAR): (**a**) the distribution of the red band points of forest; (**b**) the distribution of the NIR band points of forest; (**c**) the distribution of the red band points of farmland; (**d**) the distribution of the NIR band points of farmland.

### *3.3. NBAR Time Series Analysis*

Using the HJ-1 NBAR data, we can see the time series line graph, as shown in Figure 11. The reflectivity change in forestland is in accordance with the change in most of the vegetation reflectivity. At around the beginning of March each year, the vegetation starts to grow, and the NIR reflectivity significantly increases, remaining at around 0.25. Upon the arrival of winter, the vegetation withers and the NIR reflectivity decreases rapidly. The red band reflectivity is significantly lower than the NIR reflectivity, at around 0.05, and the fluctuation is small, increasing before spring and decreasing slowly during spring. As in farmland, the difference from most of the vegetation is due to human influence, as the surface is covered by bare soil before the growth period. The reflectivity changes in farmland are relatively stable. In the early stages of growth, the NIR band reflectivity increases, then decreases during the growth period, whereas the red band reflectivity decreases at the beginning of the growing season, and then increases during the growing season. In sum, the seasonal difference of forest and farmland albedo is mainly affected by the seasonal difference in plant growth.

This time series provided comparable dynamic data, avoiding the angular impact, to reflect the dynamic variation patterns of vegetated target areas. Therefore, we used the data to do the forecast analysis for disaster prediction, environmental control, etc.

We used the HJ-1 NBAR data and MODIS NBAR data of interest areas 1 and 2 as the raw data. For the HJ-1 NBAR data in the regions of interest, according to the average, we set a threshold of 5% and selected the pixels within the threshold to the new average, excluding mixed pixel effects. We implemented the SARIMA method using Eviews, which is freely available for data analysis, regression analysis, and forecasting. Observing the ACF and PACF figures, we selected the intervals of coefficients. After the seasonal difference and AIC screening, the model with the smallest AIC value was selected as the best, and the resulting model was determined in Table 5.

The white noise test of the model was effective. The residual sequences are purely random and the autocorrelation coefficients fall into the random interval, which indicates that the models are feasible and the prediction results are meaningful. The Shapiro normality test for the residuals produced $p$-values of 0.372 and 0.386, showing that the original hypothesis of following the normal distribution is rational, that is, the residuals exhibit a normal distribution.

For each land cover type and band, the predicted time series and the corresponding measured NBAR data are displayed in Figure 11. It can be seen from the figures that the HJ-1 predicted values and the measured values are very close. The calculated root mean square error is 8.36%, 7.58%, 8.87%, and 9.05%, indicating that the match is good. However, for the MODIS predicted values, especially in the forest, under the influence of mixed pixels, the predicted values are significantly higher than the measured values; there are large differences in small fluctuations. The HJ-1 NBAR predicted time series matches the measured NBAR values and reflects the seasonal changes in vegetation properties better; however, it is just an approximation of the MODIS NBAR predicted time series. In general, the predicted time series NBAR can represent the variation of interest area with a more reliable value compared with MODIS NBAR predicted data. The recursive estimation method is robust and reliable.

We used RMSE, MAPE, AAD, and R to evaluate the results, as shown in Table 6. The accuracy test values indicate that the HJ-1 predictions are closer to the real measured values than the MODIS predictions. Figure 12 shows scatter plots of the HJ-1 predicted and actual observations. In the scatter plots, the points are distributed on both sides of the $y = x$ line. For the MODIS predicted values, the MAPE values are within 25% and the $R$ value is 0.75. For the HJ-1 predicted values, the MAPE values are within 20%, the $R$ value is 0.91, and the HJ-1 predicted value is highly correlated with the actual observed value. The HJ-1 predicted values match the measured data very well. Thus, the proposed HJ-1 inversion data and time series methods perform well for NBAR recursive estimation.

From these results, we found that (1) the fusion data had advantages of high spatial and temporal resolution, and matched HJ-1 data well; (2) the HJ-1 NBAR data adjusted by BRDF model were close to MOD43 data, and can reflect the annual variation characteristics of surface vegetation growth well; (3) the HJ-1 data had high-resolution features compared with the MOD43 NBAR data, reflecting the regional scale reflectivity spatial changes more accurately, and making related research and analysis more sophisticated; (4) compared to the forecast results of MODIS NBAR, the HJ-1 NBAR matched the surface measured data better, and can reflect the seasonal changes in vegetation properties better.

**Table 5.** SARIMA coefficients.

| Band | HJ-1 Equation | MODIS Equation |
|---|---|---|
| R-forest | SARIMA(0,1,3) × $(0,1,1)_{45}$ | SARIMA(0,1,2) × $(0,1,2)_{45}$ |
| NIR-forest | SARIMA(0,0,2) × $(1,1,1)_{45}$ | SARIMA(0,1,2) × $(0,1,1)_{45}$ |
| R-farmland | SARIMA(0,0,3) × $(0,1,1)_{45}$ | SARIMA(0,0,1) × $(0,1,1)_{45}$ |
| NIR-farmland | SARIMA(0,0,1) × $(0,1,2)_{45}$ | SARIMA(0,1,1) × $(0,1,2)_{45}$ |

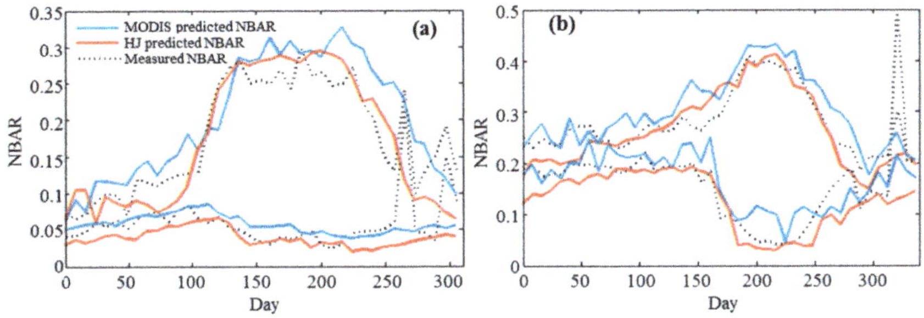


**Figure 11.** Comparison between predicted and observed NBAR time series of the red and near-infrared band data: (**a**) forest; (**b**) farmland. Each pair of the black, red, and blue curves in the subfigures correspond to red and near-infrared bands of the measured NBAR, HJ-1 predicted NBAR, and MODIS predicted NBAR, in which the red band curves have a relatively low value and the near-infrared curves have a relatively high value.

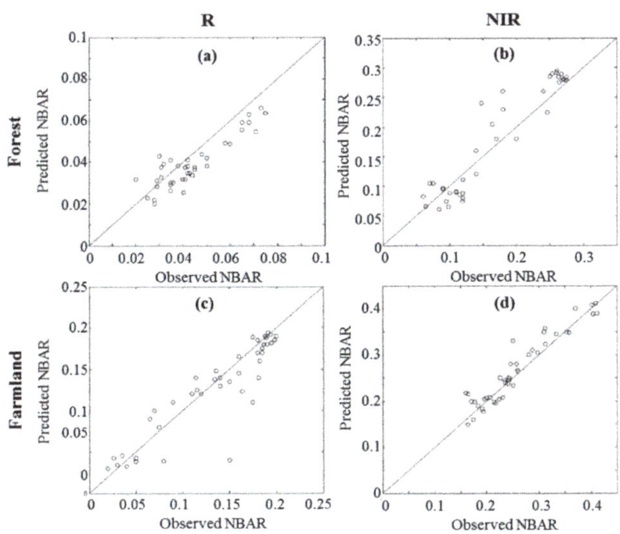


**Figure 12.** Four scatter plots of HJ-1 predicted reflectivity and observed reflectivity (x, observed; y, predicted): (**a**) the distribution of the red band points of forest; (**b**) the distribution of the NIR band points of forest; (**c**) the distribution of the red band points of farmland; (**d**) the distribution of the NIR band points of farmland.

**Table 6.** HJ-1 and MODIS NBAR prediction result accuracy test.

| Band | | RMSE | MAPE | AAD | $R$ | $R^2$ |
|---|---|---|---|---|---|---|
| R-forest | HJ-1 | 0.007 | 11.8% | 0.005 | 0.862 | 0.743 |
| | MODIS | 0.037 | 20.55% | 0.021 | 0.676 | 0.457 |
| NIR-forest | HJ-1 | 0.029 | 18.9% | 0.022 | 0.943 | 0.889 |
| | MODIS | 0.040 | 23.69% | 0.038 | 0.873 | 0.762 |
| R-farmland | HJ-1 | 0.018 | 14% | 0.013 | 0.946 | 0.895 |
| | MODIS | 0.056 | 23.51% | 0.034 | 0.695 | 0.483 |
| NIR-farmland | HJ-1 | 0.024 | 8.5% | 0.017 | 0.948 | 0.899 |
| | MODIS | 0.038 | 13.75% | 0.032 | 0.906 | 0.821 |

## 4. Conclusions

The inconsistent data quality in RS images, especially HJ-1 CCD data, is mainly influenced by environmental factors, and creates problems in the application of these data. In this paper, the mountain valley with farmland and forestland in North China is selected as the experimental area, and an improved method is presented to obtain the HJ-1 inversion NBAR data using linear matching of the RTLSR model, and then to predict reflectivity using the SARIMA model. The HJ-1 data and MODIS data are fused to obtain ample available data for inversion. The fusion data have advantages of high spatial and temporal resolution, as well as ensuring the requirements of high quality and quantity of small-scale regional data in the case study.

Using these data, the prediction of HJ-1 NBAR is found to be more similar with observed values than MODIS. Compared with MODIS NBAR products, the HJ-1 NBAR data reduce the error due to mixed pixels, and are suitable for small-scale RS surface monitoring. This provides a reference for dynamic information acquisition and application of small satellite data, and contributes to the improvement in RS estimation of surface environment variables. Moreover, the NBAR time series is a special sequence of surface observations, which contains important information about the four seasons and solar radiation. Thus, the predicted HJ-1 NBAR in the case study can contribute to the early warning of pests and phonological changes, and it can also be used to calculate other surface parameters and analyze the variation in different time scales awaiting further study, through detailed understanding and the application of inverse methods.

**Author Contributions:** Conceptualization, Z.H. and H.L.; methodology, H.L. and H.W.; formal analysis and investigation, H.L. and Z.H.; validation, H.W.; data curation, H.L.; Writing—original draft, H.L. and Z.H.; writing—review and editing, H.L., Z.H. and H.W.; funding acquisition, H.W. All authors have read and agreed to the published version of the manuscript.

**Funding:** The study was supported by National Natural Science Foundation of China (Grant No. 42006153), Young Elite Scientists Sponsorship Program by CAST (Grant No. 2021QNRC001), the Research and Innovation Fund of Tianjin Research Institute for Water Transport Engineering, M.O.T., China (Grant No. TKS20220203).

**Institutional Review Board Statement:** Not applicable.

**Informed Consent Statement:** Not applicable.

**Data Availability Statement:** Not applicable.

**Acknowledgments:** We owe great appreciation to the anonymous reviewers for their critical, helpful, and constructive comments and suggestions.

**Conflicts of Interest:** The authors declare no conflict of interest.

## References


1. Bai, Z. China's constellation of small satellites for environment and disaster monitoring and forecasting. *Aerosp. China* **2009**, *5*, 10–15. (In Chinese)
2. Wu, C.; Wang, Q.; Zhang, Y.; Li, J.; Wu, Y.; Zhu, L.; Yao, Y. Remote sensing application system for water environments developed for Environment Satellite 1. *Sci. China Earth Sci.* **2010**, *53*, 45–50. [CrossRef]
3. Zhong, B.; Ma, P.; Nie, A.; Yang, A.; Yao, Y.; Lü, W.; Zhang, H.; Liu, Q. Land cover mapping using time series HJ-1/CCD data. *Sci. China Earth Sci.* **2014**, *57*, 1790–1799. [CrossRef]
4. Ji, J.; Sha, J. Wetland change detection in Longxiang Island area based on object-based classification method and multi-source remote sensing images. *J. Fujian Norm. Univ. (Nat. Sci. Ed.).* **2017**, *33*, 78–86. (In Chinese)
5. Li, Q.; Wang, H.; Liu, W.; Yan, C. Evaluation on net primary productivity of alpine meadow based on HJ-1 satellite data: A case study in the zoige grassland. *J. Desert Res.* **2013**, *33*, 1250–1255. (In Chinese)
6. Liu, Z.; Hu, M.; Hu, Y.; Wang, G. Estimation of net primary productivity of forests by modified CASA models and remotely sensed data. *Int. J. Remote Sens.* **2018**, *39*, 1092–1116. [CrossRef]
7. Wang, X.; Meng, J. Mapping soil organic matter content in field using HJ-1 satellite image. *Trans. Chin. Soc. Agricul. Eng.* **2014**, *30*, 101–108. (In Chinese)
8. Ren, C.; Zhang, B.; Wang, Z.; Li, L.; Jia, M. Mapping forest cover in northeast China from Chinese HJ-1 satellite data using an object-based algorithm. *Sensors* **2018**, *18*, 4452. [CrossRef]
9. Bai, L.; Wang, C.; Zang, S.; Wu, C.; Luo, J.; Wu, Y. Mapping soil alkalinity and salinity in northern Songnen Plain, china with the HJ-1 hyperspectral imager data and partial least squares regression. *Sensors* **2018**, *18*, 3855. [CrossRef]
10. Sun, R.; Rong, Y.; Su, H.; Chen, S. NDVI time-series reconstruction based on MODIS and HJ-1 data spatial-temporal fusion. *J. Remote Sens.* **2016**, *20*, 361–373.
11. Roy, D.P.; Ju, J.; Lewis, P.; Schaaf, C.; Gao, F.; Hansen, M.; Lindquist, E. Multi-temporal MODIS-Landsat data fusion for relative radiometric normalization, gap filling, and prediction of Landsat data. *Remote Sens. Environ.* **2008**, *112*, 3112–3130. [CrossRef]
12. Walker, J.J.; de Beurs, K.M.; Wynne, R.H.; Gao, F. Evaluation of Landsat and MODIS data fusion products for analysis of dryland forest phenology. *Remote Sens. Environ.* **2012**, *117*, 381–393. [CrossRef]
13. Zhao, J.; Li, J.; Liu, Q.; Fan, W.; Zeng, Y.; Xu, B.; Yin, G. Leaf area index inversion combining with HJ-1/CCD and Landsat 8/OLI data in the middle reach of the Heihe river basin. *J. Remote Sens.* **2015**, *19*, 733–749. (In Chinese)
14. Wang, J.; Wang, J.; Shi, Y.; Zhou, H.; Liao, L. A recursive update model for estimating high-resolution LAI based on the NARX neural network and modis times series. *Remote Sens.* **2019**, *11*, 219. [CrossRef]
15. Loew, A.; Govaerts, Y. Towards multi-decadal consistent meteosat surface albedo time series. *Remote Sens.* **2010**, *2*, 957–967. [CrossRef]
16. Ju, J.; Roy, D.P.; Shuai, Y.; Schaaf, C. Development of an approach for generation of temporally complete daily nadir MODIS reflectance time series. *Remote Sens. Environ.* **2010**, *114*, 1–20. [CrossRef]
17. Yan, L.; Roy, D.P. Large-area gap filling of Landsat reflectance time series by spectral-angle-mapper based spatial-temporal similarity (samsts). *Remote Sens.* **2018**, *10*, 609. [CrossRef]
18. Li, T.; Wang, J.; Zhou, H. Modeling MODIS NBAR time series of vegetated surfaces and its use in LAI recursive estimation. In *IEEE Geoscience and Remote Sensing Symposium*; IEEE: Piscataway, NJ, USA, 2014; Volume 2014, pp. 2166–2169.
19. Li, T.; Wang, J.; Zhou, H.; Xiao, Z. MODIS NBAR time series modeling with two statistical methods and application to leaf area index recursive estimation. *IEEE J.-Stars.* **2015**, *8*, 1–9.
20. Mahiny, A.S.; Turner, B.J. A comparison of four common atmospheric correction methods. *Photogramm. Eng. Remote Sens.* **2007**, *73*, 361–368. [CrossRef]
21. Zhang, H.; Jiao, Z.; Li, X.; Huang, X.; Dong, Y. A priori knowledge application in the retrieval of surface albedo using hj-1 ccd data. *J. Remote Sens.* **2013**, *17*, 286–305. (In Chinese)
22. Zhu, X.; Chen, J.; Gao, F.; Chen, X.; Masek, J.G. An enhanced spatial and temporal adaptive reflectance fusion model for complex heterogeneous regions. *Remote Sens. Environ.* **2010**, *114*, 2610–2623. [CrossRef]
23. Hilker, T.; Wulder, M.A.; Coops, N.C.; Linke, J.; McDermid, G.; Masek, J.G.; Gao, F.; White, J.C. A new data fusion model for high spatial- and temporal-resolution mapping of forest disturbance based on landsat and modis. *Remote Sens. Environ.* **2009**, *113*, 1613–1627. [CrossRef]
24. Zhao, Y.; Huang, B.; Song, H. A robust adaptive spatial and temporal image fusion model for complex land surface changes. *Remote Sens. Environ.* **2018**, *208*, 42–62. [CrossRef]
25. Ying, H.; Leung, Y.; Cao, F.; Fung, T.; Xue, J. Sparsity-based spatiotemporal fusion via adaptive multi-band constraints. *Remote Sens.* **2018**, *10*, 1646. [CrossRef]
26. Heimhuber, V.; Tulbure, M.G.; Broich, M. Addressing spatio-temporal resolution constraints in Landsat and MODIS-based mapping of large-scale floodplain inundation dynamics. *Remote Sens. Environ.* **2018**, *211*, 307–320. [CrossRef]
27. Sun, C.; Liu, Q.; Wen, J.; Li, D.; Yu, K.; Zhang, Z. An algorithm for retrieving land surface albedo from hj-1 ccd data. *Remote Sens. Land Resour.* **2013**, *25*, 58–63.
28. Wanner, W.; Li, X.; Strahler, A.H. On the derivation of kernels for kernel-driven models of bidirectional reflectance. *J. Geophys. Res.* **1995**, *100*, 21077–21089. [CrossRef]

29. Roujean, J.L.; Leroy, M.; Deschamps, P.Y. A bidirectional reflectance model of the earth's surface for the correction of remote sensing data. *J. Geophys. Res.* **1992**, *97*, 20455–20468. [CrossRef]
30. Ebhuoma, O.; Gebreslasie, M.; Magubane, L. A seasonal autoregressive integrated moving average (SARIMA) forecasting model to predict monthly malaria cases in KwaZulu-Natal, South Africa. *S. Afr. Med. J.* **2018**, *108*, 573–578. [CrossRef]
31. Ben Abbes, A.; Bounouh, O.; Farah, I.R.; de Jong, R.; Martínez, B. Comparative study of three satellite image time-series decomposition methods for vegetation change detection. *Eur. J. Remote Sens.* **2018**, *51*, 607–615. [CrossRef]
32. Li, X.; Zhang, C.; Li, W.; Anyah, R.O.; Tian, J. Exploring the trend, prediction and driving forces of aerosols using satellite and ground data, and implications for climate change mitigation. *J. Clean. Prod.* **2019**, *223*, 238–251. [CrossRef]
33. Adeola, A.M.; Botai, J.O.; Mukarugwiza Olwoch, J.; De W. Rautenbach, H.C.J.; Adisa, O.M.; De Jager, C.; Botai, C.M.; Aaron, M. Predicting malaria cases using remotely sensed environmental variables in Nkomazi, South Africa. *Geospatial Health* **2019**, *14*, 1. [CrossRef] [PubMed]
34. Xia, D.; Wang, B.; Li, H.; Li, Y.; Zhang, Z. A distributed spatial–temporal weighted model on mapreduce for short-term traffic flow forecasting. *Neurocomputing* **2016**, *179*, 246–263. [CrossRef]

applied sciences

MDPI

Article

# Selective Search Collaborative Representation for Hyperspectral Anomaly Detection

**Chensong Yin, Leitao Gao, Mingjie Wang and Anni Liu ***

The 54th Research Institute of China Electronics Technology Group Corporation, Shijiazhuang 050050, China
* Correspondence: anniliu@bupt.edu.cn

**Abstract:** As an important tool in hyperspectral anomaly detection, collaborative representation detection (CRD) has attracted significant attention in recent years. However, the lack of global feature utilization, the contamination of the background dictionary, and the dependence on the sizes of the dual-window lead to instability of anomaly detection performance of CRD, making it difficult to apply in practice. To address these issues, a selective search collaborative representation detector is proposed. The selective search is based on global information and spectral similarity to realize the flexible fusion of adjacent homogeneous pixels. According to the homogeneous segmentation, the pixels with low background probability can be removed from the local background dictionary in CRD to achieve the purification of the local background and the improvement of detection performance, even under inappropriate dual-window sizes. Three real hyperspectral images are introduced to verify the feasibility and effectiveness of the proposed method. The detection performance is depicted by intuitive detection images, receiver operating characteristic curves, and area under curve values, as well as by running time. Comparison with CRD proves that the proposed method can effectively improve the anomaly detection accuracy of CRD and reduce the dependence of anomaly detection performance on the sizes of the dual-window.

**Keywords:** hyperspectral image; anomaly detection; selective search; collaborative representation

## 1. Introduction

Hyperspectral images (HSI) [1] have attracted widespread attention in many applications [2], such as geological exploration, environmental monitoring, and target detection [3] because they possess both spatial information and rich spectral information, which makes them appropriate for material identification and object detection. Among them, the utilization of HSIs for anomaly detection [4], which is a type of target detection without prior knowledge of the target [5,6], has practical significance for both civilian and military purposes.

The anomaly [7] refers to minority pixels with significantly different spectral characteristics from the majority pixels, which are referred to the background of the whole HSI. Desirable anomaly detection methods require appropriate background modeling [8] without prior knowledge to suppress the background and make the anomaly prominent [9]. Therefore, an accurate and effective model of the complex background [10] is critical in hyperspectral anomaly detectors [11], which can be generally categorized into two major types: statistical theory-based methods [12–14] and representation-based methods [15–17].

Statistical theory-based methods typically hypothesize that the background obeys a certain distribution and use statistical methods to estimate the likelihood that a pixel is attached to the background. As a representative of statistical theory-based methods, the Reed-Xiaoli (RX) detector [12] describes the background via a statistical multivariate Gaussian distribution model. However, this may not be correct for real HSI, leading to a higher false alarm rate. A series of improved algorithms have been proposed to improve the efficiency of the RX detector [18–20]. For example, the covariance matrix of the background is standardized in a regularized-RX detector [18], which can avoid the pathological state of matrix inversion. A nonlinear RX algorithm named kernel RX (KRX) [19] utilizes the kernel theory to project linearly inseparable HSI data to a high-dimensional feature space, where the differences between anomaly and background are strengthened. However, statistical theory-based methods suffer in situations with many kinds of background objects with irregular distributions. Given that these objects together do not necessarily follow some statistical distribution, it is difficult to correctly model many different background objects.

The rationale for the representation-based methods is that each background pixel in the HSI can be well expressed as a linear combination of a background dictionary by a function model, whereas the abnormal pixel cannot be similarly expressed. The abnormal degree of the pixel is evaluated by the representation residual; the larger the representation residual of a pixel, the higher the probability that the pixel will be attached to the anomaly. The representation-based methods mainly include sparse representation [21,22] and collaborative representation [23–25] according to different regularization constraints and different background dictionary constructions. Sparse representation holds that background pixels can be linearly and sparsely represented by a few atoms in an overcomplete dictionary, whereas anomaly pixels cannot. Therefore, sparse model optimization and complete dictionary construction are two key research aspects of sparse representation hyperspectral anomaly detectors. In [22], a low-rank and sparse representation (LRASR) detector applied a low-rank representation to model the background, while a sparsity-induced regularization term was employed to constrain the representation coefficient matrix. K-means was used to select the background dictionary atoms from the HSI data cube. Although sparse representation methods can detect anomalous targets to some extent, it is difficult to construct an over-complete background dictionary without abnormal pixel pollution when prior information is absent.

In addition to the sparse representation method, the collaboration representation detector (CRD) [23] has been proven to be simpler and more efficient, directly employing the neighboring pixels of the testing pixel within a sliding dual-window as the background dictionary and paying close attention to the collaboration among dictionary atoms. An $l_2$-norm-constrained coefficient vector is imposed to allow all of the atoms in the dictionary to participate in representation. However, the performance of CRD is unsatisfactory when the dual-window sizes are inappropriate, which are heavily dependent on the size of abnormal targets [23]. In the absence of prior information, it is difficult to set appropriate dual-window sizes [26] to ensure the purity of the background dictionary [27]. Furthermore, only local information attributes to the constraint weight vector and the global structure information of the image are disregarded, which limits the performance of the CRD as well. To improve detection accuracy, a collaborative-representation-based with outlier removal anomaly detector (CRBORAD) [28,29] has been proposed for purifying the background dictionary by eliminating some pixels with a small probability in the Gaussian distribution of neighborhood pixels. However, it is necessary to consider the actual situation that may not conform to statistical laws. The kernel method is also applied in CRD to improve its detection performance by projecting linearly inseparable data into a high-dimensional feature space in which those data become more separable [30,31]. Through the recursion between the kernel covariance matrices at the last and the current moments, as well as the detection values directly imposed on the regularized matrix instead of the kernel operations, repeated calculation is avoided and the overall computational complexity reduces, which was proposed in [32] and named real-time kernel CRD (RT-KCRD). A morphology-based collaborative representation detector (MCRD) [33] has been proposed to increase the incorporation of global spatial information of HSI by applying morphological filters. Each morphological profile (MP) is repeatedly acquired from each principal component of the data, which greatly increases computational complexity and running time. In [25], the various local spatial distribution information of the neighboring pixels of a test pixel were considered by adding a summation strategy to the local window in the CRD algorithm, which improved the accuracy of the linear representation and detection performance. However, this came at the cost of increased computational complexity and time, which is called local summation anomaly detection based on collaborative representation and inverse distance weight. Furthermore, the importance uniformity of different bands was broken in [34] by a self-weighted collaborative representation-based detector (SWCRD). Weight learning and collaborative representation are combined into a joint objective function to assign suitable weights to each band and simultaneously achieve collaborative representation, resulting in the improvement of anomaly detection of CRD.

In this paper, a selective search collaborative representation detector (SSCRD) is proposed to improve the anomaly detection capability and robustness of CRD, which not only effectively improves anomaly detection performance, but also avoids the sensitivity of detection accuracy on the size of the dual-window by taking global spatial information into consideration. First, a selective search is applied to HSI, in which adjacent pixels with similar spectral features are fused into several regions of variable size. The regions that occupy most of the pixels can be considered as the background, and the remaining small regions have a high probability of being abnormal. A selective search realizes the preliminary judgment of whether the pixels in an HSI are background or abnormal. Then, the result of the selective search is cleverly combined with CRD to effectively purify the local background defined by the dual-window, greatly reducing the possibility of background contamination. We evaluated and compared the proposed SSCRD using three real HSIs. The results show that the proposed SSCRD improves the detection accuracy of CRD. Moreover, superior detection performance can be maintained even when dual-window sizes are inappropriate.

## 2. Materials and Methods

### 2.1. Selective Search

As a region possesses richer information than a single pixel, a selective search aims to make full use of global spatial information and spectral information in an his, and to achieve homogeneous segmentation of the HSI. The selective search process contains two main phases: over-segmentation and region merging. First, a graph-based image segmentation technique [35] is applied to the HSI to divide the image into a series of initial regions as the processing unit. The similarities between all neighboring regions are then calculated and recorded. Subsequently, the greedy algorithm is applied to iteratively merge the regions until the separation of different objects is achieved. The results of the selective search will be regarded as the basic judgment in CRD for more precisely representing background characteristics.

Consider a 3-D HSI cube by $HSI = (X, E)$ in $R^d$, which is constituted by pixel spectra $X = \{x_i\}_{i=1}^{N}$ and edges $(x_i, x_j) \in E$ representing the connection of the pairs of neighboring pixels. $N$ is the total number of pixels in the image and $d$ is the number of spectral bands. Each edge $(x_i, x_j) \in E$ has a corresponding weight $w(x_i, x_j)$ which is defined by the spectral difference of the two neighboring pixels $x_i$ and $x_j$, connected by the edge. To achieve homogeneous over-segmentation of the HSI, the edges between two contiguous pixels in the same region should have relatively low weights, meaning that they belong to the same class. Therefore, the weight is defined as the following dynamic similarity operator for comprehensively evaluating the difference degree of the pixel spectra, which simultaneously considers the amplitude and shape similarity of pixel spectral curves:

$$w = Ent \times (1 - r) + LD \tag{1}$$

where $Ent$ is the information entropy of the spectral difference curve [35], usually defined as

$$Ent = -\sum_{t=1}^{d} P_t \lg(P_t) \tag{2}$$

and $P_t = x_t / \sum_{t=1}^{d} x_t$ is the probability of the $t^{th}$ value in the spectral difference vector $x$. $r$ is the correlation coefficient, a common criterion to describe the shape similarity of two spectral vectors, which is generally defined as:

$$r = \frac{\sum_{t=1}^{d} (x_{it} - \overline{x_i}) \cdot \sum_{t=1}^{d} (x_{jt} - \overline{x_j})}{\sqrt{\sum_{t=1}^{d} (x_{it} - \overline{x_i})^2 \cdot \sum_{t=1}^{d} (x_{jt} - \overline{x_j})^2}} \tag{3}$$

$\overline{x_*}$ represents the mean of the spectral vectors in all bands. $LD$ represents Lance distance [36], which is a typical numerical index of spectral similarity featuring insensitivity to singular values and a high ability of noise suppression. The $LD$ value ranges from 0 to 1, which is convenient for combining with the shape index $r$. The Lance distance is denoted as:

$$LD = \frac{1}{d} \sum_{t=1}^{d} \frac{|x_{it} - x_{jt}|}{x_{it} + x_{jt}} \tag{4}$$

Specifically, the smaller the $LD$ and the closer $r$ to 1 mean that the two spectra are more similar. When the shapes of the two spectra greatly differ, $Ent$ is larger and the weight of $1 - r$ is larger at this time, which means that the shape index plays a greater role in measuring the similarity of the two spectral curves. When $Ent$ is equal to 0, the shapes of the spectral curves are completely consistent. At this time, the difference in the spectral curve is mainly reflected in the numerical value, which is measured by $LD$. Generally, the smaller the weight, the more similar the two pixels are, and the sooner the two pixels are combined into a component.

In addition to the accurate description of spectral similarity, the criterion for determining whether two components can be the same is also critical. In general, two connected components can be considered the same when the minimum weight connecting the two components is less than the minimum internal difference for these two components. The minimum internal difference is defined as the minimum value of the largest weight of the minimum spanning tree of the two components. To make this criterion applicable to the case in which the two connected components are two single pixels, a threshold function $\tau(C) = K/[C]$ is appended to control the degree of difference between two different components to be greater than their respective internal differences, in which $[C]$ represents the number of pixels in the component and $K$ is a constant parameter that depends on the size of HSI and is preset to control the size of the formed components. Given that anomalies can occur at any location and in any shape, this over-segmentation method is opportunely applicable to hyperspectral anomaly detection, which is completely data-driven and can divide the image into a series of initial regions with freewill shape and size. Compared with some other over-segmentation algorithms, the graph-based over-segmentation method does not need control over the number of generated initial regions and the assignment for seed pixels of the regions, which is more convenient.

After the initial segmentation [37] of HSIs is achieved, it is important to further coalesce the minor regions to ensure that adjacent regions with the similar spectral properties are not discrete. Therefore, the spectrum similarity computation between all neighboring regions is then implemented to judge whether the regions belong to the same class or object. The average spectrum of all pixels in each region is utilized to represent the spectral characteristics of regions, and thereby, to calculate the similarities of adjacent regions. The similarity is computed by the linear combination of the spectral angle $sa$ and the correlation coefficient $r$, given by:

$$simi = \left(1 - \frac{sa}{\pi}\right) + \frac{(1+r)}{2} \tag{5}$$

Spectral angle (SA) mapper is a widely used criterion to evaluate the shape similarity of two spectral vectors by calculating the angle between two spectral vectors $x_i$ and $x_j$, and the angle can be calculated by:

$$sa = \arccos\left(\frac{x_i \cdot x_j}{\|x_i\| \cdot \|x_j\|}\right) \tag{6}$$

The two neighboring regions are more similar when the *simi* is larger, meaning the sooner the two regions will be merged into one region. In detail, the smaller the $sa$ and the closer $r$ to 1 mean that the two spectra are more similar. Furthermore, $1 - sa/\pi$ is in the range [0, 1] and $1 + r$ is in the range [0, 2] ($sa/\pi$ is in the range [0, 1] and $r$ is in the range $[-1, 1]$). Therefore, we divide $1 + r$ by 2 to ensure equal weights for $sa$ and $r$. Any reasonable combination of similarity criteria is allowed, and the definition we proposed is one of the more effective solutions.

The similarities of all pairs of initial neighboring regions are calculated and recorded in a similarity matrix. The greedy algorithm is used to iteratively merge the two adjacent regions with the highest similarity. By comparing the maximum of the spectral similarity of all adjacent regions with the pre-set threshold $T$, two involved regions will be merged into the same region when the similarity maximum is greater than the threshold. The pre-set threshold $T$ is used for constraining the degree of region, and its value is in the range of [0, 2], determined by the minimum and the maximum of the similarity matrix. The similarity values related to the two merged regions in the original similarity matrix are deleted. The similarity matrix is updated by computing the similarities between the new region and its neighboring regions. This process is looped until the similarity maximum is smaller than the threshold and the merging cannot be continued. Finally, the overall HSI is merged into a few regions. The regions that occupy most of the pixels are considered the background and these pixels are labeled as one. Regions with relatively few pixels have a high probability of anomaly and the corresponding pixels are labeled as zero. The result of the selective search is a 2D matrix $P$ (of size $n \times m$) with zero and one elements, where $n$ and $m$ are the width and height of the HSI, respectively.

### 2.2. Selective Search Collaborative Representation Detector

For each pixel $y$ in HSI, it has a local background composed of its own neighboring pixels, which are falling in a dual-window with the inner window $w_{in} \times w_{in}$ and the outer window $w_{out} \times w_{out}$ centered at the pixel $y$, as shown in Figure 1. Then, a 2D matrix $X_{lb} = \{x_i\}_{i=1}^{s}$ can be used to represent the local background, where $s$ is the number of local background pixels calculated via $s = \omega_{out} \times \omega_{out} - \omega_{in} \times \omega_{in}$. In addition, the local background $X_{lb} = \{x_i\}_{i=1}^{s}$ for each pixel has its corresponding pixel label matrix $P_{lb} = \{0,1\}_{i=1}^{s}$ (of size $1 \times s$) from the result of the selective search $P$. Dot-multiplying the local neighborhood matrix $X_{lb} = \{x_i\}_{i=1}^{s}$ and the pixel label matrix $P_{lb} = \{0,1\}_{i=1}^{s}$ can effectively remove those pixels with higher anomaly probability in the neighborhood matrix, thereby obtaining a purer local background description.

The proposed SSCRD is based on the principle that the background pixels can be approximately represented by their adjacent pixels but the anomaly pixels cannot, which is in accordance with that of CRD. This principle can be interpreted into mathematical language, which is used to solve the weight vector $\alpha$, so that $\|y - (P_{lb} * X_{lb})\alpha\|_2^2$ is minimized under the constraint that $\|\alpha\|_2^2$ is also minimized for obtaining the optimal solution, where $y$ is the pixel under test. The objective function of SSCRD can be expressed as:

$$\underset{\alpha}{\operatorname{argmin}} \|y - (P_{lb} * X_{lb})\alpha\|_2^2 + \lambda\|\alpha\|_2^2 \tag{7}$$

where $\lambda$ is the Lagrange multiplier. Equation (7) can be expanded as:

$$\underset{\alpha}{\operatorname{argmin}}\left[\alpha\left((P_{lb} * X_{lb})^T(P_{lb} * X_{lb}) + \lambda I\right)\alpha - 2\alpha^T(P_{lb} * X_{lb})^T y\right] \tag{8}$$

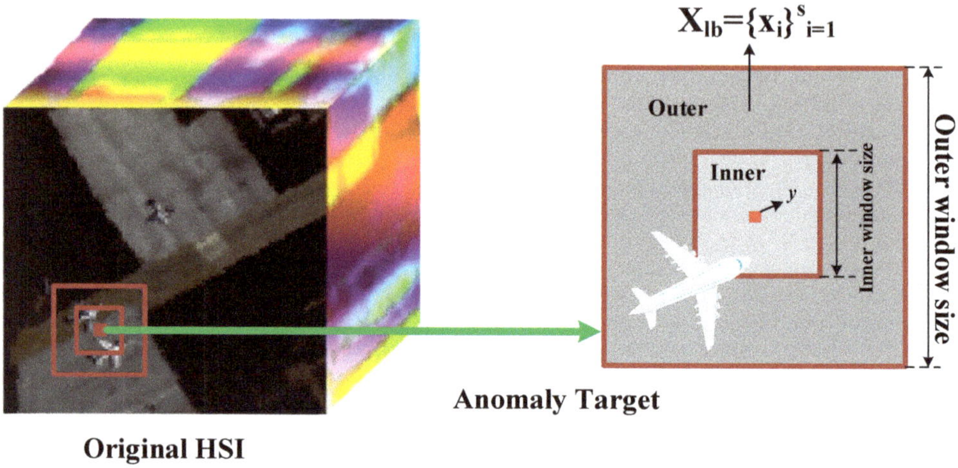


**Figure 1.** Sketch of the dual-window in CRD with the San Diego-II data set. Some aircraft pixels in the image may fall into the outer window.

Then, taking the partial derivative of $\alpha$ and setting the resultant equation to zero can obtain:

$$\hat{\alpha} = \left((P_{lb} * X_{lb})^T(P_{lb} * X_{lb}) + \lambda I\right)^{-1}(P_{lb} * X_{lb})^T y \tag{9}$$

where $I$ is the identity matrix of $s \times s$. For surrounding pixels that are quite different from the center pixel, the coefficients should be small. Therefore, a distance-related weight to each pixel of $P_{lb} * X_{lb}$ is assigned by introducing the Tikhonov regularization matrix $\Gamma_s$ used in [38,39]. The objective function of SSCRD can be expressed as:

$$\underset{\alpha}{\operatorname{argmin}}\|y - (P_{lb} * X_{lb})\alpha\|_2^2 + \lambda\|\Gamma_s\alpha\|_2^2 \tag{10}$$

where $\Gamma_s$ is defined as:

$$\Gamma_s = \begin{bmatrix} \|y - P_{lb_1} * X_{lb_1}\|_2 & & 0 \\ & \ddots & \\ 0 & & \|y - P_{lb_s} * X_{lb_s}\|_2 \end{bmatrix} \tag{11}$$

Similarly, taking the partial derivative of $\alpha$ and setting the resultant equation to zero can obtain:

$$\hat{\alpha} = \left((P_{lb} * X_{lb})^T(P_{lb} * X_{lb}) + \lambda\Gamma_s^T\Gamma_s\right)^{-1}(P_{lb} * X_{lb})^T y \tag{12}$$

Then, the residual image can be obtained by subtracting the original hyperspectral data and the predicted background data, expressed as:

$$\gamma = \|y - (P_{lb} * X_{lb})\hat{\alpha}\|_2 \tag{13}$$

If $\gamma$ is larger than a threshold, then the corresponding pixel is claimed to be an anomalous pixel. The flow of the proposed selective search collaborative representation anomaly detection algorithm is summarized in Algorithm 1.

In CRD, the background dictionary of the pixels to be measured framed by the dual-window will be seriously polluted by abnormal pixels under the condition of an inappropriate dual-window size, which leads to the difficulty of effectively separating abnormal and background pixels, resulting in unsatisfactory anomaly detection effectiveness of CRD. The SSCRD proposed in this article can effectively reduce the abnormal pollution of local background, improving the robustness of CRD and obtaining satisfactory detection capability under different dual-window sizes.

**Algorithm 1:** Selective Search Collaborative Representation Anomaly Detector

**Input** (1). A hyperspectral image *HSI*
(2). The parameter K for the creation of initial region
(3). The threshold T for the region Merging
(4). The dual-window sizes
**Output** A two-dimensional map recording the detection result *Img-R*
**Procedure** For the input hyperspectral image data *HSI*
(1). Obtain initial region $R = \{r_1, \ldots, r_n\}$ using graph-based image segmentation;
(2). Further region merging via greedy algorithm
(a) Calculate similarity matrix *Simi*
(b) Get highest similarity $maxSimi = max(Simi) = simi(r_i, r_j)$;
(c) **While** $maxSimi \geq T$ **do**
Merge corresponding regions $r_t = r_i \cup r_j$;
Update similarity set *Simi*: $Simi = Simi \cup Simi_t$;
Update highest similarity $maxSimi = max(Simi) = simi(r_i, r_j)$;
(d) Obtain the label matrix *P*;
(3). **For** each pixel $y$ **do**
(a) Getting$X_{lb} = \{x_i\}_{i=1}^{s}$ and $P_{lb} = \{0,1\}_{i=1}^{s}$ from the dual-window;
(b) Calculating weight vector $\hat{\alpha}$ via the $l_2$-regularized minimization with Equation (12);
(c) Obtaining the final residual $\gamma$ via Equation (13);

## 3. Results

To evaluate the detection performance of the proposed SSCRD, experiments on three typical HSIs were conducted. The SSCRD was compared with CRD [23], CRBORAD [29], and LRSRD [22] to verify its effectiveness. The receiver operating characteristic (ROC) [40], the area under the curve (AUC) [41], and running time were applied to evaluate the performances. All the experiments were implemented in MATLAB on a Dell XPS13 personal laptop with an Intel Core i5-8250U CPU with 8 GB of RAM.

### 3.1. Dataset

(1) San Diego: Two images of size 100 × 100 are extracted from this dataset, mainly covering the urban scene where San Diego Airport is located in California, USA https://www.erdc.usace.army.mil/ (accessed on 1 May 2020). A total of 189 bands from the raw dataset are reserved for the experiments. The aircraft in each image are regarded as anomalies.
(2) ABU Airport: One image is derived from this dataset obtained from the airborne visible/infrared imaging spectrometer (AVIRIS) sensor, http://aviris.jpl.nasa.gov/ (accessed on 1 May 2020), with 3.5-m spatial resolution. The size of the image is 100 × 100 with 191 spectral bands. Three aircraft in the image are regarded as anomalies.

### 3.2. Detection Performance

The segmentation parameters $K$ for three images are all set as 2 since the sizes of them are the same. The regularization parameter $\lambda$ for CRD, CRBORAD, and SSCRD are all set as $10^{-6}$ according to the original literature [23]. All of the settings for parameter $T$ correspond to the optimal label map $P$ for each image, which are obtained by traversal of the range of parameter values. In addition, pixels with a higher background probability are represented by black, and pixels with a higher abnormal probability are represented by white to more prominently display the label map $P$.

The colour detection graphs for the San Diego-I image, the corresponding false color image, the ground-truth map, the result of graph-based over-segmentation, and the corresponding label map $P$ obtained from the selective search are shown in Figure 2. The threshold $T$ in the region fusion is set at 1.15 for the San Diego-I image because of the optimal regional fusion effect. The color detection graphs of CRD-based methods are shown in Figure 2, and are obtained when the sizes of the dual-window are (7,11), which is a typical sample of unbefitting dual-window sizes for CRD in the San Diego-I image. Since the proposed SSCRD aims to make up for the defect that the local background in CRD is abnormally polluted, and then to improve the anomaly detection accuracy of CRD, the sizes of the dual-window for the representation of detection results are set to present the intuitive and effective improvement of SSCRD compared with CRD. Compared with the ground-truth map, the anomaly detection performance of CRD is unsatisfactory because of the blurred aircraft shapes and the difficulty in distinguishing the aircraft from pixels misjudged as anomalies. As shown in Figure 2, the detection graph of CRBORAD is slightly improved compared with that of CRD, and more pixels of three aircraft are identified. However, the shapes of the aircraft remain unidentifiable, and some background pixels are misjudged as abnormal, which seriously affects the correct judgment of the anomaly. By contrast, SSCRD successfully detects more abnormal pixels compared with CRD and CRBORAD, making the shape of the aircraft more complete and clearer. Furthermore, the detection performance of LRSRD is passable because of the recognizable locations and shapes of the three aircraft. However, some background pixels on the bottom left edge of the image have extremely high scores to be misjudged as anomalies. This could be attributed to an inaccurate and incomplete background dictionary.

The ROC curves of the anomaly detectors for the San Diego-I image are shown in Figure 3 for the quantitative comparison of detection performance under the situation that the sizes of the dual-window are (7,11) as well. As shown in Figure 3, the probability of detection for CRD is less than 0.3 when the false alarm rate is 0.1, while the probability of detection for CRD is 0.78 when the false alarm rate rises to 0.5. In general, the ROC curve of the detection results of the CRD on the San Diego-I image shows a large gap from the upper left corner of the coordinate axis, meaning that the CRD fails to accurately detect the anomalies. The corresponding AUC value (in percent) of the CRD is 71.81. Compared with the ROC curve of CRD, the ROC curve of CRBORAD is closer to the top left corner, indicating a slight improvement in anomaly detection performance. Concretely, the probability of detection reaches 0.55 when the false alarm rate is 0.1, while the probability of detection is 0.8 when the false alarm rate is 0.25. The AUC value (in percent) of CRBORAD is 85.48. Based on the ROC curve of SSCRD, the detection effect is significantly improved by the selective search compared with CRD and CRBORAD. When the false alarm rate is 0.1, the probability of detection reaches 0.87, and the curve is closest to the upper left corner of the coordinate axis. Moreover, the detection probability of SSCRD is always higher than that of CRD and CRBORAD when the false-alarm rate changes from 0 to 1, showing a considerable improvement compared to CRD. The RCO curves of LRSRD and SSCRD are slightly different, and the AUC value (in percent) of SSCRD and LRSRD are 94.77 and 92.90, respectively, indicating the superior detection performance of SSCRD for the San Diego-I image.

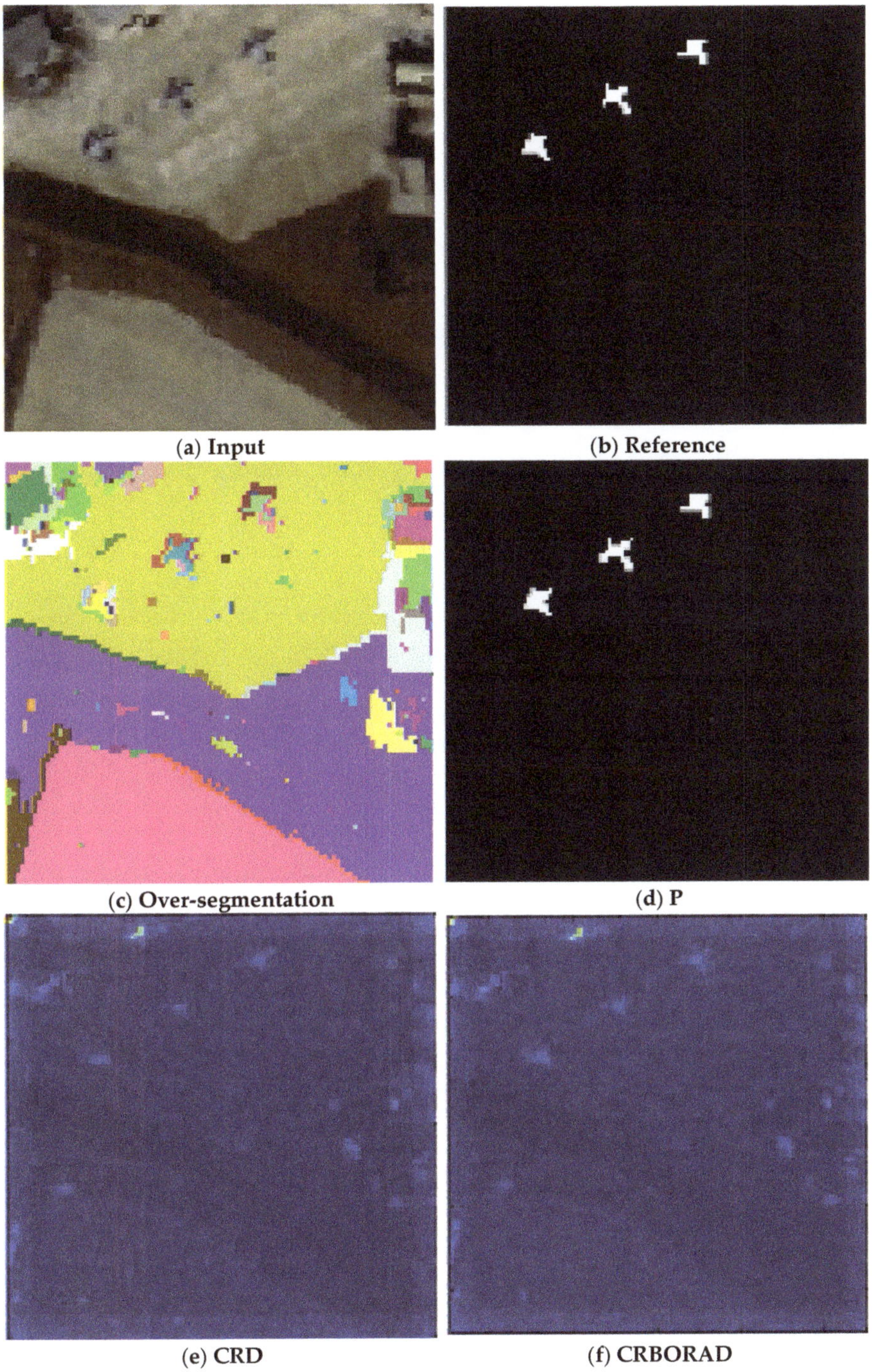


**Figure 2.** *Cont.*

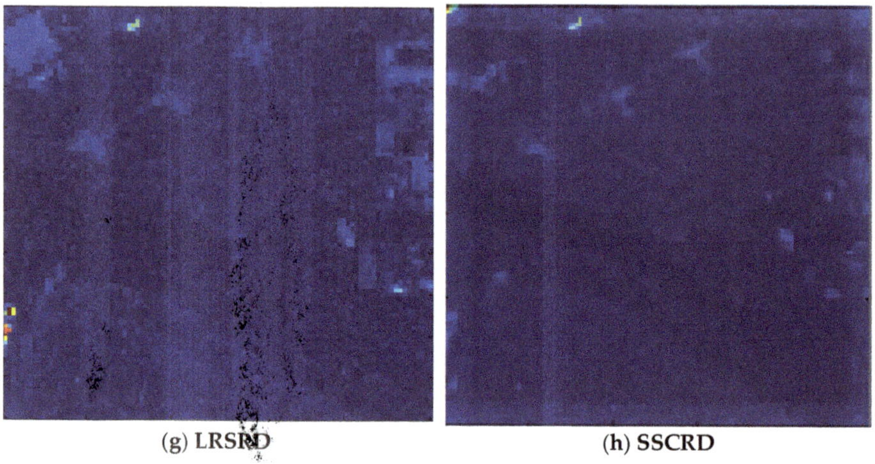


**Figure 2.** Color detection maps for San Diego-I image with dual-window sizes (7,11). (**a**) False color image; (**b**) ground-truth map; (**c**) over-segmentation; (**d**) label map P; (**e**) the detection result of CRD; (**f**) the detection result of CRBORAD; (**g**) the detection result of LRSRD; (**h**) the detection result of SSCRD.

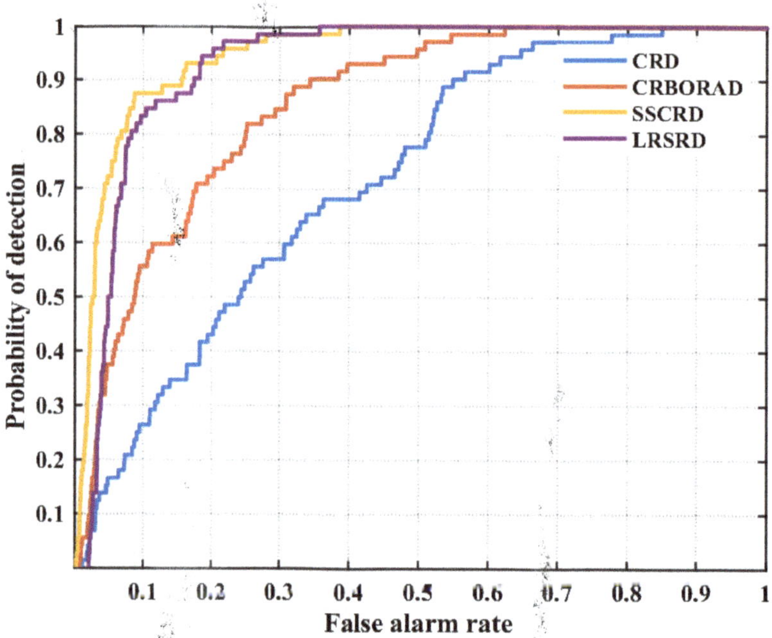


**Figure 3.** ROC curves for San Diego-I dataset with dual-window sizes (7,11) for CRD, CRBORAD, and SSCRD.

For the San Diego-II image, the false color image, ground-truth maps, the result of over-segmentation, the label map $P$ obtained from the selective search, and the corresponding color detection graphs are shown in Figure 4. The threshold $T$ in the region fusion is set at 1.40 for the San Diego-II image because of the optimal regional fusion effect. To present the intuitive and effective improvement of SSCRD compared with CRD, the sizes of the dual-window for the San Diego-II image are set at (9,13) for CRD, CRBORAD, and SSCRD, which is a typical sample of unbefitting dual-window sizes for CRD in the San Diego-II image. As shown in Figure 4e, the anomaly detection effectiveness of CRD is very general. First, the shapes of the two airplanes as the abnormal target to be identified at the lower left corner are relatively obscure, and it is difficult to distinguish them from the background pixels misjudged as abnormal. Second, the position and shape of the other airplane as the abnormal target to be identified are completely indistinguishable. Compared with CRD, the anomaly detection capacity of CRBORAD improves to some extent, as shown in Figure 4f. Not only are the positions and shape of the two airplanes at the lower left corner generally discernible, but also some anomaly pixels belonging to another smaller airplane also stand out. However, the shape of this anomaly target is still too fuzzy to recognize it as an airplane. Due to the selective search, both the positions and shapes of the three aircraft are clearly presented in the color detection graph of SSCRD. The scores of those anomaly pixels are high enough to highlight them from a complex background, thereby achieving effective anomaly detection. Thus, SSCRD greatly improves the hyperspectral anomaly detection capability of CRD. As for LRSRD, the two airplanes in the lower left corner are clearly detected, but the other airplane is vague and unrecognizable. Many background pixels in the upper-right and lower-right corners of the image possess high scores, which disrupts the effective detection of anomalies.

The ROC curves of the four anomaly detectors under the dual-window sizes of (9,13) for the San Diego-II image are shown in Figure 5 for a quantitative comparison of detection performance. As shown in Figure 5, the probability of detection for CRD is less than 0.6 when the false alarm rate is 0.1, while the probability of detection for CRBORAD is about 0.78, and that for SSCRD is already close to 1. The furthest away from the upper left corner of the coordinate axis indicates that the anomaly detection accuracy of CRD needs to improve. The ROC curve of SSCRD is much closer to the upper left corner of the coordinate system, indicating a great improvement compared with CRD and a better detection performance than that of CRBORAD and LRSRD. The corresponding AUC values (in percent) are 85.19, 88.69, 94.30, and 98.68 for CRD, CRBORAD, LRSRD, and SSCRD, respectively.

The false color image, ground-truth maps, the result of over segmentation, the label map $P$ obtained from the selective search, and the corresponding color detection graphs for the ABU Airport image are shown in Figure 6. The threshold $T$ in the region fusion is set at 1.95 for the ABU Airport image because of the optimal regional fusion effect. The sizes of the dual-window for the ABU Airport image are set at (11,15) for CRD, CRBORAD, and SSCRD, which is a typical sample of unbefitting dual-window sizes for CRD in the ABU Airport image. As can be seen from the CRD detection results (Figure 6e), except that the position of the left aircraft is vaguely detected, the rest of the anomalies are not detected. The shapes and outlines of the aircraft are blurred. By comparing the detection result of CRBORAD (Figure 6f) with the ground-truth map (Figure 6b), the position of the aircraft on the left side is detectable and its shape is roughly distinguished. The positions of the two smaller aircraft on the right side are indistinct. The shape and outer profile are totally unidentifiable. In the detection results of SSCRD (Figure 6h), the position and shape of the aircraft on the left side can be clearly detected, and the position of the two aircraft on the right side are clearly detected, but their shapes and contours cannot be successfully detected. Moreover, the misjudgement of the detection graph is similar to that of CRD. There is some misjudgement at the bottom left of the image, in which some background pixels are judged to be abnormal, which may be attributed to the basic principle of CRD. The extremely high scores of those with misjudged backgrounds are responsible for the ambiguity of anomaly detection. Compared with CRD, CRBORAD, and SSCRD, the score of the larger aircraft on the left obtained from LRSRD is sufficiently high to make this anomaly clearly visible. However, the two smaller aircraft on the right side are still indistinct. LRSRD has a significantly higher false alarm rate. Many background pixels are displayed in the anomaly detection image.

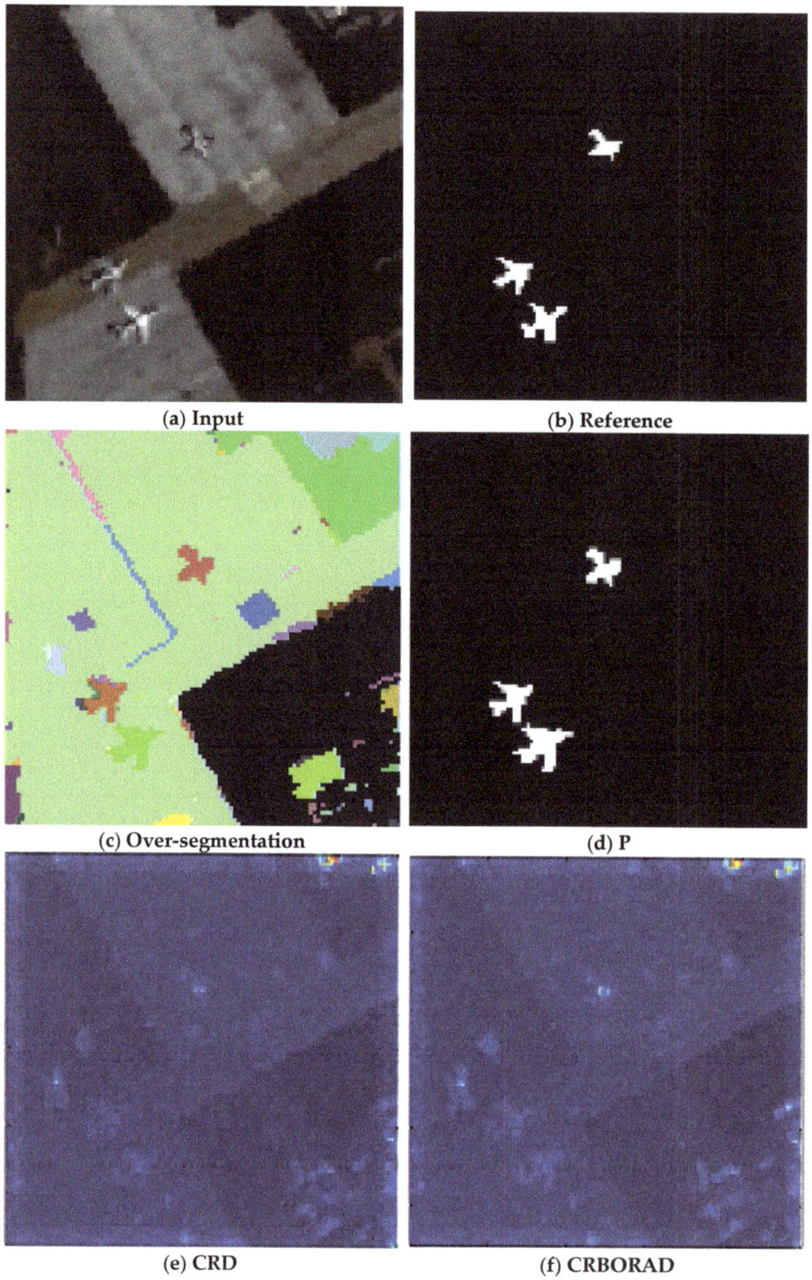


**Figure 4.** *Cont.*

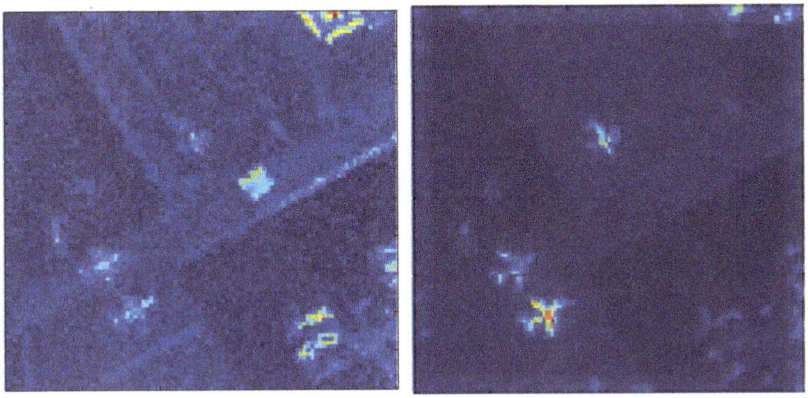

(g) LRSRD (h) SSCRD

**Figure 4.** Color detection maps for San Diego-II image with dual-window sizes (9,13). (**a**) False color image; (**b**) ground-truth map; (**c**) over-segmentation; (**d**) label map P; (**e**) the detection result of CRD; (**f**) the detection result of CRBORAD; (**g**) the detection result of LRSRD; (**h**) the detection result of SSCRD.

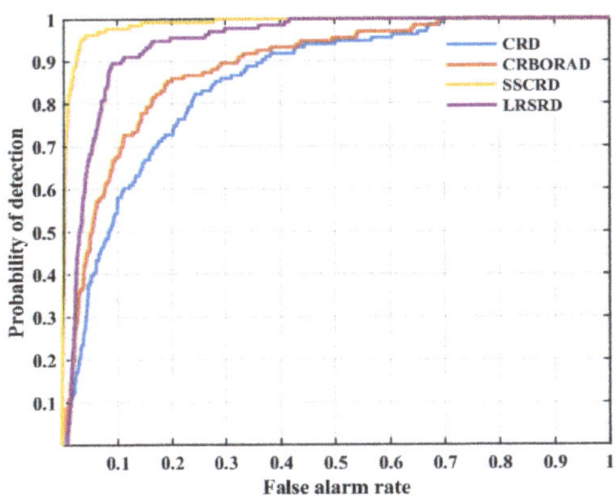


**Figure 5.** ROC curves for San Diego-II dataset with dual-window sizes (9,13) for CRD, CRBORAD, and SSCRD.

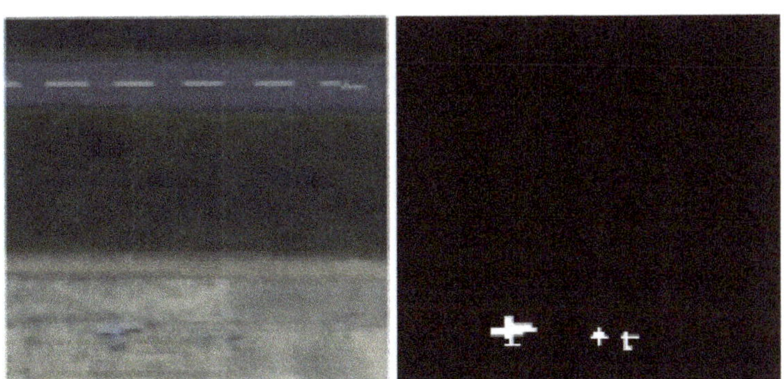

(a) Input (b) Reference

**Figure 6.** *Cont.*

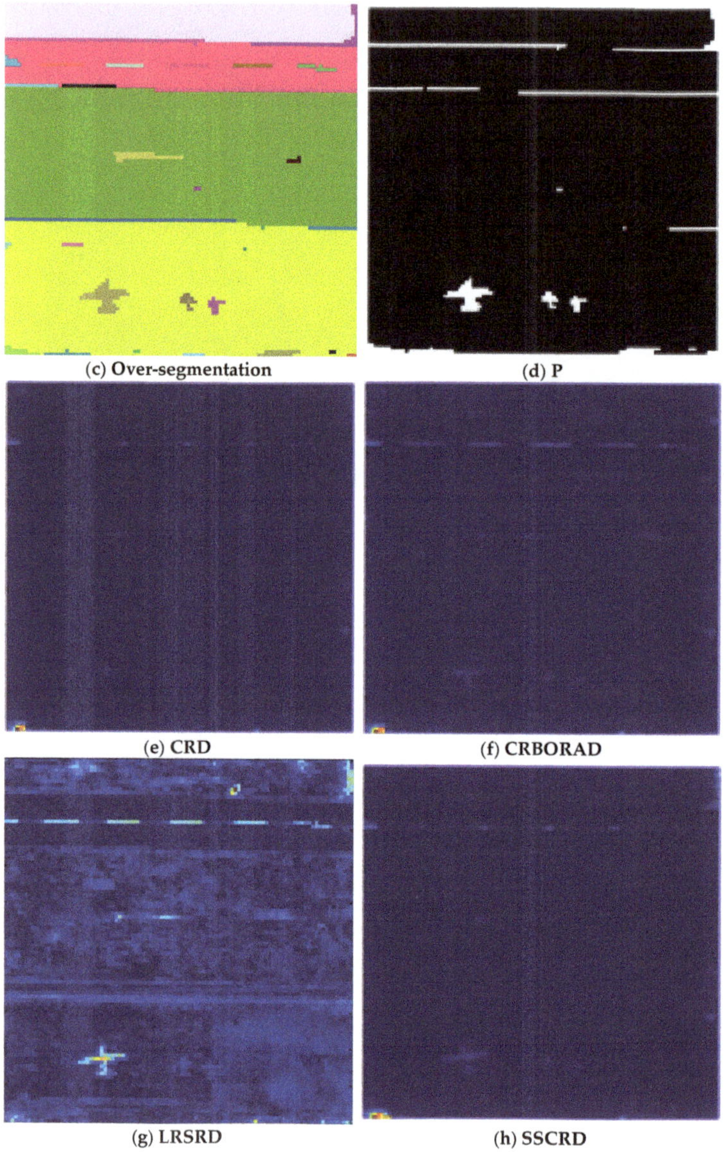


**Figure 6.** Color detection maps for ABU Airport dataset with dual-window sizes (11,15). (**a**) False color image; (**b**) ground-truth map; (**c**) over-segmentation; (**d**) label map P; (**e**) the detection result of CRD; (**f**) the detection result of CRBORAD; (**g**) the detection result of LRSRD; (**h**) the detection result of SSCRD.

The ROC curves of the four anomaly detectors for the ABU Airport image are shown in Figure 7 for the quantitative comparison of detection performance under the situation that the sizes of the dual-window are (11,15). Figure 7 shows that the detection effect of CRD is very poor. The curve is at a significant distance from the upper left corner of the coordinate axis. Moreover, when the probability of detection is greater than 0.8, the corresponding false alarm rate is as high as 0.35. The corresponding AUC value (in percent) of the CRD is 80.88. Compared with CRD, the ROC curve of CRBORAD is significantly closer to the upper left corner of the coordinate axis. When the false alarm rate is 0.1, the probability of detection reaches 0.8. The AUC value (in percent) of CRBORAD is 93.78. Compared with CRBORAD, the ROC curve of SSCRD is closer to the upper left corner of the coordinate axis, and the probability of detection reaches 0.9 when the false alarm rate is 0.1. The ROC curve of SSCRD is much closer to the upper left corner of the coordinate axis than that of CRD and CRBORAD, and the corresponding AUC value of SSCRD is 96.74, showing a considerable improvement to CRD and superior detection performance. Due to the failure detection of two small aircraft, the ROC curve (in percent) of LRSRD is slightly lower than CRBORAD and SSCRD, and the AUC value (in percent) of LRSRD is 92.99.

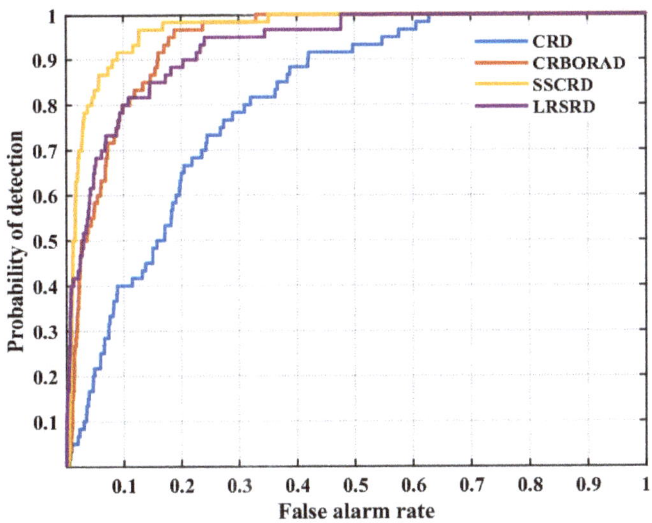


**Figure 7.** ROC curves for ABU Airport image dataset with dual-window sizes (11,15) for CRD, CRBORAD, and SSCRD.

## 4. Discussion

The sizes of the dual-window are two important parameters of CRD, which depend on the size of the anomaly in the image of each dataset and determine the anomaly detection availability of CRD. It is necessary to discuss the effect of dual-window sizes on the anomaly detection performance. Tables 1–3 display the AUC values of CRD, CRBORAD, and SSCRD, respectively, with varying window sizes on the three images.

**Table 1.** AUC values (in %) and computing times (in seconds) for San Diego-I image.

| Win_Size | | CRD | | CRBORAD | | SSCRD | |
|---|---|---|---|---|---|---|---|
| $W_{out}$ | $W_{in}$ | AUC | Time | AUC | Time | AUC | Time |
| 15 | 13 | 96.90 | 12.58 | **98.38** | 15.50 | 98.19 | 29.15 |
| | 11 | 91.38 | 26.71 | **95.94** | 25.85 | 95.73 | 42.80 |
| 13 | 11 | 94.10 | 9.61 | **97.49** | 11.01 | 97.41 | 25.96 |
| | 9 | 84.07 | 20.36 | 93.26 | 20.77 | **94.88** | 36.88 |
| 11 | 9 | 88.51 | 7.60 | 94.99 | 8.96 | **97.12** | 23.58 |
| | 7 | 71.81 | 14.88 | 85.48 | 16.74 | **94.77** | 33.61 |
| 9 | 7 | 80.90 | 4.97 | 89.03 | 7.32 | **96.73** | 23.65 |
| | 5 | 64.35 | 10.31 | 71.46 | 12.55 | **96.47** | 29.05 |
| 7 | 5 | 78.76 | 4.05 | 80.12 | 5.56 | **98.59** | 21.07 |
| | 3 | 62.42 | 6.81 | 64.22 | 8.56 | **99.13** | 22.74 |

**Table 2.** AUC values (in %) and computing times (in seconds) for San Diego-II image.

| Win_Size | | CRD | | CRBORAD | | SSCRD | |
|---|---|---|---|---|---|---|---|
| $W_{out}$ | $W_{in}$ | AUC | Time | AUC | Time | AUC | Time |
| 19 | 17 | 94.51 | 17.21 | 95.00 | 18.93 | **97.43** | 26.28 |
| | 15 | 88.14 | 42.22 | 89.86 | 33.10 | **95.86** | 42.09 |
| 17 | 15 | 94.90 | 15.37 | 95.25 | 16.30 | **98.19** | 22.05 |
| | 13 | 89.42 | 35.65 | 91.66 | 31.64 | **97.59** | 40.00 |
| 15 | 13 | 94.29 | 12.44 | 95.50 | 14.07 | **98.48** | 20.25 |
| | 11 | 86.88 | 29.10 | 90.38 | 28.39 | **98.11** | 36.41 |
| 13 | 11 | 91.62 | 10.22 | 93.54 | 11.89 | **98.49** | 18.12 |
| | 9 | 85.19 | 21.10 | 88.69 | 22.05 | **98.68** | 29.05 |
| 11 | 9 | 90.23 | 8.14 | 92.52 | 9.72 | **99.07** | 17.16 |
| | 7 | 84.35 | 16.68 | 88.84 | 16.95 | **99.27** | 24.02 |

**Table 3.** AUC values (in %) and computing times (in seconds) for ABU Airport image.

| Win_Size | | CRD | | CRBORAD | | SSCRD | |
|---|---|---|---|---|---|---|---|
| $W_{out}$ | $W_{in}$ | AUC | Time | AUC | Time | AUC | Time |
| 19 | 17 | 94.91 | 15.94 | 96.35 | 18.09 | **97.03** | 22.14 |
| | 15 | 86.29 | 39.43 | 91.75 | 31.95 | **93.38** | 38.14 |
| 17 | 15 | 93.15 | 15.80 | 96.71 | 18.26 | **97.49** | 19.54 |
| | 13 | 81.10 | 36.83 | 94.03 | 33.00 | **95.59** | 38.95 |
| 15 | 13 | 90.98 | 13.15 | 96.76 | 14.23 | **97.96** | 17.24 |
| | 11 | 80.88 | 30.09 | 93.78 | 29.40 | **96.74** | 31.98 |
| 13 | 11 | 89.19 | 10.06 | 95.80 | 12.47 | **98.20** | 14.98 |
| | 9 | 77.68 | 21.23 | 89.21 | 22.93 | **97.96** | 25.50 |
| 11 | 9 | 84.37 | 7.84 | 88.90 | 10.16 | **98.89** | 12.66 |
| | 7 | 70.64 | 16.06 | 79.80 | 17.45 | **98.73** | 21.14 |

For the San Diego-I image, the AUC values (in percent) of CRD range from 62.42 to 96.90, and the AUC values (in percent) of CRBORAD range from 64.22 to 98.38 when the size of the dual-window varies. Moreover, the AUC values (in percent) of the proposed SSCRD range from 94.88 to 99.13 as the size of the dual-window changes. When the size of the dual-window is set to (15,13), (15,11), or (13,11), the AUC values of the three algorithms are not significantly different, and all are at a high level. This is because the inner window with a relatively large size covers those nearby anomaly pixels when the pixel to be measured is abnormal, so there are almost no abnormal pixels in the local background framed by the dual-window. In this case, purification processing of the local background is superfluous. However, improper selection of the dual-window sizes results in a lower AUC value, meaning the deterioration of detection performance of CRD and CRBORAD. As the size of the dual-window decreases, the AUC values of CRD and CRBORAD both decrease. For example, the AUC values (in percent) of CRD and CRBORAD are 64.35 and 71.16 when the size of the dual-window is (5,9). The AUC value of CROBORAD deteriorates slightly less than that of CRD. The small size of the inner window could not frame the nearby abnormal pixels when the size of the anomaly to be detected is relatively large, resulting in some abnormal pixels within the outer window and the impurity of the local background. Due to the selective search process, the AUC of SSCRD has a stable high-level value, meaning that the dependence of the CRD anomaly detection performance on the size of the dual window is greatly reduced. When the size of the dual-window is (5,9), the AUC value (in percent) of SSCRD is 96.47, which is more than 25% higher than that of CRD and CROBORAD. In addition, different sizes of dual-window correspond to different local background dictionaries and slightly different detection performances of SSCRD. From Table 1, we calculate the average improvement effect of SSCRD on CRD to be 21.8% under different dual-window sizes, while the average improvement effect of CRBORAD on CRD is only about 7.3%.

A similar phenomenon can be observed for San Diego-II, as shown in Table 2. The AUC values of SSCRD are larger than those of CRD and CRBORAD under different dual-window sizes. Furthermore, the relatively large inner window size enables CRD and CRBORAD to obtain larger AUC values when the size of the outer window is consistent. The highest AUC value of SSCRD is 99.27, which is close to 100%, while the highest AUC values of CRD and CRBORAD are 94.09 and 95.50, respectively. Furthermore, the AUC values of CRD and CRBORAD under different dual-window sizes fluctuate in the range of nearly 10% and 6%, whereas the range of AUC value fluctuation of SSCRD is less than 4%. The minimum AUC value of SSCRD is 95.86, which is even greater than the maximum AUC values of CRD and CRBORAD, indicating a significant improvement in the detection performance of CRD via applying selective search.

For the ABU Airport image, as shown in Table 3, the AUC values (%) of CRD range from 70.64 to 94.91, and the AUC values (%) of CRBORAD range between 79.80 and 96.76. For comparison, the AUC values (%) of SSCRD range from 93.38 to 98.89. On the one hand, SSCRD has a better effect on improving the anomaly detection performance of CRD compared with CRBORAD. On the other hand, SSCRD effectively enhances the robustness of CRD anomaly detection performance under different dual-window sizes. Specifically, we deduce that the average improvement effect of CRBORAD on CRD is 9.05%, and the average improvement effect of SSCRD on CRD is about 15.34%, according to Table 3.

Lastly, the running times of these three local anomaly detectors are listed in Tables 1–3, indicating that the running times of the three algorithms are not much different and that the more pixels in the local background, the longer the running time.

## 5. Conclusions

In this study, we proposed a selective search collaborative representation detector to improve the widely-studied collaborative representation detector. Due to a selective search, not only is the anomaly detection performance of CRD improved, but the sensitivity of the detection accuracy to the size of the dual-windows is also avoided. The selective search cleverly combines the global structure of the HSI with the spectral similarity of adjacent pixels to divide the whole image into several homogeneous regions, so as to achieve the initial separation of background pixels and highly likely abnormal pixels. This preliminary judgment of whether the pixels in an HSI are background or abnormal can effectively purify the local background defined by a dual-window, and then greatly reduce the possibility of background contamination. Three HSIs were introduced to verify the feasibility and effectiveness of the proposed SSCRD. Compared with CRD and CRBORAD, the proposed SSCRD maintained superior detection performance under different sizes of dual-windows.

**Author Contributions:** Conceptualization, C.Y. and A.L.; Methodology, L.G.; Validation, C.Y., M.W. and A.L.; Formal Analysis, L.G.; Resources, C.Y.; Data Curation, M.W.; Writing—Original Draft Preparation, C.Y.; Writing—Review and Editing, C.Y. and A.L.; Visualization, L.G.; Supervision, A.L. All authors have read and agreed to the published version of the manuscript.

**Funding:** This research received no external funding.

**Institutional Review Board Statement:** Not applicable.

**Informed Consent Statement:** Not applicable.

**Data Availability Statement:** Not applicable.

**Acknowledgments:** The authors thank Editors and Reviewers for the careful editorial processing and the constructive suggestions.

**Conflicts of Interest:** The authors declare no conflict of interest.

## References


1. Khan, M.J.; Khan, H.S.; Yousaf, A.; Khurshid, K.; Abbas, A. Modern trends in hyperspectral image analysis: A review. *IEEE Access* **2018**, *6*, 14118–14129. [CrossRef]
2. Paris, S.; Mary, D.; Ferrari, A. Detection Tests Using Sparse Models, With Application to Hyperspectral Data. *IEEE Trans. Signal Process.* **2013**, *61*, 1481–1494. [CrossRef]
3. Stein, D.W.J.; Beaven, S.G.; Hoff, L.E.; Winter, E.M.; Schaum, A.P.; Stocker, A.D. Anomaly detection from hyperspectral imagery. *IEEE Signal Process. Mag.* **2002**, *19*, 58–69. [CrossRef]
4. Matteoli, S.; Diani, M.; Corsini, G. A tutorial overview of anomaly detection in hyperspectral images. *IEEE Aerosp. Electron. Syst. Mag.* **2010**, *25*, 5–28. [CrossRef]
5. Merrill, N.; Olson, C.C. Unsupervised Ensemble-Kernel Principal Component Analysis for Hyperspectral Anomaly Detection. In Proceedings of the IEEE/CVF Conference on Computer Vision and Pattern Recognition Workshops, Seattle, WA, USA, 13–19 June 2020; pp. 112–113.
6. Feng, Y.; Liu, J.; Liu, W. Coincidence of the Rao Test, Wald Test and GLRT for anomaly detection in hyperspectral imagery. *Signal Process.* **2020**, *169*, 107416. [CrossRef]
7. Su, H.; Wu, Z.; Zhu, A.X.; Du, Q. Low rank and collaborative representation for hyperspectral anomaly detection via robust dictionary construction. *ISPRS J. Photogramm.* **2020**, *169*, 195–211. [CrossRef]
8. Matteoli, S.; Diani, M.; Theiler, J. An overview of background modelling for detection of targets and anomalies in hyperspectral remotely sensed imagery. *IEEE J. Sel. Top. Appl. Earth Obs. Remote Sens.* **2014**, *7*, 2317–2336. [CrossRef]
9. Nasrabadi, N.M. Hyperspectral target detection: An overview of current and future challenges. *IEEE Signal Process. Mag.* **2014**, *31*, 34–44. [CrossRef]
10. Borghys, D.; Kåsen, I.; Achard, V.; Perneel, C. Comparative evaluation of hyperspectral anomaly detectors in different types of background. *Proc. SPIE* **2012**, *8390*, 83902J.
11. Manolakis, D.; Shaw, G. Detection algorithms for hyperspectral imaging applications. *IEEE Signal Process. Mag.* **2002**, *19*, 29–43. [CrossRef]
12. Reed, I.S.; Yu, X. Adaptive multiple-band CFAR detection of an optical pattern with unknown spectral distribution. *IEEE Trans. Acoust. Speech Signal Process.* **1990**, *38*, 1760–1770. [CrossRef]
13. Guo, Q.; Zhang, B.; Ran, Q.; Gao, L.; Li, J.; Plaza, A. Weighted-RXD and linear filter-Based RXD: Improving background statistics estimation for anomaly detection in hyperspectral imagery. *IEEE J. Sel. Top. Appl. Earth Obs. Remote Sens.* **2014**, *7*, 2351–2366. [CrossRef]
14. Du, B.; Zhao, R.; Zhang, L.; Zhang, L. A spectral-spatial based local summation anomaly detection method for hyperspectral images. *Signal Process.* **2016**, *124*, 115–131. [CrossRef]
15. Huyan, N.; Zhang, X.; Zhou, H.; Jiao, L. Hyperspectral anomaly detection via background and potential anomaly dictionaries construction. *IEEE Trans. Geosci. Remote Sens.* **2019**, *57*, 2263–2276. [CrossRef]
16. Ling, Q.; Guo, Y.; Lin, Z.; An, W. A constrained sparse representation model for hyperspectral anomaly detection. *IEEE Trans. Geosci. Remote Sens.* **2019**, *57*, 2358–2371. [CrossRef]
17. Ma, N.; Peng, Y.; Wang, S. A fast recursive collaboration representation anomaly detector for hyperspectral image. *IEEE Geosci. Remote Sens. Lett.* **2019**, *16*, 588–592. [CrossRef]
18. Nasrabadi, N.M. Regularization for spectral matched filter and RX anomaly detector. *Proc. SPIE* **2008**, *6966*, 696604.
19. Kwon, H.; Nasrabadi, N.M. Kernel RX-algorithm: A nonlinear anomaly detector for hyperspectral imagery. *IEEE Trans. Geosci. Remote Sens.* **2005**, *43*, 388–397. [CrossRef]
20. Tao, R.; Zhao, X.; Li, W.; Li, H.; Du, Q. Hyperspectral anomaly detection by fractional Fourier entropy . *IEEE J. Sel. Top. Appl. Earth Observ. Remote Sens.* **2019**, *12*, 4920–4929. [CrossRef]
21. Chen, Y.; Nasrabadi, N.M.; Tran, T.D. Sparse representation for target detection in hyperspectral imagery. *IEEE J. Sel. Top. Signal Process.* **2011**, *5*, 629–640. [CrossRef]
22. Xu, Y.; Wu, Z.; Li, J.; Plaza, A.; Wei, Z. Anomaly detection in hyperspectral images based on low-rank and sparse representation. *IEEE Trans. Geosci. Remote Sens.* **2016**, *54*, 1990–2000. [CrossRef]
23. Wei, L.; Qian, D. Collaborative representation for hyperspectral anomaly detection . *IEEE Trans. Geosci. Remote Sens.* **2015**, *53*, 1463–1474.
24. Zhu, D.; Bo, D.; Zhang, L. Binary-class collaborative representation for target detection in hyperspectral images. *IEEE Geosci. Remote Sens. Lett.* **2019**, *16*, 1100–1104. [CrossRef]
25. Tan, K.; Hou, Z.; Wu, F.; Du, Q.; Chen, Y. Anomaly detection for hyperspectral imagery based on the regularized subspace method and collaborative representation. *Remote Sens.* **2019**, *11*, 1318. [CrossRef]
26. Tu, B.; Yang, X.; Zhou, C.; He, D.; Laza, A.P. Hyperspectral anomaly detection using dual window density. *IEEE Trans. Geosci. Remote Sens.* **2020**, *58*, 8503–8517. [CrossRef]
27. Zhang, Y.; Fan, Y.; Xu, M. A background-purification-based framework for anomaly target detection in hyperspectral imagery. *IEEE Geosci. Remote Sens. Lett.* **2020**, *17*, 1238–1242. [CrossRef]
28. Vafadar, M.; Ghassemian, H. Anomaly detection of hyperspectral imagery using modified collaborative representation. *IEEE Trans. Geosci. Remote Sens.* **2018**, *15*, 577–581. [CrossRef]

29. Su, H.; Wu, Z.; Du, Q.; Du, P. Hyperspectral anomaly detection using collaborative representation with outlier removal. *IEEE J. Sel. Topics Appl. Earth Observ. Remote Sens.* **2018**, *11*, 5029–5038. [CrossRef]
30. Roweis, S.T.; Saul, L.K. Nonlinear dimensionality reduction by locally linear embedding. *Science* **2000**, *290*, 2323–2326. [CrossRef]
31. Li, W.; Prasad, S.; Fowler, J.E. Decision fusion in kernel-induced spaces for hyperspectral image classification. *IEEE Trans. Geosci. Remote Sens.* **2014**, *52*, 3399–3411. [CrossRef]
32. Zhao, C.; Li, C.; Yao, X. Real-time kernel collaborative representation-based anomaly detection for hyperspectral imagery. *Infrared Phys. Technol.* **2020**, *107*, 103325. [CrossRef]
33. Imani, M. Anomaly detection using morphology-based collaborative representation in hyperspectral imagery. *Eur. J. Remote Sens* **2018**, *51*, 457–471. [CrossRef]
34. Wang, R.; Hu, H.; He, F.; Nie, F.; Cai, S.; Ming, Z. Self-weighted collaborative representation for hyperspectral anomaly detection. *Signal Process.* **2020**, *177*, 107718. [CrossRef]
35. Felzenszwalb, P.; Huttenlocher, D. Efficient Graph-Based Image Segmentation. *Int. J. Comput. Vis.* **2004**, *59*, 167–181. [CrossRef]
36. Shannon, C.E. A mathematical theory of communication. *Bell Syst. Tech. J.* **1948**, *27*, 379–423. [CrossRef]
37. Lance, G.N.; Williams, W.T. Computer programs for hierarchical polythetic classification ("similarity analyses"). *Comput. J.* **1966**, *9*, 60–64. [CrossRef]
38. Li, W.; Du, Q. Unsupervised nearest regularized subspace for anomaly detection in hyperspectral imagery. In Proceedings of the 2013 IEEE International Geoscience and Remote Sensing Symposium, IGARSS 2013, Melbourne, Australia, 21–26 July 2013; pp. 1055–1058.
39. Li, W.; Tramel, E.W.; Prasad, S.; Fowler, J.E. Nearest regularized subspace for hyperspectral classification. *IEEE Trans. Geosci. Remote Sens.* **2014**, *52*, 477–489. [CrossRef]
40. Kerekes, J. Receiver operating characteristic curve confidence intervals and regions. *IEEE Geosci. Remote Sens. Lett.* **2008**, *5*, 251–255. [CrossRef]
41. Ferri, C.; Hernández-Orallo, J.; Flach, P. A coherent interpretation of AUC as a measure of aggregated classification performance. In Proceedings of the International Conference on Machine Learning, Bellevue, WA, USA, 28 June–2 July 2011; pp. 657–664.

*applied sciences*

MDPI

*Article*

# An Automatic Geometric Registration Method for Multi Temporal 3D Models

**Haixing Shang [1,*], Guanghong Ju [1], Guilin Li [2], Zufeng Li [1] and Chaofeng Ren [3,*]**

1 Northwest Engineering Corporation Limited, Power China Group, Xi'an 710065, China
2 Tibet Electric Power Corporation Limited of CHN ENERGY Group, Linzhi 860019, China
3 College of Geological Engineering and Geomatics, Chang'an University, Xi'an 710054, China
* Correspondence: shang_hx@nwh.cn (H.S.); ren_cf@163.com (C.R.)

**Abstract:** The application research of ground change detection based on multi-temporal 3D models is attracting more and more attention. However, the conventional methods of using UAV GPS-supported bundle adjustment or measuring ground control points before each data collection are not only economically costly, but also have insufficient geometric accuracy. In this paper, an automatic geometric-registration method for multi-temporal 3D models is proposed. First, feature points are extracted from the highest resolution texture image of the 3D model, and their corresponding spatial location information is obtained based on the triangular mesh of the 3D model, which is then converted into 3D spatial-feature points. Second, the transformation model parameters of the 3D model to be registered relative to the base 3D model are estimated by the spatial-feature points with the outliers removed, and all the vertex positions of the model to be registered are updated to the coordinate system of the base 3D model. The experimental results show that the position measurement error of the ground object is less than 0.01 m for the multi-temporal 3D models obtained by the method of this paper. Since the method does not require the measurement of a large number of ground control points for each data acquisition, its application to long-period, high-precision ground monitoring projects has great economic and geometric accuracy advantages.

**Keywords:** unmanned aerial vehicles (UAVs); photogrammetry; 3D model; geometric registration; accuracy; feature points

## 1. Introduction

The combined optical camera system technology of unmanned aerial vehicles (UAVs) provides the potential to acquire high-spatial-resolution images of large areas [1]. Based on images at different viewing angles obtained by UAVs, digital orthophoto models (DOMs), digital surface models (DSMs), dense 3D points, and 3D models are created via photogrammetry. The results are then used to evaluate the monitored ground objects' displacement rate and topographic surface changes [2–6]. The 3D model is widely used in several fields due to its ability to both directly display the structural details of the scene and to perform high precision geometric measurements, such as heritage-building documentation [7], landslide monitoring [8,9], glacial geomorphology [10], building modeling [11], and change detection [12]. Due to the small image size of the UAV camera, especially for oblique photogrammetry engineering projects, the number of images may be as many as tens of thousands or even hundreds of thousands, which greatly increases the data-processing time and economic cost [13]. In order to obtain the dynamic displacement process of the ground objects in the monitoring area, it is necessary to obtain the UAV image data at a certain frequency. However, it is a challenging task to ensure that the 3D model generated from the multi-temporal UAV images has a unified spatial reference.

Based on the UAV optical image data, the technique integrating the structure-from-motion (SfM) and multi-view stereo (MVS) is used to generate a detailed 3D model [14–17].

Normally, UAV imagery is geo-referenced via ground control points (GCPs), which are placed on the study area and measured using global navigation satellite system (GNSS) receivers or total stations [18]. In order to ensure that the 3D model generated from UAV images of different time sequences have a unified spatial reference, the conventional processing method is to survey a certain number of GCPs before each data acquisition. For small- or medium-sized UAV photogrammetry projects, five GCPs can meet the geometric requirements [19]. However, more GCPs are needed for large engineering projects, especially those that collect image data by terrain-following flights [20]. Arranging and measuring a large number of GCPs for construction sites, or dangerous areas that are difficult for people to reach, is costly and lacks the flexibility. As the manufacturing cost of airborne GNSS-RTK (real time kinematic) receivers decreases, it has become possible to integrate them into low-cost UAVs [21]. The camera position with centimeter accuracy can be obtained by GNSS-RTK equipment, which can be introduced into the global-position-system-supported (GPS-supported) bundle-adjustment (BA) optimization to significantly reduce the number of GCPs required, or even eliminate the GCPs completely [22,23]. Other research results show that the accuracy of the UAV GPS-supported BA method cannot obtain centimeter-level accuracy; in particular, there may be systematic errors in the vertical direction [24,25]. It can be seen from this research that high-precision geometric positioning for ground objects based on UAV photogrammetry technology cannot yet completely abandon the use of GCPs. In fact, the absolute position of an object is not the user's concern in many cases, but rather relative changes. It is more valuable to obtain the spatial position of the ground object on the 3D model obtained from different temporal data. Obviously, it is not economical to measure a large number of GCPs before each flight mission, and it is impossible to obtain a high-precision unified space benchmark for the multi-temporal 3D models. For historical 3D model data especially, it is no longer possible to supplement surveying GCPs.

The goal of this paper was to complete the geometric registration of multi-period 3D models with a uniform spatial reference. The basis of geometric registration between models is to obtain the spatial coordinate information of the ground point of different multi-temporal 3D models. The conventional method can obtain the spatial position of corresponding points by manual measurement, and then estimate the geometric transformation model parameters of the two-period 3D model based on the position difference. Due to the influence of human factors, the efficiency of this method is too low and the reliability of the results is insufficient. So, it is necessary to propose a method that can automatically match the corresponding points of the multi-temporal 3D models and calculate their positions with high precision. There are inevitably incorrect corresponding points after spatial-feature-point extraction, which are considered as outliers. The conventional method of eliminating outliers based on two-view geometry verification is not effective, so the robust and high-precision solution of the relative geometric transformation parameters of the models is another problem.

In this paper, a method for automatic geometric registration of multi-temporal 3D models is presented. In detail, it consists of three parts. Firstly, the spatial-feature points are extracted from the 3D model, and the initial matching between the spatial-feature points of different time models is completed. Secondly, based on the corresponding spatial positions of the initial spatial-feature points, the relative geometric transformation model parameters of the multi-period 3D model are calculated. Finally, all the vertex positions of the registered 3D model are transformed based on the calculated model parameters, so as to keep them unified with the spatial reference of the baseline model. The rest of the paper is organized as follow. Section 2 briefly introduces the study area and the study-data-acquisition process. Section 3 introduces the proposed methodology and key technologies used in this study. Section 4 describes the experimental objectives, experimental process and results, and provides a detailed analysis based on the experimental results. Section 5 offers the conclusions of the study.

## 2. Study Area and Data

### *2.1. Study Area*

The study area is located within the Xinjiang province (79°43′26″ E and 39°28′57″ N), 81 km east north from Akesu in China, as shown in Figure 1a. The average annual precipitation is 80.4 mm, and the main rainfall period is concentrated in May to September. The annual evaporation is 1643–2202 mm, which is 27 times the average annual precipitation. The dry climate results in low vegetation coverage and a large number of stones on the surface of the study area. As shown in Figure 1a,b the study site is located on the Akesu River and is crossed by a road. There is a steep rock with a height difference of 100 m in the west of the study area, as shown in Figure 1c. The stones often become loose from the steep rock and fall to the road, which causes economic loss, property damage, and loss of life [26]. The goal of our team was to rely on UAV technology to periodically acquire UAV images of the study site and to generate 3D models by photogrammetry. The multi-temporal 3D models identified the rocks that were at risk of falling, or had moved. To achieve this goal, the first problem to be solved was how to make multi-temporal 3D models have a unified spatial reference.

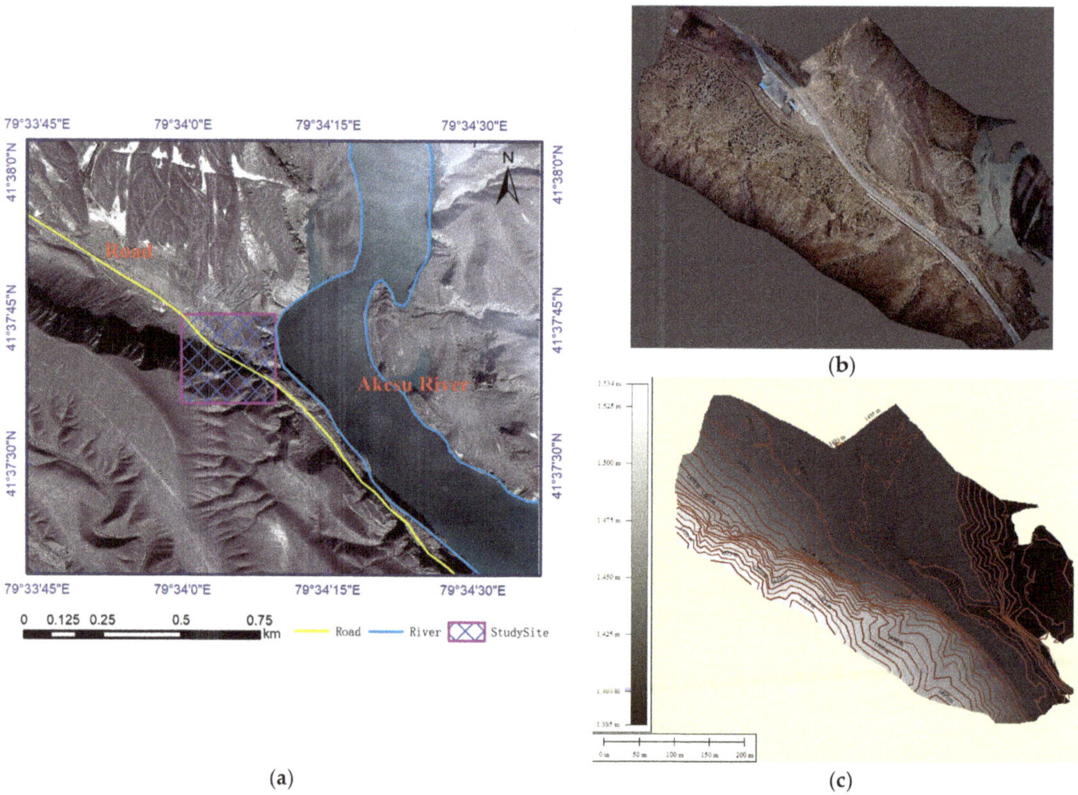


**Figure 1.** An illustration of the location and topography of the study site. (**a**) The location of the study site. (**b**) The digital ortho model of the study site. (**c**) The digital elevation model and contour of the study site.

### *2.2. Data Acquisition*

The flight platform was a six-rotor UAV, as shown in Figure 2c, equipped with a full-frame Sony ILCE-7RM4 camera. The detailed parameters of the flight platform and camera system are shown in Table 1. According to the technical requirement that the ground sampling distance (GSD) of the UAV image is 2 cm, the designed flight routes are shown in Figure 2a, where the relative flight altitude was 130 m, and the overlap of heading and side direction were 80% and 60%, respectively. During the experiment period, the same flight routes were used twice to collect UAV data. The first acquisition was on 22 June 2021, and a total of 182 images were acquired. The second collection time was 15 July 2021, and a total of 184 images were obtained.

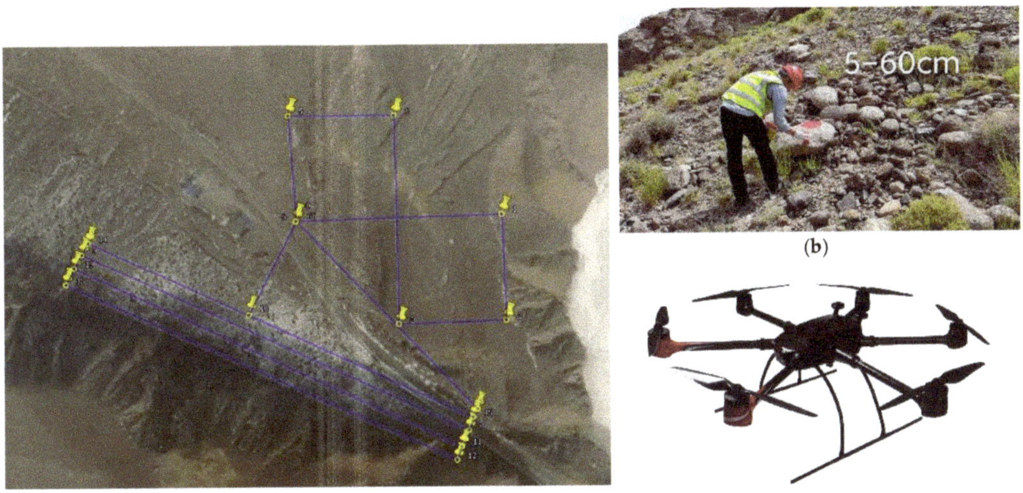


(a) (b) (c)

**Figure 2.** UAV data acquisition. (**a**) The flights of the study site. (**b**) The spray-painted stone. (**c**) The six-rotor flight platform.

**Table 1.** The parameters of the flight platform and camera system.

| Hovering Time (min) | Load Capacity (kg) | Maximum Working Altitude (m) | Sensor Size (pixel) | Focal Length (mm) | Pixel Size (mm) |
|---|---|---|---|---|---|
| 75 | 15 | 5000 | 9504 × 6336 | 25 | 0.0037 |

In order to evaluate the difference between the two cycle models, we selected stones of different sizes within the study site to be painted red, and measured their spatial positions using GNSS-RTK, as shown in Figure 2b.

## 3. Methodology

The geometric registration of multi-temporal 3D models is to define and solve the transformation model parameters from the vertices of the registered 3D models to the spatial location of the based 3D models. The main process consists of three steps:

(1) The 3D model feature points are extracted from the texture image by the interest point operator, and their position information is computed.
(2) The feature vector of spatial-feature points consists of the image information around the feature points, which are used to match the corresponding points.
(3) According to the corresponding position of the spatial-feature points, the transformation model parameters between models are estimated, than the new transformation model is used to update the spatial location information of the 3D model to be registered.

### 3.1. Spatial-Feature-Point Extraction

The research object of this paper was a 3D model, which contained data including vertices, triangulated-mesh connecting vertices, and texture-image data corresponding to vertices, as shown in Figure 3a.

(a) (b)

**Figure 3.** The composition of the 3D model. (**a**) The vertices of the 3D model and the triangulation connecting the vertices. (**b**) The texture image of the 3D model and the texture coordinates corresponding to the vertices.

In Figure 3a, the yellow circles are vertices in the 3D model data, which are connected by black lines to form a triangulation. Each vertex corresponds to a unique normalized 2D texture coordinate, as shown by the white origin in Figure 3b. Then, the corresponding 2D texture coordinates, which are the normalized coordinates of the feature points in the texture image, and 3D vertices are rendered to form a regular 3D model.

The prerequisite of 3D model registration is to obtain high-precision spatial-feature points and their corresponding spatial positions. However, the conventional method of extracting image feature points directly extracts the coordinates of 2D image points, and then recovers their spatial positions based on the SfM pipeline algorithm [27]. Since texture images of 3D model data are irregular images, as shown in Figure 3b, the direct transformation relationship between texture coordinates and their spatial coordinates cannot be directly formed. Therefore, we propose a spatial feature-point-extraction method that can calculate the corresponding spatial positions of feature points on texture images.

Firstly, feature points are extracted from texture images. The scale-invariant feature transform (SIFT) feature has been widely used in the field of feature extraction because of its advantages of invariant to lighting, scale, and rotation changes and can overcome affine image changes [28]. In this paper, SIFT feature points are first extracted from texture images, and their image coordinates are normalized to texture coordinates, as shown in Equation (1).

$$\begin{cases} t_x = \frac{x}{wid} \\ t_y = 1 - \frac{y}{hei} \end{cases} \quad (1)$$

where $wid$ and $hei$ represent the width and height of the texture image (in pixels), respectively. $t_x$ and $t_y$ represent the normalized texture coordinates.

Secondly, the plane coordinates of feature points $\begin{bmatrix} X_w & Y_w \end{bmatrix}^T$ are interpolated. The normalized texture coordinates $\begin{bmatrix} t_x & t_y \end{bmatrix}^T$ of each triangle were searched to obtain the texture vertex coordinates of the triangles. A schematic diagram of a triangle and model triangle is shown in Figure 4.

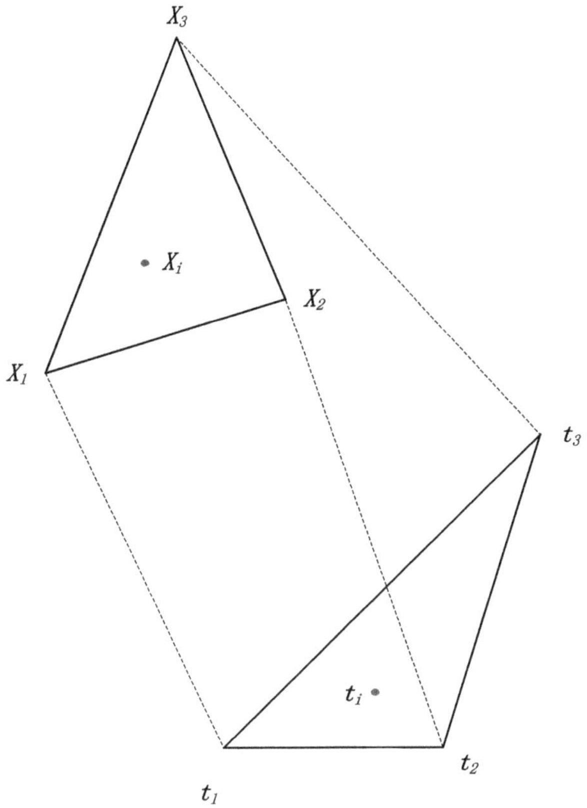


**Figure 4.** The mapping relationship between texture triangles and model triangles.

In Figure 4, $t_1$, $t_2$, and $t_3$ are the three normalized vertex coordinates of the texture triangle, respectively. $X_1$, $X_2$, and $X_3$ are the spatial plane coordinates of the model triangle, respectively. $t_i$ is the normalized texture position corresponding to the extracted feature point. $X_i = \begin{bmatrix} X_w & Y_w & Z_w \end{bmatrix}^T$ is the spatial position corresponding to $t_i$. According to the corresponding relationship between the model triangle and the texture triangle, the plane coordinates of texture triangle and model triangle are defined as a 2D affine transformation, which can be expressed as Equation (2).

$$\begin{bmatrix} X_w \\ Y_w \end{bmatrix} = \begin{bmatrix} a & b & c \\ d & e & f \end{bmatrix} \begin{bmatrix} t_x \\ t_y \\ 1 \end{bmatrix} \quad (2)$$

where $(a, b, c, d, e, f)$ are 2D affine-transformation parameters.

The three normalized texture coordinates of the texture triangle and the vertex plane coordinates of the corresponding model triangle are put into Equation (2). After finishing, which can be expressed as:

$$\mathbf{A} \cdot \mathbf{X} = \mathbf{L} \quad (3)$$

where $\mathbf{A} = \begin{bmatrix} tx_1 & ty_1 & 1 & 0 & 0 & 0 \\ 0 & 0 & 0 & tx_1 & ty_1 & 1 \\ tx_2 & ty_2 & 1 & 0 & 0 & 0 \\ 0 & 0 & 0 & tx_2 & ty_2 & 1 \\ tx_3 & ty_3 & 1 & 0 & 0 & 0 \\ 0 & 0 & 0 & tx_3 & ty_3 & 1 \end{bmatrix}, \mathbf{X} = \begin{bmatrix} a \\ b \\ c \\ d \\ e \\ f \end{bmatrix}, \mathbf{L} = \begin{bmatrix} Xw_1 \\ Yw_1 \\ Xw_2 \\ Yw_2 \\ Xw_3 \\ Yw_3 \end{bmatrix}$.

Based on the least-squares adjustment method, the affine-transformation parameters in Equation (3) can be obtained from Equation (4).

$$\mathbf{X} = \left(\mathbf{A}^{\mathrm{T}}\mathbf{P}\mathbf{A}\right)^{-1}\left(\mathbf{A}^{\mathrm{T}}\mathbf{P}\mathbf{L}\right) \tag{4}$$

All the spatial-feature points obtained by matching are considered as equal-weighted observations, and their corresponding stochastic models are defined as an identify matrix, as shown in Equation (4) for **P**. Affine-transformation parameters can be estimated based on Equation (4). Then, according to affine-transformation parameters and Equation (2), the corresponding space plane coordinates $\left[X_w \quad Y_w\right]^{\mathrm{T}}$ of feature point $t_i$ can be calculated.

Thirdly, the feature-point elevation coordinate $Z_w$ is fitted. The vertex coordinates of the model triangle are defined as planes, and the elevation coordinates of the feature point can be obtained by interpolation in the plane after fitting. If the spatial coordinates of the model triangle vertices are $\left[X_1 \quad Y_1 \quad Z_1\right]^{\mathrm{T}}$, $\left[X_2 \quad Y_2 \quad Z_2\right]^{\mathrm{T}}$, and $\left[X_3 \quad Y_3 \quad Z_3\right]^{\mathrm{T}}$, respectively, and the spatial coordinates of the feature points are $\left[X_w \quad Y_w \quad Z_w\right]^{\mathrm{T}}$, then the above four points are in the same plane, which can be expressed as follows:

$$\begin{vmatrix} X_w & Y_w & Z_w & 1 \\ X_1 & Y_1 & Z_1 & 1 \\ X_2 & Y_2 & Z_2 & 1 \\ X_3 & Y_3 & Z_3 & 1 \end{vmatrix} = 0 \tag{5}$$

Equation (5) can be expressed as:

$$Z_w = Z_1 - \frac{(X_W - X_1)(Y_{21}Z_{31} - Y_{31}Z_{21}) + (Y_w - Y_1)(Z_{21}X_{31} - Z_{31}X_{21})}{X_{21}Y_{31} - X_{31}Y_{21}} \tag{6}$$

After obtaining the plane coordinates $\left[X_w \quad Y_w\right]^{\mathrm{T}}$ of the feature point, the elevation coordinates $Z_w$ of the feature point can be fitted by Equation (6).

### 3.2. Spatial-Feature-Point Matching

When the feature points of all 3D-model texture images are extracted and their spatial positions are calculated, it is necessary to establish the corresponding relationship between the spatial-feature points of the base 3D model and the spatial-feature points of the 3D model to be registered. Based on the 128-dimension normalized feature vector of SIFT points, the similarity value of the feature vector is obtained after calculating the Euclidean distance, and then it is determined whether the two feature points are corresponding points [28]. In conventional image-matching, the feature points and feature vectors of image A and B are obtained, respectively, and then the similarity between the feature points set of image A and image B are determined successively to complete the two-view image matching. After the feature points and feature vectors of a batch 3D model are completely acquired, the exhaustive search and matching method cannot be realized due to the large amount of data. Therefore, this paper adopts the axis-aligned bounding-box (AABB) search method to match the corresponding points between the feature points of the base 3D model and the feature points of the 3D model to be registered.

After extracting the feature points of the base 3D models, the set of their spatial-feature points is generated, as shown by the blue dots in Figure 5a. In the same way, the spatial-feature-point set of the model to be registered is generated, as shown in Figure 5b. The set of spatial-feature points in Figure 5a is evenly divided into AABB bounding boxes, so that the number of feature points contained in each bounding box is less than a threshold (value 3000), as shown in the bounding box set $\{B_i | i = 1, 2, \cdots 6\}$ in Figure 5a. Then, according to the prior knowledge r of the positioning error between the base model and the model to be registered, the bounding box is expanded by r to form the search bounding box $\{S_i | i = 1, 2, \cdots 6\}$ for the set of feature points to be registered. After the similarity matching between the bounding box and the searching bounding box is completed, a complete list of corresponding points between the base feature points and the feature points to be registered can be obtained.

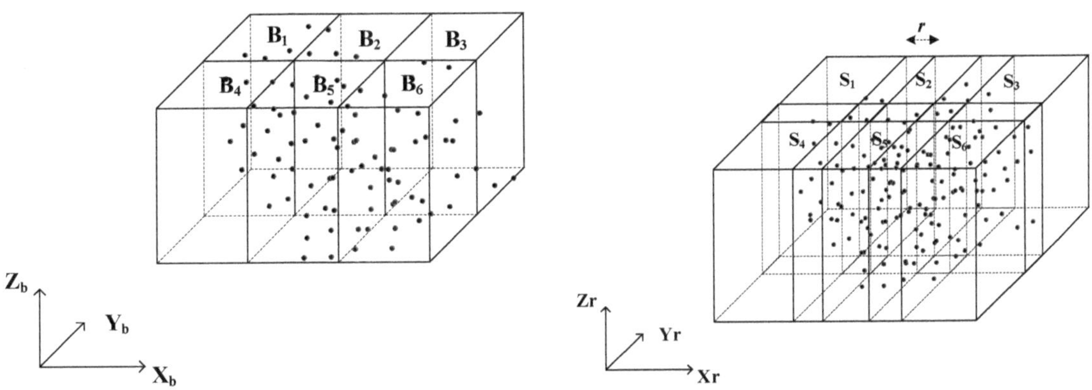


(a) (b)

**Figure 5.** The AABB bounding box of the search range. (**a**) The bounding box of the base model space feature points. (**b**) The searching bounding box of the model space feature points to be registered.

### 3.3. Registration Model Solution

The geometric transformation model from the vertex of the registered 3D model to the vertex of the base 3D model can be defined as 3D similarity transformation, as shown in Equation (7).

$$\begin{bmatrix} X_b \\ Y_b \\ Z_b \end{bmatrix} = \lambda \cdot \mathbf{R} \cdot \begin{bmatrix} X_r \\ Y_r \\ Z_r \end{bmatrix} + \begin{bmatrix} r_x \\ r_y \\ r_z \end{bmatrix} \tag{7}$$

where $\begin{bmatrix} X_r & Y_r & Z_r \end{bmatrix}^T$ is the vertex position of the model to be registered and $\begin{bmatrix} X_b & Y_b & Z_b \end{bmatrix}^T$ is the corresponding vertex position of the base model. $\mathbf{R}$ and $\begin{bmatrix} r_x & r_y & r_z \end{bmatrix}^T$ represents the rotation matrix and the translation from the vertex of the model to be registered to the base model, respectively. $\lambda$ represents the scale factor from the vertex of the model to be registered to the base model, and when $\lambda = 1$, Equation (7) becomes a 3D rigid transformation model.

The detailed definition of rotation matrix $\mathbf{R}$ in Equation (7) is shown in Equation (8).

$$\mathbf{R} = \mathbf{R}_\varphi \mathbf{R}_\omega \mathbf{R}_\kappa = \begin{bmatrix} \cos\varphi & 0 & -\sin\varphi \\ 0 & 1 & 0 \\ \sin\varphi & 0 & \cos\varphi \end{bmatrix} \cdot \begin{bmatrix} 1 & 0 & 0 \\ 0 & \cos\omega & -\sin\omega \\ 0 & \sin\omega & \cos\omega \end{bmatrix} \cdot \begin{bmatrix} \cos\kappa & -\sin\kappa & 0 \\ \sin\kappa & \cos\kappa & 0 \\ 0 & 0 & 1 \end{bmatrix} = \begin{bmatrix} \cos\varphi\cos\kappa - \sin\varphi\sin\omega\sin\kappa & -\cos\varphi\sin\kappa - \sin\varphi\sin\omega\cos\kappa & -\sin\varphi\cos\omega \\ \cos\omega\sin\kappa & \cos\omega\cos\kappa & -\sin\omega \\ \sin\varphi\cos\kappa + \cos\varphi\sin\omega\sin\kappa & -\sin\varphi\sin\kappa + \cos\varphi\sin\omega\cos\kappa & \cos\varphi\cos\omega \end{bmatrix} \tag{8}$$

where $(\varphi, \omega, \kappa)$ are the rotation angle of the rotation matrix $\mathbf{R}$.

When a sufficient number of corresponding spatial points are obtained, the Levenberg–Marquardt (LM) algorithm [29] is used for solving non-linear squares problems, and the seven transformation parameters of Equation (7) can be obtained. The open source library "levmar" [30] of the LM algorithm is used to solve the transformation parameters. However, the initial corresponding spatial-feature points obtained by the similarity of feature vectors inevitably contain incorrect correspondences, which are usually referred to as outliers. Therefore, the random sampling consistency algorithm (RANSAC) is adopted to eliminate the outliers and robustly obtain the transformation model parameters [31]. Finally, the estimated model parameters are used to transform all vertices of the registered model to complete the geometric registration.

### 3.4. Data Processing

After acquiring multi-temporal UAV images and position and orientation system (POS) data, the software of "Context Capture V10.16.0.75" was used for 3D reconstruction to obtain 3D models. Then the feature points of the 3D model were extracted and their corresponding spatial positions were calculated based on the method in Section 3.1. Thirdly, the spatial-feature points were initially matched to obtain the initial correspondence based on the method in Section 3.2. Fourthly, the model transformation parameters from the registration model to the base 3D model were estimated based on the initial correspondence, and all vertices of the registered 3D model were updated based on the method in Section 3.3. The workflow of the data processing is shown in Figure 6.

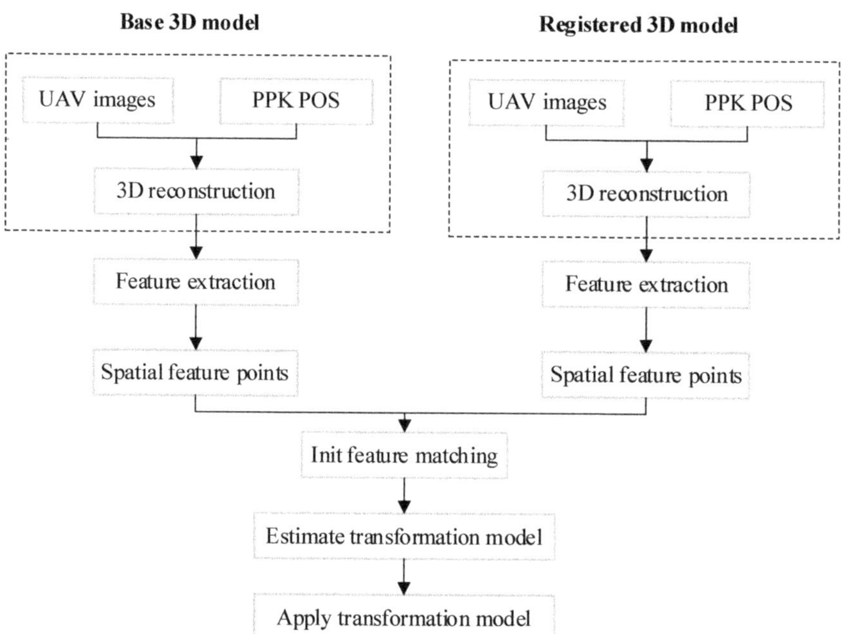


**Figure 6.** The workflow of data processing.

## 4. Experiment Results and Discussions

To verify the effectiveness of the algorithm, we selected seven stones of different sizes in the study area and applied red paint to their surfaces. The distribution of the stones is shown in Figure 7.

C1–C7 in Figure 7 represent the numbering of the seven stones. The positions of the seven stones were accurately measured by GNSS-RTK equipment. Before the second period of data acquisition, the positions of the seven stones were changed and their spatial positions were measured again, and then the UAV image data were collected. The purpose of the experiment was to evaluate the positioning accuracy of multi-temporal 3D models geometric registration by the moving positions of the seven stones.

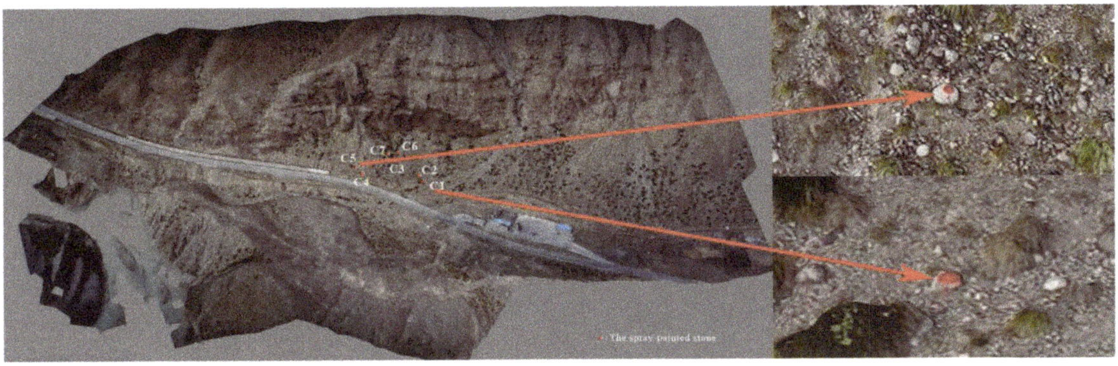


**Figure 7.** The 3D model of the study site and the spray-painted stones.

### *4.1. Distribution and Correspondence Analysis of Spatial-Feature Points*

The number and distribution of spatial-feature points are the basic conditions for 3D model geometric registration. Firstly, the feature points were extracted from the texture images in the 3D model, and then the spatial positions of the feature points were calculated based on the triangulation of the 3D model to generate the corresponding spatial-feature points. Figure 8 shows the feature-point-extraction results of two texture images.

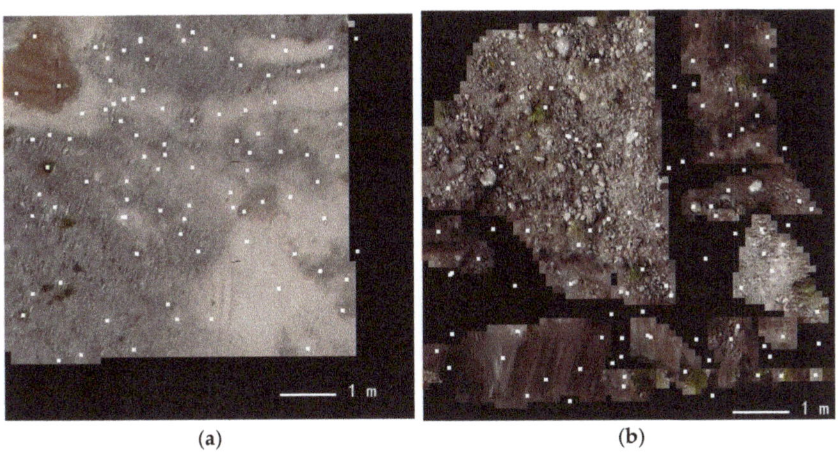


**Figure 8.** The distribution of feature points in texture images. (**a**) A continuous texture image. (**b**) A fragmented texture image.

The white dots in Figure 8 represent the extracted feature points, of which the feature points extracted in Figure 8a were evenly distributed and not significantly different from those of the conventional image. In Figure 8b, due to the fragmentation of texture images, many feature points were distributed in the non-textured image area. Obviously, these feature points were invalid and needed to be eliminated in the feature-matching process. According to the analysis of feature-point-extraction results in Figure 8, although 3D texture images are irregular and fragmented, a sufficient number of feature points could still be extracted. After extracting all the feature points of the lowest-level model, the spatial positions of the feature points were calculated, and the planar feature points were converted into spatial-feature points. The spatial-feature points were superimposed and displayed with the 3D grid model, as shown in Figure 9.

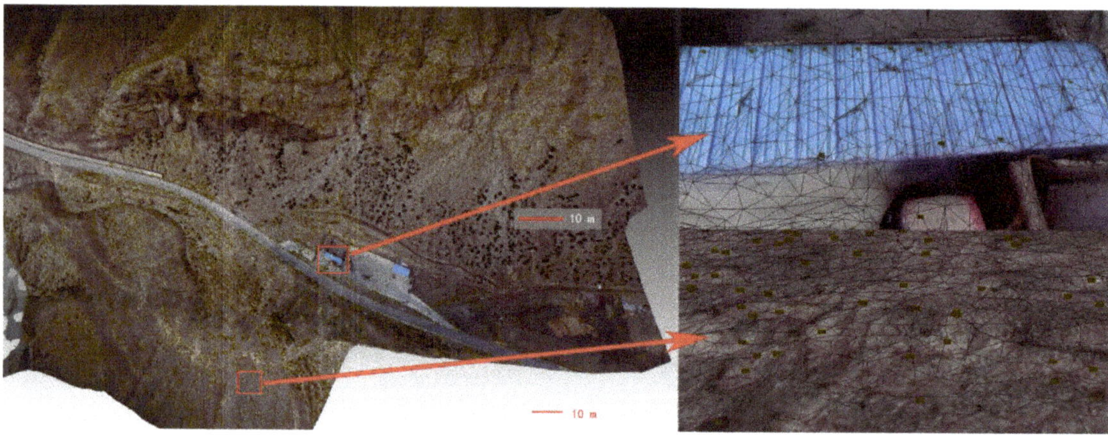


**Figure 9.** The distribution of feature points in the mesh model.

The green dots in the left of Figure 9 are the extracted spatial-feature points, which were uniformly distributed on the surface of the mesh model. The right side of Figure 9 shows the enlarged display of the red box. Analysis from the visual performance showed that the extracted spatial-feature points were strictly fitted to the surface of the mesh, and their planar positions and elevation interpolation results were consistent with the actual situation, and were reasonable. After obtaining the spatial-feature points, the initial similarity determination and 3D model geometric registration were performed to obtain the corresponding points in different stages, as shown in Figure 10.

The yellow dots in Figure 10a,b represent the distribution of spatial-feature points extracted from the 3D model in two periods, respectively, which were evenly distributed but had no correspondence between spatial-feature points. Figure 10c,d shows the initial matching correspondence of the spatial-feature points for the two periods' data, from which it can be seen that it still contained a certain number of outlier points, as shown in the red-boxed area. Figure 10e,f shows the final correspondence of the spatial-feature points obtained after model registration. Compared with Figure 10c,d, the outliers existing in the initial corresponding spatial-feature points have been eliminated. After statistics, the number of spatial-feature points extracted from the data in the two periods was 100,834 and 93,751, respectively. After similarity-determination, a total of 4914 initial corresponding points were obtained. A total of 2816 final corresponding spatial points were obtained after the two periods of 3D model registration. Based on the analysis of the above results, it is possible to obtain sufficient and evenly distributed corresponding points by extracting spatial-feature points from the multi-period 3D model.

### *4.2. Geometric-Registration-Accuracy Analysis of Spatial-Feature Points*

A total of three methods are used to generate 3D models for evaluating the relative geometric positioning accuracy of multi-temporal 3D models. The first method is to generate 3D models based on low-precision UAV airborne POS data. The second method uses post-processed kinematic (PPK) technology to obtain high-precision position at the time of image acquisition, and uses a GPS-support method to complete 3D reconstruction to obtain 3D models. The third method uses the 3D-model alignment method proposed in this paper, and performs vertex updates on the 3D model to be aligned based on the model-alignment parameters. The spatial-feature points of the above three types of 3D models were extracted, respectively, and the corresponding points of the final spatial features were obtained. The positions of the spatial-feature points in the base 3D model were subtracted from the positions of those in the 3D model to be aligned to obtain any residual positioning errors, and the results are shown in Figure 11.

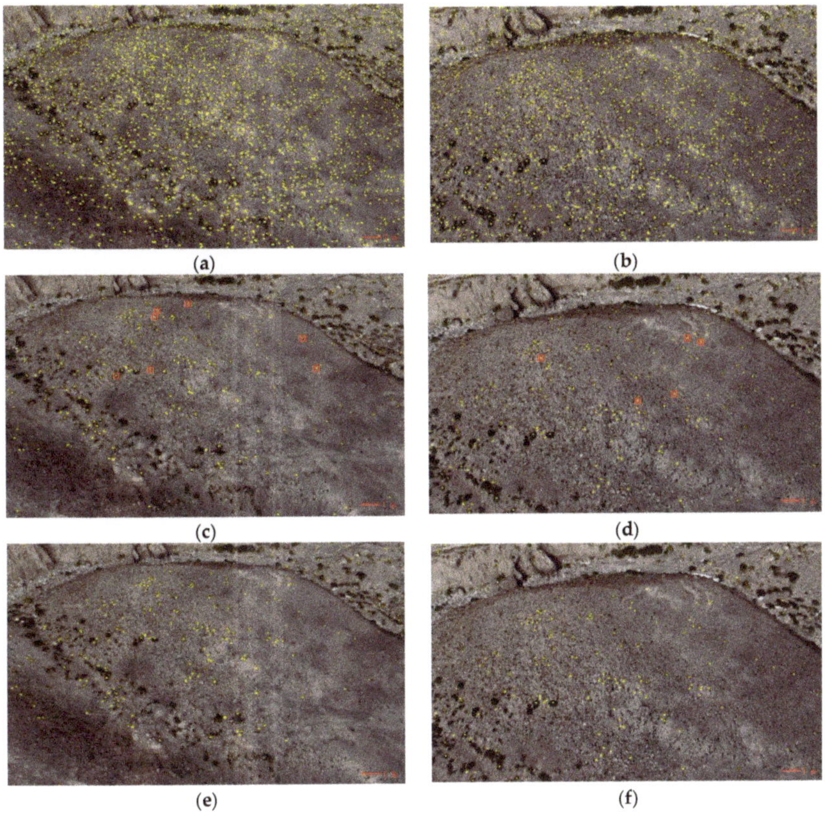

**Figure 10.** The corresponding distribution of spatial-feature points, the yellow dots in the figure are the spatial-feature points. (**a**) The initial spatial-feature points extracted. (**c**) The initial correspondence of spatial-feature points, with incorrect correspondence in the red box. (**e**) The fine correspondence of spatial-feature points after registration. (**b**,**d**,**f**) corresponds to (**a**,**c**,**e**) and are the results of the second-cycle data.

DX, DY, and DZ in Figure 11 represent the position residual errors in X, Y, and Z directions, respectively. The residual error distribution in Figure 11a shows that the positioning accuracy of conventional UAV-borne POS was low, and there were obvious systematic errors among the generated multi-temporal 3D models. When high-precision PPK data was used to generate 3D models, the positioning accuracy of its multi-period 3D models was greatly improved. The relative positioning accuracy of the two temporal 3D models was less than 0.2 m, as shown in Figure 11b. However, the average values in the Y and Z directions were not zero, indicating that a small amount of systematic positioning error still existed in these two directions. The residual errors of the corresponding points of the 3D model after using the geometric alignment of this paper were all close to 0 m, showing a random distribution in Figure 11c. The residual-error accuracy of spatial-feature points obtained by the three methods was calculated, and the results are shown in Table 2.

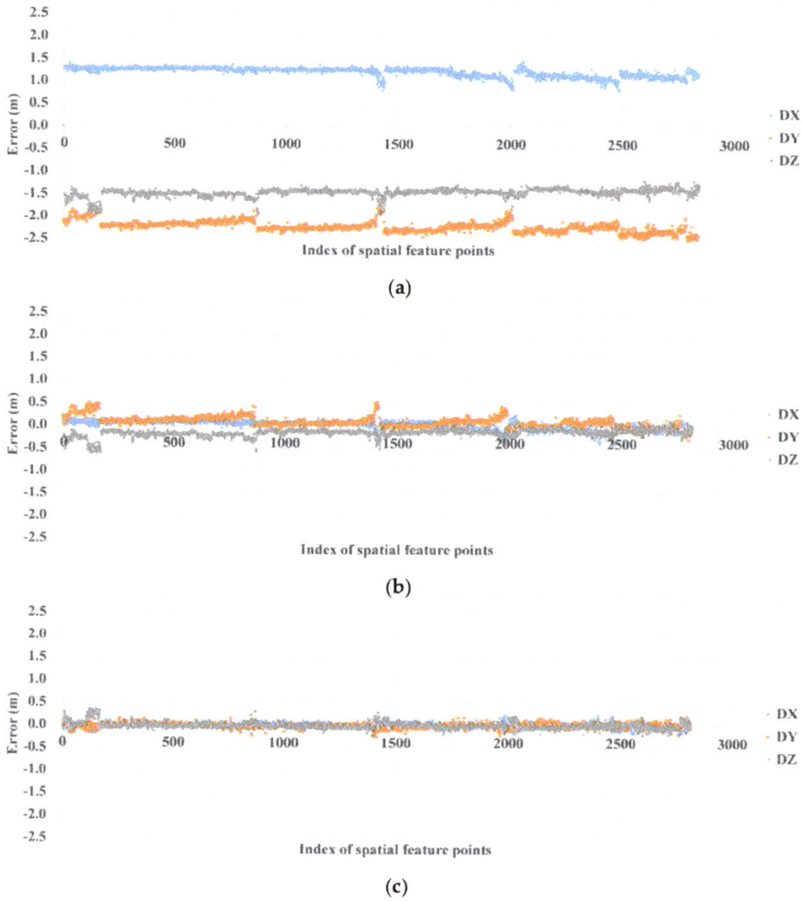


**Figure 11.** The corresponding point residual errors of the spatial-feature points. (**a**) Results obtained based on UAV airborne POS system. (**b**) Results obtained based on PPK POS data. (**c**) Results obtained based on 3D-model automatic registration.

**Table 2.** Comparison of the positioning accuracy of corresponding points for different 3D models.

| Method | AVG X (m) | AVG Y (m) | AVG Z (m) | STD X (m) | STD Y (m) | STD Z (m) |
|---|---|---|---|---|---|---|
| POS | 1.21 | −2.21 | −1.46 | 0.12 | 0.12 | 0.13 |
| PPK | 0.01 | 0.08 | −0.16 | 0.09 | 0.11 | 0.09 |
| REG | $5.37 \times 10^{-5}$ | $-2.38 \times 10^{-5}$ | $7.10 \times 10^{-7}$ | 0.05 | 0.06 | 0.07 |

In Table 2, AVG represents the average error of corresponding points of spatial features, which reflects the systematic error of relative positioning accuracy of multi-temporal 3D models. STD represents the standard deviation, which reflects the internal aggregation of corresponding points of spatial features. Based on the results in Table 2, it can be seen that the overall average error of the 3D model generated from conventional UAV airborne POS data was 2.91 m. The accuracy was low and cannot be directly used for change detection in multi-temporal 3D models. The overall average error of the multi-temporal 3D models generated based on PPK data was better than 0.18 m, but there were still a few systematic errors in the Y and Z directions, which were consistent with the results in Figure 11b. The overall average error of the 3D-model alignment method proposed in this paper was almost equal to 0 m. After compensation by the transformation parameters, the overall STD of the corresponding points was reduced from 0.36 m to 0.14 m. In contrast, the geometric registration of multi-temporal 3D models based on spatial-feature points proposed in this paper can effectively eliminate the spatial location differences among 3D models and obtain a unified spatial reference.

### *4.3. Accuracy Analysis of 3D Model after Registration*

In order to further evaluate the positioning error between the registered 3D model and the base 3D model, we accurately measured the displacement of seven spray-painted stones between the two sequential 3D models.

Figure 12 shows the screenshots of the sprayed stones numbered C1 and C5 in the two sequential 3D models. The GNSS-RTK equipment was used to measure the spatial position of the sprayed stones before the two UAV data collections, and their spatial displacement was calculated and used as the true value. Using the same three 3D model data as in Figure 11, the displacement of the sprayed stones in the sequential 3D model was calculated after manually measuring their spatial position in the 3D model. The actual displacement obtained from GNSS-RTK equipment was subtracted from the model displacement measured manually to obtain the difference in displacement of the seven sprayed stones, and the results are shown in Table 3.

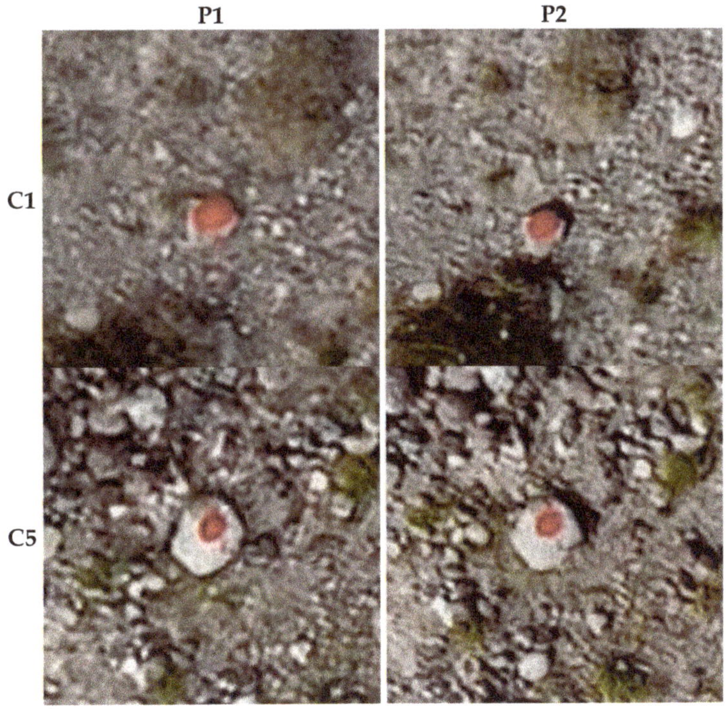


**Figure 12.** The sprayed stones in the two sequential 3D model. P1 represents the 3D of the first period, and P2 represents the second period.

**Table 3.** The displacement and difference of the sprayed stones in the two sequential 3D models.

| Name | Displacement (m) | | | | Difference (m) | | |
|---|---|---|---|---|---|---|---|
| | GPS | POS | PPK | REG | POS | PPK | REG |
| C1 | 0.49 | 3.13 | 0.41 | 0.49 | −2.64 | 0.07 | −0.01 |
| C2 | 0.62 | 2.87 | 0.58 | 0.61 | −2.25 | 0.04 | 0.01 |
| C3 | 0.88 | 3.03 | 0.83 | 0.87 | −2.15 | 0.06 | 0.01 |
| C4 | 0.63 | 3.14 | 0.53 | 0.61 | −2.51 | 0.10 | 0.02 |
| C5 | 0.37 | 2.97 | 0.24 | 0.33 | −2.61 | 0.13 | 0.04 |
| C6 | 0.73 | 3.20 | 0.69 | 0.77 | −2.46 | 0.04 | −0.04 |
| C7 | 0.47 | 3.14 | 0.34 | 0.45 | −2.67 | 0.13 | 0.02 |
| AVG | - | | | | −2.47 | 0.08 | 0.01 |

The GPS column in Table 3 shows the actual spatial displacement of the sprayed stones. Based on analysis of the results in Table 3, the sequential 3D model generated by conventional POS data shows that the average value of the difference in spatial displacements was −2.47 m, far exceeding the actual spatial displacements of the sprayed stones. So, 3D models generated based on conventional POS data cannot be directly used for ground object change detection. The average value of the spatial-displacement difference was reduced to 0.08 m after using PPK data, which is about three times the GSD. The results show that the multi-period 3D model generated by PPK data can be localized and discovered when the spatial-location-change value of ground objects exceeds three times the GSD. In contrast, the average of the spatial displacement difference was as low as 0.01m for the model-registration method in this paper. The results in Table 3 show that the relative distance measurement error of the aligned 3D model was less than 0.01 m. The results show that the 3D model generated based on the method in this paper has a highly uniform spatial reference for the multi-temporal 3D models and can be directly applied to the change detection of the multi-temporal 3D models.

## 5. Conclusions

In this paper, we propose a fully automatic 3D model registration method, which converts 2D feature points extracted from the highest resolution texture image of the model into 3D spatial-feature points, and solves the model-transformation parameters of the 3D model to be aligned with respect to the base 3D model, and then establishes a unified spatial reference for multi-temporal 3D models. The experimental results show that the multi-temporal 3D models generated by high-precision PPK data combined with GPS-support method can locate ground object displacement that is more than three times that of the GSD of the UAV image. In comparison, the average value of the difference between the measured ground object displacement and the actual object displacement was 0.01 m for the 3D model generated by the method in this paper. This method does not need to measure GCPs at each data acquisition to make a multi-time series 3D model with a uniform spatial reference. For periodic dynamic inspection tasks, it has the great advantage of saving project cost and economic cost.

**Author Contributions:** Conceptualization, H.S.; methodology, H.S. and C.R.; software, C.R.; validation, G.L. and Z.L.; formal analysis, G.J.; investigation, G.L.; resources, G.L.; data curation, H.S.; writing—original draft preparation, C.R.; writing—review and editing, C.R.; visualization, H.S.; supervision, Z.L.; project administration, H.S.; funding acquisition, C.R. All authors have read and agreed to the published version of the manuscript.

**Funding:** This research was funded by the Northwest Engineering Corporation Limited Major Science and Technology Projects, grant number XBY-ZDKJ-2020-08, and the National Natural Science Foundation of China, grant number (41801383). The APC was funded by 41801383.

**Institutional Review Board Statement:** Not applicable.

**Informed Consent Statement:** Not applicable.

**Data Availability Statement:** The data presented in this study are available on request from the corresponding author.

**Conflicts of Interest:** The authors declare no conflict of interest.

## References

1. Mathews, A.J. A practical UAV remote sensing methodology to generate multispectral orthophotos for vineyards: Estimation of spectral reflectance using compact digital cameras. *Int. J. Appl. Geospat. Res.* **2015**, *6*, 65–87. [CrossRef]
2. Lu, P.; Catani, F.; Tofani, V.; Casagli, N. Quantitative hazard and risk assessment for slow-moving landslides from Persistent Scatterer Interferometry. *Landslides* **2014**, *11*, 685–696. [CrossRef]
3. Abdulla, A.R.; He, F.; Adel, M.; Naser, E.S.; Ayman, H. Using an unmanned aerial vehicle-based digital imaging system to derive a 3D point cloud for landslide scarp recognition. *Remote Sens.* **2016**, *8*, 95. [CrossRef]
4. Conforti, M.; Mercuri, M.; Borrelli, L. Morphological changes detection of a large earthflow using archived images, LiDAR-derived DTM, and UAV-based remote sensing. *Remote Sens.* **2020**, *13*, 120. [CrossRef]
5. Komarek, J.; Klouček, T.; Prošek, J. The potential of Unmanned Aerial Systems: A tool towards precision classification of hard-to-distinguish vegetation types? *Int. J. Appl. Earth Obs. Geoinf.* **2018**, *71*, 9–19. [CrossRef]
6. Klouek, T.; Komárek, J.; Surov, P.; Hrach, K.; Vaíek, B. The use of UAV mounted sensors for precise detection of bark beetle infestation. *Remote Sens.* **2019**, *11*, 1561. [CrossRef]
7. Castilla, F.J.; Ramón, A.; Adán, A.; Trenado, A.; Fuentes, D. 3D sensor-fusion for the documentation of rural heritage buildings. *Remote Sens.* **2021**, *13*, 1337. [CrossRef]
8. Lucieer, A.; Jong, S.; Turner, D. Mapping landslide displacements using structure from motion (SfM) and image correlation of multi-temporal UAV photography. *Prog. Phys. Geogr.* **2014**, *38*, 97–116. [CrossRef]
9. Guzzetti, F.; Mondini, A.C.; Cardinali, M.; Fiorucci, F.; Santangelo, M.; Chang, K.T. Landslide inventory maps: New tools for an old problem. *Earth-Sci. Rev.* **2012**, *112*, 42–66. [CrossRef]
10. Ewertowski, M.; Tomczyk, A.; Evans, D.; Roberts, D.; Ewertowski, W. Operational framework for rapid, very-high resolution mapping of glacial geomorphology using low-cost unmanned aerial vehicles and structure-from-motion approach. *Remote Sens.* **2019**, *11*, 65. [CrossRef]
11. Rakha, T.; Gorodetsky, A. Review of unmanned aerial system (UAS) applications in the built environment: Towards automated building inspection procedures using drones. *Autom. Constr.* **2018**, *93*, 252–264. [CrossRef]
12. Kohv, M.; Sepp, E.; Vammus, L. Assessing multitemporal water-level changes with uav-based photogrammetry. *Photogramm. Rec.* **2017**, *32*, 424–442. [CrossRef]
13. Huang, F.; Yang, H.; Tan, X.; Peng, S.; Tao, J.; Peng, S. Fast reconstruction of 3D point cloud model using visual SLAM on embedded UAV development platform. *Remote Sens.* **2020**, *12*, 3308. [CrossRef]
14. Smith, M.; Carrivick, J.; Quincey, D. Structure from motion photogrammetry in physical geography. *Prog. Phys. Geogr.* **2015**, *40*, 247–275. [CrossRef]
15. Westoby, M.J.; Brasington, J.; Glasser, N.F.; Hambrey, M.J.; Reynolds, J.M. 'Structure-from-Motion' photogrammetry: A low-cost, effective tool for geoscience applications. *Geomorphology* **2012**, *179*, 300–314. [CrossRef]
16. Stathopoulou, E.K.; Welponer, M.; Remondino, F. Open-source image-based 3D reconstruction pipelines: Review, comparision and evaluation. *ISPRS-Int. Arch. Photogramm. Remote Sens. Spat. Inf. Sci.* **2019**, *XLII-2/W*, 331–338. [CrossRef]
17. Ren, C.; Zhi, X.; Pu, Y.; Zhang, F. A multi-scale UAV image matching method applied to large-scale landslide reconstruction. *Math. Biosci. Eng.* **2021**, *18*, 2274–2287. [CrossRef]
18. Le, V.C.; Cao, X.C.; Long, N.Q.; Le, T.; Anh, T.T.; Bui, X.N. Experimental investigation on the performance of DJI Phantom 4 RTK in the PPK Mode for 3D mapping open-pit mines. *Inz. Miner.* **2020**, *1*, 65–74.
19. Long, N.; Goyal, R.; Bui, L.; Cao, C.; Canh, L.; Quang Minh, N.; Bui, X.-N. Optimal choice of the number of ground control points for developing precise DSM using light-weight UAV in small and medium-sized open-pit mine. *Arch. Min. Sci.* **2021**, *66*, 369–384.
20. Ren, C.; Shang, H.; Zha, Z.; Zhang, F.; Pu, Y. Color balance method of dense point cloud in landslides area based on UAV images. *IEEE Sens. J.* **2022**, *22*, 3516–3528. [CrossRef]
21. Troner, M.; Urban, R.; Seidl, J.; Reindl, T.; Brouek, J. Photogrammetry using UAV-mounted GNSS RTK: Georeferencing strategies without GCPs. *Remote Sens.* **2021**, *13*, 1336. [CrossRef]
22. Gong, Y.; Meng, D.; Seibel, E.J. Bound constrained bundle adjustment for reliable 3D reconstruction. *Opt. Express* **2015**, *23*, 10771–10785. [CrossRef] [PubMed]
23. Triggs, B. Bundle Adjustment—A modern synthesis. In *Proceedings of International Workshop on Vision Algorithms: Theory & Practice*; Springer: Berlin/Heidelberg, Germany, 1999; pp. 298–372.
24. Troner, M.; Urban, R.; Reindl, T.; Seidl, J.; Brouek, J. Evaluation of the georeferencing accuracy of a photogrammetric model using a quadrocopter with onboard GNSS RTK. *Sensors* **2020**, *20*, 2318. [CrossRef]
25. Taddia, Y.; Stecchi, F.; Pellegrinelli, A. Coastal mapping using DJI Phantom 4 RTK in post-processing kinematic mode. *Drones* **2020**, *4*, 9. [CrossRef]
26. Corominas, J.; Westen, C.V.; Frattini, P.; Cascini, L.; Smith, J.T. Recommendations for the quantitative analysis of landslide risk. *Bull. Eng. Geol. Environ.* **2014**, *73*, 209–263. [CrossRef]

27. James, M.R.; Robson, S.; D"Oleire-Oltmanns, S.; Niethammer, U. Optimising UAV topographic surveys processed with structure-from-motion: Ground control quality, quantity and bundle adjustment. *Geomorphology* **2017**, *280*, 51–66. [CrossRef]
28. Lowe, D. Distinctive image features from scale-invariant key points. *Int. J. Comput. Vis.* **2003**, *60*, 91–110. [CrossRef]
29. Nocedal, J.; Wright, S.J.; Mikosch, T.V.; Resnick, S.I.; Robinson, S.M. *Numerical Optimization*; Springer: Berlin/Heidelberg, Germany, 1999.
30. Levmar: Levenberg-Marquardt Nonlinear Least Squares Algorithms in C/C++. Available online: http://users.ics.forth.gr/~{}lourakis/levmar/ (accessed on 22 October 2020).
31. Fischler, M.A.; Bolles, R.C. Random sample consensus: A paradigm for model fitting with applications to image analysis and automated cartography. *Read. Comput. Vis.* **1981**, *24*, 381–395. [CrossRef]

applied sciences

MDPI

Article

# Target Identification via Multi-View Multi-Task Joint Sparse Representation

**Jiawei Chen [1], Zhenshi Zhang [2] and Xupeng Wen [3,*]**

[1] School of Computational Science and Engineering, Georgia Institute of Technology, Atlanta, GA 30332, USA
[2] Undergraduate School, National University of Defense Technology, Changsha 410073, China
[3] School of Traffic and Transportation Engineering, Central South University, Changsha 410073, China
* Correspondence: wenxupeng@csu.edu.cn; Tel.: +86-181-0125-1215

**Abstract:** Recently, the monitoring efficiency and accuracy of visible and infrared video have been relatively low. In this paper, we propose an automatic target identification method using surveillance video, which provides an effective solution for the surveillance video data. Specifically, a target identification method via multi-view and multi-task sparse learning is proposed, where multi-view includes various types of visual features such as textures, edges, and invariant features. Each view of a candidate is regarded as a template, and the potential relationship between different tasks and different views is considered. These multiple views are integrated into the multi-task spare learning framework. The proposed MVMT method can be applied to solve the ship's identification. Extensive experiments are conducted on public datasets, and custom sequence frames (i.e., six sequence frames from ship videos). The experimental results show that the proposed method is superior to other classical methods, qualitatively and quantitatively.

**Keywords:** target identification; multi-view multi-task; spare representation; feature extraction

## 1. Introduction

With the emergence of various high-resolution visible light and infrared cameras, video monitoring technology has been rapidly developed and applied. Currently, many video cameras are deployed in streets, traffic, and ports, and these are mainly convenient for the staff at the monitoring center to observe and understand the site conditions of each monitoring station. However, the area covered by each monitoring point is relatively small. To expand the monitoring range, more and more video monitoring points are arranged, the workload of the personnel on duty becomes heavier, the work efficiency is relatively low, and they are no longer willing to use the monitoring video. Therefore, most video monitoring has become a kind of decoration at present. "How do I make the most of video surveillance?" It has become the focus and difficulty in the field of target identification. To solve the above problems, this paper studies the target identification method.

Recently, several computer vision and machine learning methods have shown competitive results in target identification and tracking [1,2], especially the deep learning-based method that has received extensive attention in recent years due to its excellent performance [3–7]. However, these methods rely on the parameter adjustment of the neural network model. The training time is long, and it is easy to consume a lot of computing resources [8]. Therefore, these models may be inefficient in the case of demanding tracking [9]. In addition, the visual tracking decomposition method [10] adopts the sparse principal component analysis method based on multiple features to build multiple basic trackers, but the tracking performance of this multi-feature method is not good. The sparse representation-based method [11–13] uses the particle filter framework [14] to target identification based on sparse representation. In addition, the multi-task learning [15] method is an efficient method, which learns the sparse representation of all particles in the particle filter framework [16]. Compared with the L1 method with sparse representation [17], although the multi-task method (MTT) takes advantage of the interdependence between particles, the robustness of multi-task is not good due to the existence of outliers.

To identify targets more robustly, researchers have proposed a multi-task and multi-perspective learning method to optimize the target identification problem [18,19]. In this paper, we propose a target identification method based on multi-task multi-view sparse learning (MVMT).

The main contributions of this paper are summarised as follows:

(1) A multi-view multi-task method based on sparse representation, namely MVMT, is proposed for target identification. Compared with the previous related method, the proposed MVMT can not only use the learning based on sparse representation, but also introduces a variety of complementary perspective features to enhance the expression features of the targets;
(2) Each perspective of each particle is regarded as a separate task, and the potential connections between different perspectives and different particles are considered in a multi-task learning framework;
(3) To capture the outlier tasks that frequently occur in the particle sampling process, the coefficient matrix is decomposed into two cooperative parts to enhance the robustness of multi-task learning, and the posterior probability of outliers is set to zero to ensure that samples will not be taken in the resampling process.

## 2. Related Works

Recently, researchers have proposed various target identification methods: a typical method is finding the target position through the response map generated by the online learning correlation filter (CF) and determining the target scale using a fixed scale factor. The CF based trackers include Kernelized Correlation Filters (KCF) [20], multichannel features [21], adaptive scale estimation [22], the fusion of complementary learners [23], long short-term memory [24], support vector machines [25], sparse coding [26] etc. However, these CF-based trackers have some major shortcomings: First, the tracker relies excessively on the maximum response value when determining the target position. Second, the CF-based tracker uses a fixed scale factor to determine the size of the target, which is not suitable for the actual scale change of the moving target. Therefore, when the tracking scene is too complex, the generated model is not robust enough. Besides the optical flow method, Camshift and Meanshift are also popular target tracking methods used in recent years, showing their potential in identification in tracking applications [27,28]. However, these methods identify the target according to its single feature, and it is impossible to identify a common single feature that can be applied to different situations.

To improve the robustness of prediction, various ensemble methods have been studied. The Staple tracker [29] consists of a correlation filter and a model based on color histogram, which complement each other. In [30], three different feature trackers based on support vector machines are integrated, and the trackers are adaptively selected according to the consistency of front and rear tracks. In order to eliminate redundancy between weak models, efficient diverse ensemble co-tracking [31] trains different models by generating a set of effective manual data. Multi-cue correlation filters tracker [32] assigns the suitable weak classified based on their self-wise and pairwise relationships. Due to the limited training samples, sequent training convolutional tracking (STCT) solved the problem of overfitting existing convolutional neural networks (CNNs) in visual tracking [33]. STCT uses binary masks to force basic learners to learn different features. In the correlation filter method it is proposed to fuse response maps from different convolutional neural network levels. Therefore, integrating multiple individual CF [34] has become a common technology. Under the Siamese tracking pipeline, a twofold Siamese network trains an appearance model and a semantic model, respectively, for online combination [35]. The learning part-based model of joint tracking is also discussed in the Siamese network [36].

## 3. Problem Description and Modeling

The target can be represented by multiple types of visual features, including edge [37], intensity [38], color [39], and texture [40]. Using the complementary features of these multi-sources, information can significantly improve the target identification performance [39,41–43]. These different visual features are sampled from different feature spaces. Since the view data are extracted from the same object of interest, these data are interrelated, i.e., each view datum can be obtained from the sparse representation of a part of the samples under the view. In practical applications, gradient direction histogram (HOG) [44], local binary pattern (LBP) [45] and scale-invariant feature transformation (SIFT) [46,47] are highly expressive perspective features. Therefore, the robustness of the target identification problem is enhanced by combining the above multiple perspectives and training multiple samples at the same time. To express the scheme more intuitively and conveniently, the flow chart of the multi-view and multi-task learning method is presented in Figure 1.

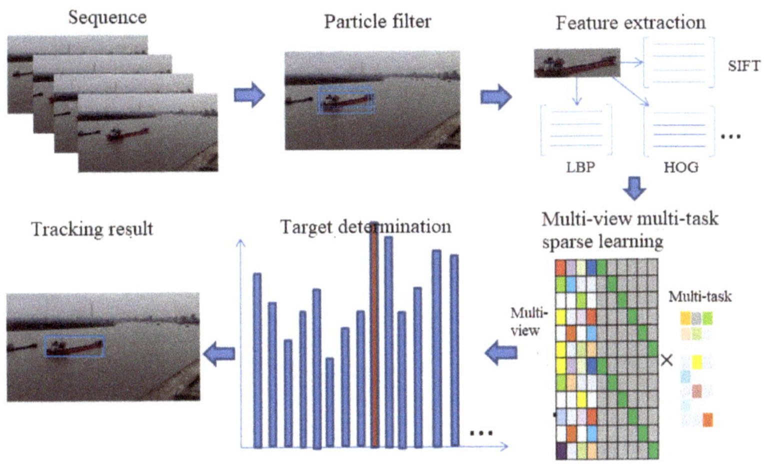


**Figure 1.** Target identification schematic of multi-view multi-task algorithm.

Assuming that $s_t$ and $x_t$ represent the state variables and observation variables at time $t$, respectively, the target identification problem can be described as the estimation of a posteriori state probability $p_n(s_t|s_{t-1})$ through a limited set of $n$ particles with importance weights $\omega_1$, which $x_{1:t} = \{x_1, \cdots, x_t\}$ are the previously observed $t$ frames of images. Particle samples $s_t^i$ are sampled from the importance distribution $q(s_t|s_{1:t}, x_{1:t})$, i.e., the simplification of state transition probability $p(s_t|s_{t-1})$. In addition, the importance weight of particle $i$ is updated through the observation probability, i.e., $w_t^i = w_{t-1}^i p(x_t|s_t^i)$

The sparse representation of feature $x$ can be reconstructed by the regularization $\ell1$ problem to minimize the error [3]:

$$\min_{w} ||MW - X||_2^2 + \lambda ||W||_1 \tag{1}$$

where $M = [D, I, -I]$ is a complete dictionary composed of target template set $D$ and positive and negative background template sets $I$ and $-I$. Each column is a target template in $D$, and the candidate region pixels are modified in the column vector. Each column in the background template set is a unit vector with one non-zero element.

$W = [a^T, e^{+T}, e^{-T}]^T$ is composed of target coefficient and positive and negative background coefficients $e^{+T}, e^{-T}$. Each column in $a^T$ represents a classification plane.

The probability of this observation is derived from the reconstruction error of $x$

$$p(x|y) = \frac{1}{Y} \exp\{-\alpha ||Da - x||^2\} \tag{2}$$

where $a$ is obtained by solving (1), $\alpha$ is constant and is used to control the shape of the Gaussian kernel, and $Y$ is the normalization parameter.

## 4. Multiple Feature Extraction Methods for Target Identification

In this paper, we introduce four typical feature extraction methods to make the advantages of complementary features, including the scale-invariant feature transformation (SIFT), the histogram of oriented gradient (HOG) [9], the local binary pattern (LBP) [47], and the Hu invariant matrix. Specifically, HOG is edge distributions of objects captured based on gradient features. LBP has a powerful function of representing object texture. Hu moment invariants are fast and can better describe larger objects in the image. In addition, to ensure the quality of extracted features, a simple and effective illumination normalization method is adopted before feature extraction [19]. The unit norm normalization is applied to extract feature vectors for each particle's viewing angle [46]. To clearly show the feature extraction process, we adopt four feature extraction methods to extract features from the same image, which are elaborated as follows. Since the ship has significant texture features and edge features, we choose the ship image as an example, and the original image of the ship is presented in Figure 2.

**Figure 2.** Original image of ship.

### 4.1. Scale-Invariant Feature Transformation (SIFT) Feature Extraction Method

Scale-invariant feature transformation (SIFT) is a machine learning method for local feature detection. The feature points of adjacent image frames are matched by the features obtained from the size and direction of the feature points in the image and their related descriptors. The extraction result of the SIFT feature on key point descriptors of the ship is shown in Figure 3.

**Figure 3.** SIFT feature on key point descriptors of ship.

From Figure 3, it can be seen that the key point descriptors show the features of the scale size and direction of the descriptors. These features have a scale variance for changing image brightness, rotation angle, or shooting angle. These quantized features can be used to match the feature points of adjacent ship image frames, i.e., when the descriptors of the two adjacent ship frames are extracted, the key descriptors between the adjacent ship frames can be matched. When the direction and module length of the 128-dimensional vector are matched, the two frames are matched.

*4.2. Gradient Direction Histogram (HOG) Feature Extraction Method*

HOG is a gradient direction histogram feature that is used to detect objects in image processing in the field of computer vision. HOG is a histogram feature formed by counting the gradient direction of each pixel in the local neighborhood of the image. The implementation method of HOG is as follows:

Firstly, Gamma space and color space are normalized. In the texture feature of a ship image, the weight of the local surface exposure factor is large. Therefore, the Gamma compression algorithm can effectively reduce the local illumination variation of the image. The ship image is transformed into a gray scale, then the gamma compression formula $I(x,y) = I(x, y)^{gamma}$ is adopted, where gamma is the compression parameter, and gamma = 0.5 is desirable. The result is shown in Figure 4, which shows the normalized Gamma space and color space of the ship images, which can reduce the influence of light factors.

**Figure 4.** Gamma compression image of the ship.

Then, the image is divided into small unit intervals that are called cell units. The direction histogram of the gradient of each pixel is obtained from the cell unit. Finally, the collected histograms are combined to form a feature descriptor. The image of ship HOG feature extraction is shown in Figure 5.

From Figure 5, it can be seen that the information of the HOG gradient mainly exists in the edges of the image, which can describe the appearance and shape of the ship target with the edge or gradient directional density distribution. Therefore, the HOG features have good invariance to the geometric and optical changes of the image. In addition, under the conditions of strong local light normalization and rough spatial sampling and fine direction sampling, fine deformation can be allowed without affecting the detection result.

**Figure 5.** HOG feature extraction image of the ship.

### 4.3. Local Binary Pattern (LBP) Feature Extraction Method

LBP is a feature extraction method for local texture features in an image. Its essence is to describe the relationship between pixels in an image and pixels in its neighborhood. LBP has the advantages of gray constancy and rotation constancy. The algorithm for extracting LBP features from the ship database is as follows:

Step 1: Divide the manually drawn object of interest box into 16 × 16 cells;

Step 2: Compare each pixel of the ship sample with the gray value of the surrounding eight pixels, and convert the binary results into decimal numbers;

Step 3: Calculate the LBP value of each pixel in each cell, count the frequency of LBP value of each pixel, and obtain the histogram;

Step 4: Splice the statistical histograms of each cell into separate feature vectors.

The LBP's texture feature of the ship is presented in Figure 6, and the histogram of the ship is shown in Figure 7.

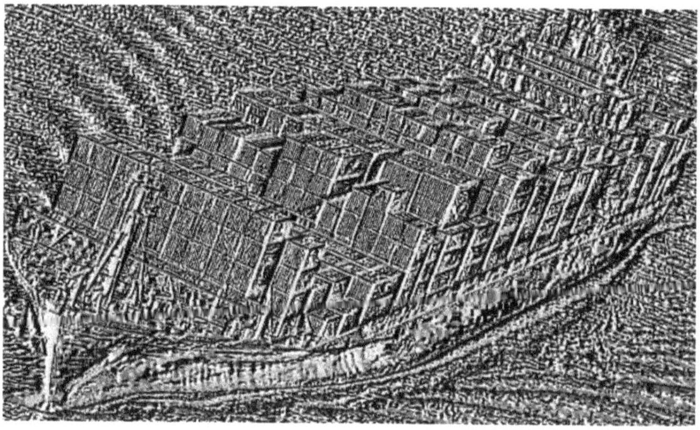

**Figure 6.** LBP texture image of the ship.

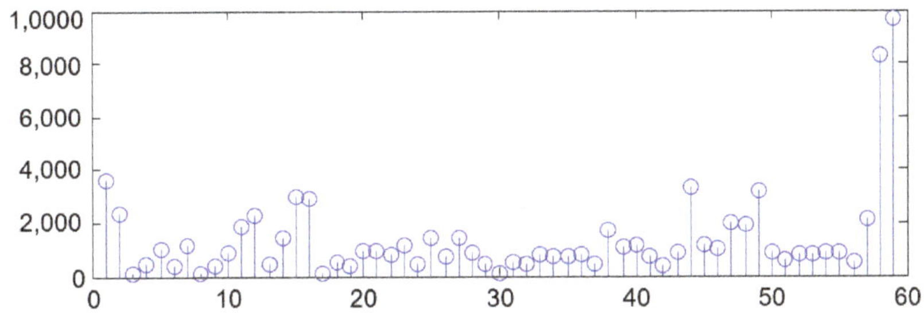


**Figure 7.** LBP histogram of ship.

In Figure 6, the principle of LBP is to compare the gray value of a pixel in the ship image with the gray value of the pixel around it, and then convert the comparison results into binary mode. From Figure 6 it can be seen that after LBP feature extraction the remarkable ship texture feature map is displayed, and it has the characteristics of gray scale constancy and rotation constancy. From Figure 7, most of the energy is distributed in 59 histograms with high probability. Taking into account the compression of ship eigenvectors, it is very reasonable for us to use these 59 histograms to describe the ship texture in this experiment.

### *4.4. Hu Invariant Matrix Feature Extraction Method*

The Hu invariant matrix is a set of characteristic quantities composed of seven invariant matrices. In 1962, it was proven that these characteristic quantities are invariant to image scaling, translation and rotation [48]. Matrix invariants can well represent the structural features of images. The advantage of Hu matrix invariants for image feature extraction is that the calculation speed is fast, but the disadvantage is that the accuracy of image recognition is low, especially for images with rich texture. Especially, Hu invariant matrices are not sensitive to texture-rich images, and generally can only recognize large objects in the image. As the ship is a large object, so the texture is relatively simple. Therefore, Hu matrix invariants are used as auxiliary features to further enhance the robustness of ship identification in combination with other visual angle features. In particular, after being processed by the Hu invariant matrix feature method, it is a vector composed of seven scalars, rather than a processed ship image.

## 5. Target Identification Method Based on Multi-View Multi-Task Sparse Learning

Several previous research works have been studied for multi-view multi-task sparse learning. Chen et al. [49] applied the multi-view sparse learning method to the field of target identification. Mei et al. [50] proposed a generative tracker based on multi-view learning, with impressive tracking performance. The tracker proposed by Hong et al. [51] cannot be directly used to perform visual ship tracking tasks because the features to be tracked are color and intensity, which may be very similar between different ships.

To address the above issues, we propose a multi-view multi-task sparse learning for target identification. Suppose there are $n$ candidates, each has $v$ different perspectives (e.g., SIFT, HOG, LBP, hu feature). Defined $X^v \in R^{dv \times n}$ as the feature matrix, it is a normalized $n$-column $d$-dimensional particle image feature vector. The $dv$ is the $v$-dimensional viewing angle and the $n$ is the number of candidates. Especially, defined $D^v \in R^{dv \times n}$ as a target dictionary, where each column is a target of the $v$-th perspective, and $n$ is the number of target templates. In this paper, it means four high-dimensional vectors of the SIFT, HOG, LBP, hu feature extraction methods. In this manner, it can effectively integrate multiple feature extraction methods. The target dictionary constructs a complete dictionary $A^v = [D^v, I_{dv}]$ by combining the background template $I_{dv}$. Each representation matrix $C^v$ is composed of two cooperative parts $P^v$ and $Q^v$. Component $P$ captures shared features among all tasks, and component $Q$ captures outliers. The multi-view sparse representation can be obtained by the following cost function.

$$\min_{W,P,Q} \frac{1}{2}||A^vC^v - X^v||_F^2 + \omega_1||P||_{1,2} + \omega_2||Q^T||_{1,2} \tag{3}$$

$$||P||_{1,2} = \sum_i (\sum_j P_{i,j}^2)^{\frac{1}{2}} \tag{4}$$

where $P_{ij}$ represents the elements in row $i$ and column $j$ of matrix $P$, $C^v = P^v + Q^v$, $P = [P^1, P^2, \ldots, P^v]$, $Q = [Q^1, Q^2, \ldots, Q^v]$, $\omega_1$ and $\omega_2$ are the parameters controlling the sparsity of $P$ and $Q$, respectively. Figure 8 illustrates the structure of the learned matrices $P$ and $Q$.

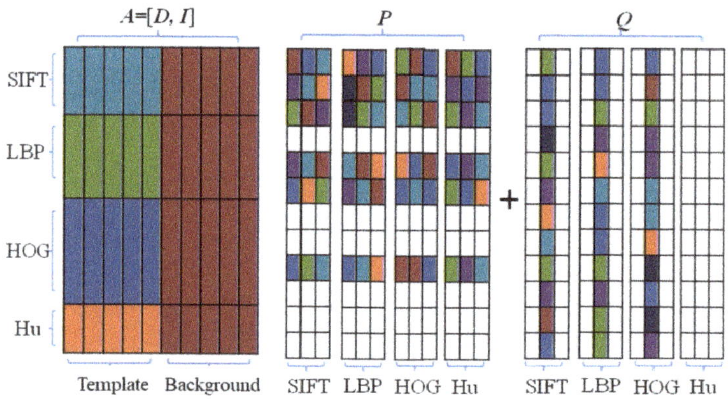


**Figure 8.** The structure of the matrices $P$, $Q$, and $A$.

It should be noted that the superposition of $P^v$ and $Q^v$ requires the same number of columns as $A^v$, so the zero matrix is used to fill the $A^v$ matrix. Definition $\hat{A}^v = [A^v, 0^v]$, where $0^v \in \Re^{dv \times (n_1 - n_v)}$, and each element of $0^v$ is zero, $n_1, \ldots, n_v$ is the dimension of each perspective. According to the target identification results, the observation probability of potential sample $i$ is defined as:

$$p_i = \frac{1}{\Gamma} \exp\{-\alpha \sum_{v=1}^{V} ||A^v C_i^v - X_i^v||^2\} \tag{5}$$

where $\Gamma$ is the normalization parameter, $C_i^v$ is the coefficient corresponding to the $v$-th angle of view of the $Q^i$ potential sample. The target identification result is the particle with the maximum observation probability. Update the target dictionary $D$ step by step and weight the template.

In the process of target identification, some outlier tasks often exist. These sample particles are far from the target and have little overlap with other particles. The outlier task is captured by introducing the coefficient matrix. In particular, if the sum of coefficients corresponding to the particle of $\ell 1$ standard is greater than the adaptive threshold $\sigma$:

$$\sum_{v=1}^{V} |Q_i^v| > \sigma, \tag{6}$$

where $Q_i^v$ is the $i$-th row of $Q^v$, it is defined as outliers, and the observation probability is set to zero, and the resampling process of outliers is ignored. Define the number of detected outlier tasks as $N_0$, and update the threshold $\sigma$ as follows:

$$\begin{cases} \sigma new = \sigma old\kappa, n0 > N0 \\ \sigma new = \sigma old/\kappa, n0 = 0 \\ \sigma new = \sigma old, 0 < n0 \leq N0, \end{cases} \quad (7)$$

where $\kappa$ is a scale factor, $N0$ is a threshold for the number of predefined outliers.

*5.1. Multi View Multi-Task Sparse Learning Target Identification Method*

In this paper, a new target identification method called multi-view multi-task learning method (MVMT), is proposed. The method mainly includes five steps: first, read the image sequence in the captured video and sample the particles according to the particle filter sampling and resampling methods; second, select candidate templates; third, extract multiple features of particles to increase the diversity of target representation from different feature spaces; fourth, use the multi-view sparse representation method for multi-task learning to train the optimal candidate template set; fifth, for each candidate template, transform the reconstruction error into a probability problem to determine the target; finally, calibrate the target and track it in a rectangular frame. To represent the process of a multi-view multi-task learning method more intuitively, the flow chart of a multi-view multi-task sparse learning target identification method is shown in Figure 9.

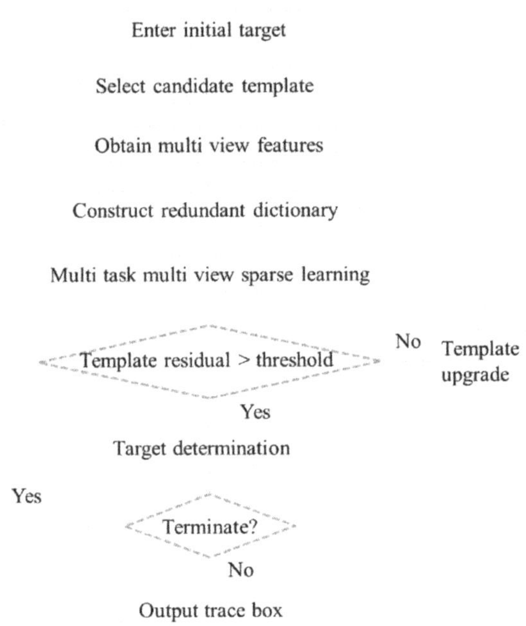


**Figure 9.** Flow chart of multi-view multi-task target identification method.

According to the flow chart in Figure 9, the target identification algorithm based on multi-task and multi-view can be divided into the following main steps:

(1) Initialize the target: Select the rectangular box of the target of interest on the first frame image;
(2) Select the candidate template: Select the candidate template according to the Gaussian distribution model and process its gray image level.

$$f(x) = \frac{1}{\sqrt{2\pi}\sigma}\exp(-\frac{(x-\mu)^2}{2\sigma^2}) \tag{8}$$

(3) Extract multi-view features: Extract the features of LBP, HOG, and SIFT view from the obtained candidate templates, and reduce the dimension of the extracted feature matrix;

(4) Construct the redundant dictionary: Assemble the obtained single column vectors of each view of a given template into a new column vector, i.e., template vector, and cycle through other candidate templates. All the template vectors obtained are arranged in columns, and the identity matrix is added to form a redundant dictionary;

(5) Multi-view multi-task sparse learning: Multi-view multi-task learning is performed on the templates in the redundant dictionary to solve the classification coefficient matrix C. For several particles selected by sampling around the target, select the candidate template with a small residual to replace the bad performance in the dictionary, and finally make the candidate template in the dictionary the optimal template set. For Equation (3), a gradient descent algorithm is proposed, so the target residual can be calculated as:

$$\cos t(W) = \frac{1}{2m}[\sum_{v=1}^{V}((A^vC^v - X^v)^2 + \omega_1||P||_{1,2} + \omega_2||Q^T||_{1,2} \tag{9}$$

For solution C, the following method is used for convergence to obtain the optimal solution:

$$c_j := c_j - \alpha\frac{1}{V}\sum_{v=1}^{V}((A^vC^v - X^v)\cdot A_j^v) \tag{10}$$

(6) Update template: if the residual error between the training sample and the label is less than the given threshold, replace the corresponding template in the dictionary, as shown in Figure 10.

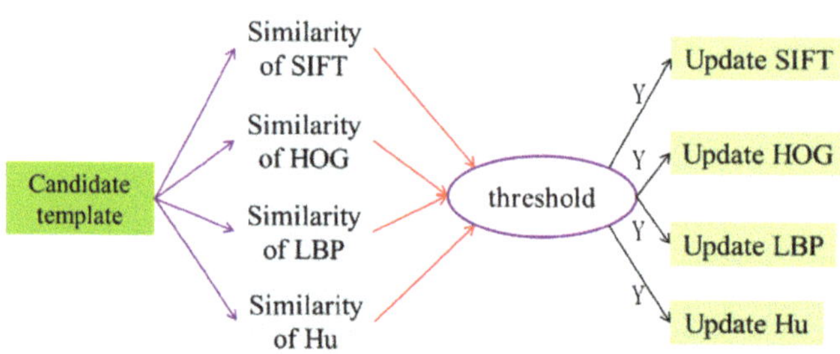


**Figure 10.** Template update schematic.

(7) Determine target: Carry out probability conversion for each classification plane (each column) in matrix C and take the classification plane with the largest probability as the final target to be tracked. For the $t$-th frame image, the determination criteria are as follows:

$$p(y_i|x_t) = \frac{1}{\Gamma}\exp\{-\alpha\sum_{v=1}^{V}||Ta(1:n,i) - Y||\} \tag{11}$$

where $\Gamma$ is the normalized factor, $N$ represents the number of candidate templates in the dictionary, and $Y$ represents the label. The flow chart of target judgment is shown in Figure 11.

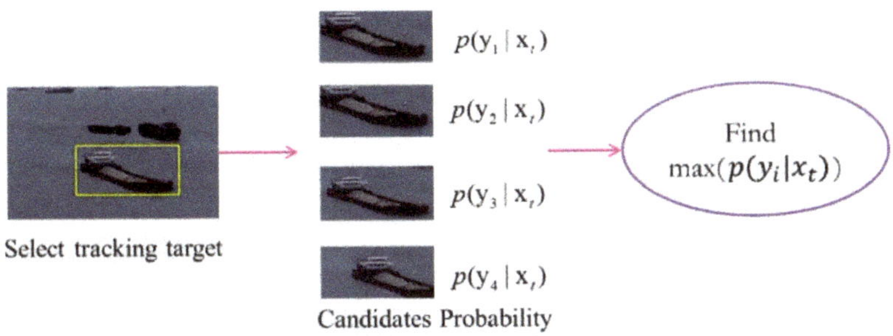


**Figure 11.** Flow chart target selection.

(8) Output box to calculate the target identification error.

*5.2. Simplified MVMT Methods*

To further verify the superiority of multi-view sparse learning performance compared with single-view single-feature tracker, the single-task single-view target identification method (STSV), single-task multi-view target identification method (STMV), and single-view multi-task target identification method (MTSV) are designed, which are three special cases of multi-view multi-task (MVMT) method.

For single-view multi-task learning algorithm (MTSV), it can be obtained from the following formula:

$$\min_{W,P,Q} ||AC - X||_2^2 + \omega_1 ||P||_{1,2} + \omega_2 ||Q^T||_{1,2} \tag{12}$$

The difference between single-view multi-task learning algorithm and multi-view multi-task learning algorithm is that when constructing a dictionary, only a single-view feature is used to describe a template and train multiple particles at the same time.

$$D = [T^1, T^2, \cdots, T^n] \tag{13}$$

where $T_k$ ($k$ = 1, 2, ..., $n$) is the extracted single-view template. By comparing Equations (3) and (12), it can be obtained that SVMT is a special case of MVMT.

## 6. Experiment and Result Analysis

To verify the effectiveness of the proposed method (MVMT), several classical trackers with excellent performance are selected as comparison methods. The experimental performance of these algorithms is compared on the image sequence frames dataset.

To make full use of the advantages of complementary features, three popular features are used in the experiment: gradient direction histogram (HOG), local binary pattern (LBP), and scale-invariant feature transformation (SIFT). HOG is edge distributions of objects captured based on gradient features. LBP has a powerful function of representing object texture. SIFT has scale invariance for changing image brightness, rotation angle or shooting angle. In addition, to ensure the quality of the extracted features, the illumination normalization method is used to extract the feature vectors from multiple perspectives of each particle before feature extraction.

MVMT (multi-view multi-task) is compared with the other three special cases: SVST, MVST, and SVMT algorithms. The source code of all algorithms runs in MTLAB. The experimental hardware environment is as follows: the CPU is i5-3612qe; the memory is 8GB; the video memory is 2GB; the operating system is windows 10; and the software development environment is matlab2018.

The number of initial candidate templates for MVMT algorithm, L1T, and MTT proposed in the algorithm parameter setting is $n$ = 10, and the number of training templates is

100. Gaussian distribution is used to sample the particles. After selecting the candidate template, the sparse learning method is used to construct the redundant dictionary for later sample training and updating. Scale-invariant feature transformation (SIFT), gradient direction histogram (HOG), local binary pattern (LBP), and Hu invariant matrix are used. The gradient descent iterative optimal algorithm is adopted to continuously update the candidate templates in the dictionary. Finally, the template with the smallest residual error is selected as the method of the next frame of image.

The experimental data are divided into two groups of data for experimental verification. One group is a public dataset, including video picture frame sequence. The other group is custom dataset, including the video sequence of ships in key waters and port.

### 6.1. Experimental Results and Analysis of Public Datasets

To evaluate the effectiveness of the proposed MVMT method, three complementary features are adopted. Four widely used sequence diagrams (car4, faceocc2, david_interior, sylv) are selected. The experimental results of the common dataset are shown in Figure 12.

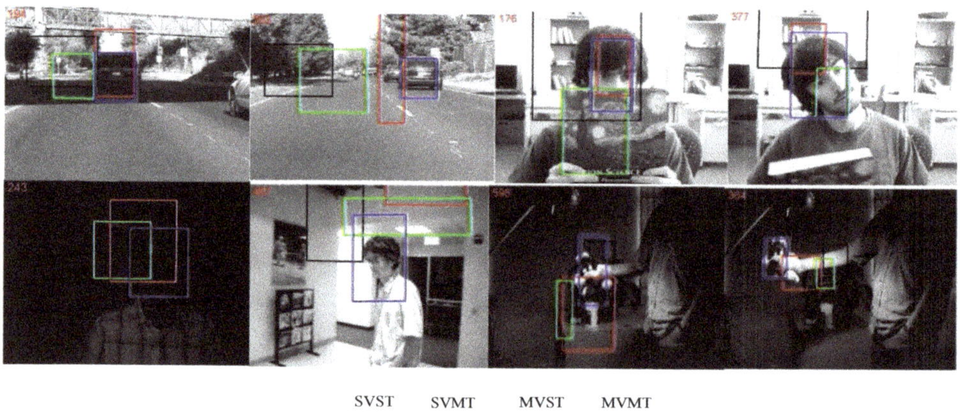


**Figure 12.** The identification results of the proposed MVMT and SVST, MVST, SVMT algorithms.

In the car4 sequence, the main influencing factors are illumination change and size change. The MVMT and the fast-moving vehicle are tracked from beginning to end, while the SVST and MVST are affected by illumination and size change of the vehicle. They contain more and more background areas, and the target is gradually lost. The target has been completely lost in 194 frames. The SVMT has good performance on the influence of illumination, but at about the 310th frame the vehicle passes the bridge and suddenly loses the target. According to the experimental results, MVMT ratio can better deal with the factors of light change and target size change.

In the faceocc2 sequence, the task is to identify the swing of David's head in the room. The main factors of this sequence are swing and occlusion. MVMT and MVST can identify the target successfully in the whole sequence, but MVMT can only identify a small part of the target after the target is occluded many times. In addition, after the head target swings, the appearance and angle change. SVST and SVMT algorithms lose some information of the target. Then after being occluded many times, SVST completely loses the target. The experimental results show that MVMT can better deal with the occlusion, appearance change and angle change in face tasks.

In the David_indoor sequence, the influence of different illumination is mainly studied. The experimental results show that in the case of dark light, that is, in the early stage of target identification, SVST has been separated from the target because the texture features or edge features are not obvious. With the significant enhancement of illumination, the

SVMT and MVST algorithms also slowly lose their target. In the whole target identification process, only MVMT algorithm has captured the target. The experimental results show that MVMT can better deal with the problems caused by light changes.

In the owl sequence, the task is to identify the owl doll. The owl sequence is relatively more fixed, and fuzzy scenes and fast motion are considered. At the early stage of the sequence frame the shape of the target changes and receives different illumination. The SVST and SVMT algorithms lose the target. With the rapid movement, the target becomes fuzzy. The MVST algorithm gradually loses the target information and cannot capture the target. In the whole owl sequence, only MVMT algorithm has captured the target. The experimental results show that MVMT can better deal with the problem of fuzzy targets.

Figure 13 shows the experimental error comparison curve. It can be seen that the identification methods of SVST and MVST are very poor, and the SVST method is the worst. The reason is that a single task lacks associated information with other tasks and is artificially divided. At the same time, the identification performance of both SVMT method and MVMT method is generally better, because the multi-task learning method cannot only capture the common features among multiple tasks but also eliminate outliers, which makes the performance of identification better. Among them, the MVMT method has the best identification performance, because the multi-view and multi-task simultaneously contain a variety of feature vectors in different state spaces, provide multiple views for the same target object, and learn together, which has very good stability and strong robustness.

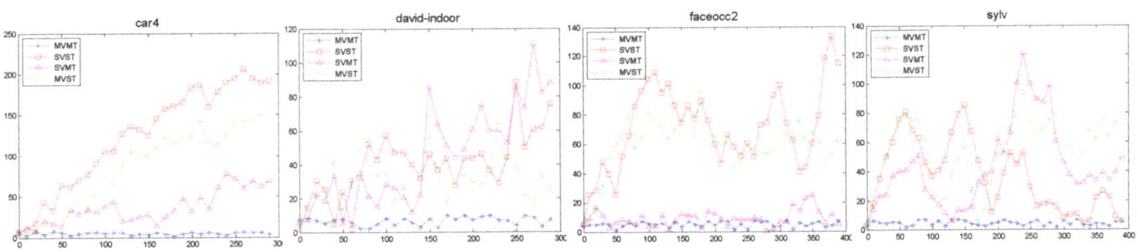


**Figure 13.** The errors of the proposed MVMT algorithm and SVST, MVST, SVMT algorithms (pixel).

### *6.2. Experiment and Analysis of Custom Datasets*

To further verify the application performance of the proposed algorithm in the actual scene, we adopt the targets from the user-defined sequence frames and compare the performance of different algorithms. The custom dataset comes from the field shooting in a port. Six sequence frames are selected from the ship video Ship-1, ship video Ship-2, ship video Ship-3, and ship video Ship-4, ship video Ship-5, and ship video Ship-6. Among them, Ship-1, Ship-2, Ship-3, and Ship-4 are dangerous watercourse scenes taken in different places during the day, Ship-5 is the scene taken at night, and Ship-6 is the video sequence obtained through infrared rays. As with the public dataset, all pictures are resized to $320 \times 240$ (pixels). In this experiment, the source code provided by L1T and MTT and the proposed MTMV source code are tested in MTLAB. Meanwhile, the above STSV (single-view single-task), STMV (multi-view single-task), and MTSV (single-view multi-task) algorithms are compared in the experiment. The identification results of these seven algorithms are shown in Figure 14.

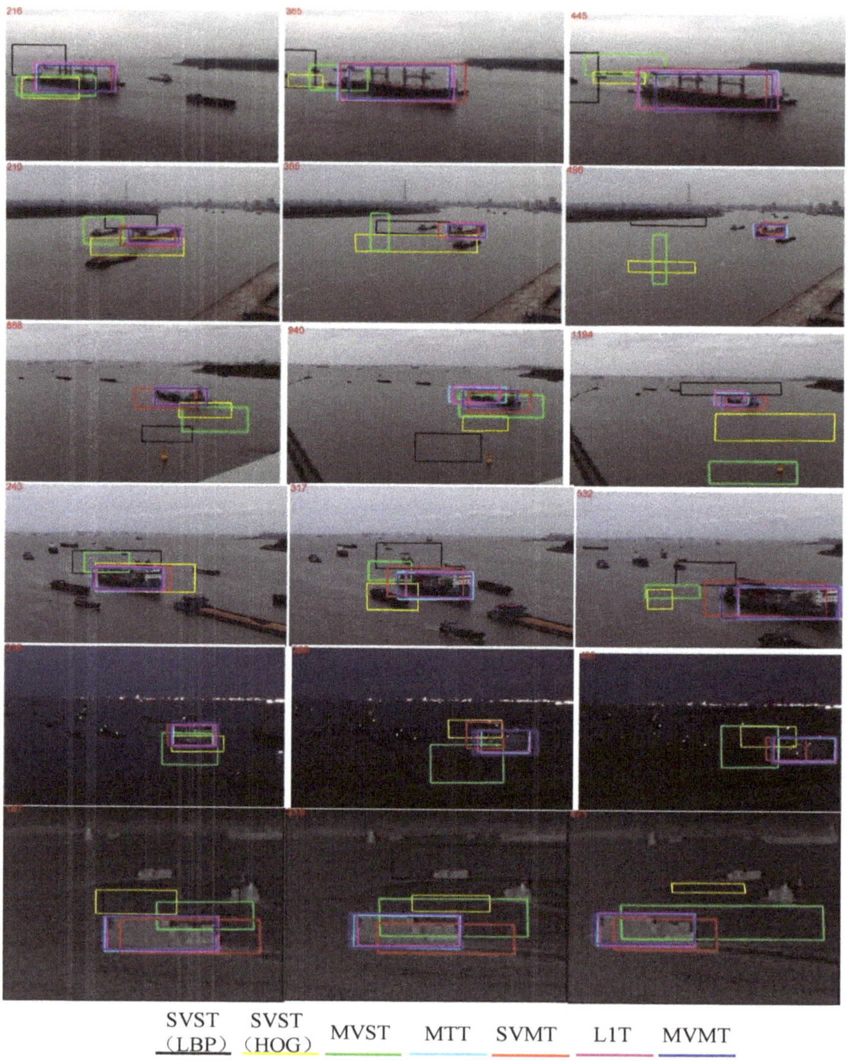


**Figure 14.** Identification results of 7 algorithms in 6 groups of ship sequence pictures.

### 6.3. *Qualitative Comparison and Analysis*

In the Ship-1 sequence, the influence of the size change of the target on the identification process is mainly considered. At the 216th frame, the SVST and MVST gradually separated from the target when the target size just began to increase. At the 365th frame, the target size has doubled from the initial target size due to the continuous increase in-variati the ship size, and the SVST and MVST have lost the target. In the whole process, although L1T, MTT and SVMT kept tracking the target, the effect was not as good as MVMT. Experiments show that Mvmt has strong robustness to target size changes.

In the Ship-2 sequence, the main interference factors are the appearance change and angle change of the target. At the 241st frame, MVST soon lost its target. At the 389th and 436th frames, MTT, SVMT, L1T, and MVMT tracked the target all the way, but SVMT only captured part of the target, which was not as effective as MVMT. The experimental results show that MVMT can deal with the problems of size change, appearance change and angle change. L1T and MVST fail to track the whole sequence, which shows that the mechanism of multi-task joint learning makes MTT, SVMT, and MVMT more robust. However, MTT and SVMT are less robust than MVMT because MVMT uses the advantages of complementary features and can detect outliers.

In the Ship-3 sequence, the main factors are fast motion and blur. At the 821st frame, the MVST quickly loses the target. At the 931st frame, after the camera lens moves rapidly, the image becomes blurred. L1T and SVMT only capture some targets, while MVMT keeps locking the targets. Compared with L1T and MVST, MVMT can fully track the target under different changes because of the robustness of additional features. In the whole database-3 sequence, only MVMT can successfully track the target in the whole sequence, while other trackers either completely lose the target or contain most of the background. It is concluded from the experiment that the single-task tracker is easily affected by the appearance change.

The Ship-4 sequence is mainly tested for occlusion factors. At the 243rd frame, the target is unobstructed, and all methods lock the target well. However, At the 317th frame, SVST and MVST have been separated from the target due to partial occlusion of the target. SVMT has included part of the background, and L1T has only captured part of the target. With the occlusion leaving, SVMT, L1T and MTT can track the target at the 532nd frame, but their performance is not as good as MVMT. This shows that MVMT can better deal with the influence of occlusion factors than other methods.

In the Ship-5 sequence, the influence of illumination change factor is mainly considered. At the 233rd frame, MVMT, MTT, and L1T can fully track the sailing ship and only MVST is separated from the target. With the gradual dimming of the light, MVST contains more and more background areas at the 346th frame. According to the experimental results, MVMT has better robustness than other methods.

In the Ship-6 sequence, it mainly detects the stability and robustness of target identification on infrared video. At the 222nd frame, MVST has lost the target. MTT, SVMT, L1T, and MVMT can keep locking the target, but SVMT is easy to contain more background. MVMT can faithfully track the sailing ship, which shows that MVMT is better than other methods and can track infrared video stably.

### 6.4. Quantitative Comparison and Analysis

To quantitatively evaluate the performance of these comparison methods, this experiment calculates the distance between the center of the result and the real position of each frame; it then draws the variation diagram of the center position error and the number of frames. Due to space limitations, only the error comparison of the six sequence diagrams provided here by the seven methods is listed. For a perfect result, the positioning error should be zero. For a more intuitive comparison, the average position errors (APEs) of the six sequences are shown in Figure 15.

To further illustrate the comparison between different video sequence diagrams, Table 1 shows the average error between the results and the real position on six sequence diagrams for the seven methods. Table 2 shows the weighted standard deviation between the results and the real position for the seven methods, and the specific weights of each particle are equal. The following two statistical tables show that the average error and standard deviation of the MVMT method are the smallest in both visible and infrared video sequences. Therefore, in general, MVMT has the best performance.

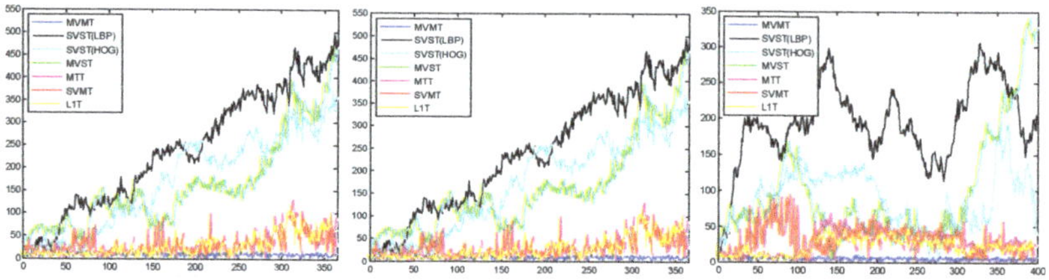


(a) APEs on Ship-1 (b) APEs on Ship-2 (c) APEs on Ship-3

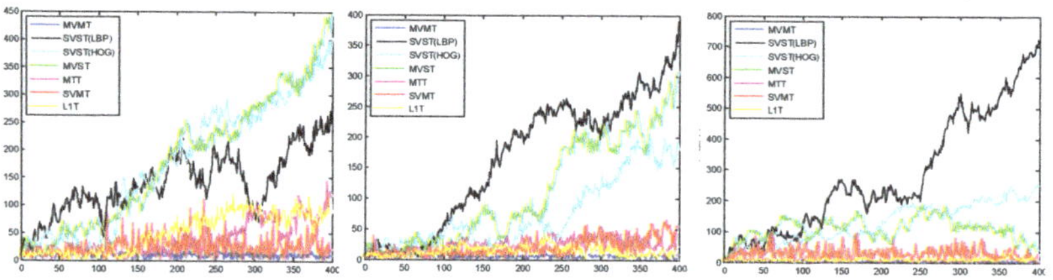


(d) APEs on Ship-4 (e) APEs on Ship-5 (f) APEs on Ship-6

**Figure 15.** 7 Average position error (pixel) of 5 comparison methods.

**Table 1.** Average error between results of five comparison methods and real position (pixels).

| Sequence | Ship-1 | Ship-2 | Ship-3 | Ship-4 | Ship-5 | Ship-6 |
|---|---|---|---|---|---|---|
| STMV | 160.4 | 280.4 | 96.1 | 179.2 | 109.1 | 110.8 |
| MTT | 28.57 | 7.12 | 25.52 | 38.06 | 20.52 | 10.42 |
| MTSV | 30.9 | 20.33 | 32.82 | 24.9 | 21.46 | 32.57 |
| L1T | 28.67 | 6.69 | 21.99 | 50.76 | 12.76 | 11.71 |
| MTMV | **9.21** | **4.56** | **7.00** | **11.04** | **3.01** | **5.99** |

The bold indicates that the experimental value of this algorithm is superior to other algorithms.

**Table 2.** Weighted standard deviation between the results of the five comparison methods and the real position (pixels).

| Sequence | Ship-1 | Ship-2 | Ship-3 | Ship-4 | Ship-5 | Ship-6 |
|---|---|---|---|---|---|---|
| STMV | 9.95 | 15.64 | 6.11 | 10.7 | 6.98 | 5.82 |
| MTT | 1.98 | 0.41 | 1.46 | 2.46 | 1.18 | 0.60 |
| MTSV | 2.01 | 1.22 | 1.89 | 1.52 | 1.31 | 1.88 |
| L1T | 1.99 | 0.39 | 1.23 | 3.08 | 0.75 | 0.66 |
| MTMV | **0.51** | **0.24** | **0.37** | **0.58** | **0.16** | **0.32** |

The bold indicates that the experimental value of this algorithm is superior to other algorithms.

## 7. Conclusions

In this paper, a multi-view multi-task target identification method based on sparse learning is proposed. Extensive experiments were conducted on public datasets and custom video datasets on ship sequence frames. Under different challenging interference factors such as size, illumination, blur, deformation, and occlusion, the experimental results show that the proposed MVMT can better identify the target in both visible and infrared video. Compared with the comparisons method, the performance of the proposed MVMT is the best. In summary, the conclusions of this paper are presented as follows:

(1) A target identification method based on machine learning is proposed. The traditional target identification method makes use of less information. In the case of many targets and complex backgrounds, tracking loss and error are common. To break through the limitations of traditional methods, we introduce a machine learning approach and provide a new idea for target identification by using sparse representation, dictionary learning method and particle filter framework;

(2) By analyzing the features of targets in the surveillance video, the texture, edge, and geometry invariant features of targets are automatically extracted, and the multi-view feature learning method is introduced to realize the information fusion of multiple features. To obtain better results, a multi-task learning method is proposed based on the multi-view feature for joint learning, which makes full use of the information transfer function between different tasks, makes up for the deficiency of the traditional single learning task, improves the classification and recognition accuracy, and provides a theoretical basis for the accurate and stable identification.

**Author Contributions:** J.C.: writing—original draft preparation, methodology, and conceptualization, and X.W.: writing—original draft preparation, visualization, validation, and software, and Z.Z.: funding acquisition, supervision and project administration. All authors have read and agreed to the published version of the manuscript.

**Funding:** This work has no funding.

**Institutional Review Board Statement:** Not applicable.

**Informed Consent Statement:** Not applicable.

**Data Availability Statement:** Not applicable.

**Conflicts of Interest:** The authors declare no conflict of interest.

## References

1. Tang, J.; Liu, F.; Zou, Y.; Zhang, W.; Wang, Y. An Improved Fuzzy Neural Network for Traffic Speed Prediction Considering Periodic Characteristic. *IEEE Trans. Intell. Transp. Syst.* **2017**, *18*, 2340–2350. [CrossRef]
2. Yan, Y.; Zhang, S.; Tang, J.; Wang, X. Understanding characteristics in multivariate traffic flow time series from complex network structure. *Phys. A Stat. Mech. Its Appl.* **2017**, *477*, 149–160. [CrossRef]
3. Chen, W.; Zhou, G.; Liu, Z.; Li, X.; Zheng, X.; Wang, L. NIGAN: A Framework for Mountain Road Extraction Integrating Remote Sensing Road-Scene Neighborhood Probability Enhancements and Improved Conditional Generative Adversarial Network. *IEEE Trans. Geosci. Remote Sens.* **2022**, *60*, 1–15. [CrossRef]
4. Chen, W.; Li, X.; Wang, L. Target Detection for Mine Remote Sensing Using Deep Learning. In *Remote Sensing Intelligent Interpretation for Mine Geological Environment*; Springer: Singapore, 2022; pp. 127–164.
5. Tong, W.; Chen, W.; Han, W.; Li, X.; Wang, L. Channel-attention-based DenseNet network for remote sensing image scene classification. *IEEE J. Sel. Top. Appl. Earth Obs. Remote Sens.* **2020**, *13*, 4121–4132. [CrossRef]
6. Liu, M.; Hu, Q.; Wang, C.; Tian, T.; Chen, W. Daff-Net: Dual Attention Feature Fusion Network for Aircraft Detection in Remote Sensing Images. In Proceedings of the IEEE International Geoscience and Remote Sensing Symposium IGARSS, Brussels, Belgium, 11–16 July 2021; pp. 4196–4199.
7. Ouyang, S.; Xu, J.; Chen, W.; Dong, Y.; Li, X.; Li, J. A Fine-Grained Genetic Landform Classification Network Based on Multimodal Feature-Extraction and Regional Geological Context. *IEEE Trans. Geosci. Remote Sens.* **2022**, *60*, 1–14. [CrossRef]
8. Craswell, N.; Mitra, B.; Yilmaz, E.; Campos, D.; Voorhees, E.M. Overview of the TREC 2019 deep learning track. *arXiv* **2020**, arXiv:2003.07820.
9. Zhang, S.; Tang, J.; Wang, H.; Wang, Y.; An, S. Revealing intra-urban travel patterns and service ranges from taxi trajectories. *J. Transp. Geogr.* **2017**, *61*, 72–86. [CrossRef]
10. Kwon, J.; Lee, K.M. Visual tracking decomposition. In Proceedings of the IEEE Conference Computer Vision and Pattern Recognition (CVPR), San Francisco, CA, USA, 13–18 June 2010; pp. 1269–1276.
11. Wang, Y.; Luo, X.; Ding, L.; Hu, S. Visual tracking via robust multi-task multi-feature joint sparse representation. *Multimed. Tools Appl.* **2018**, *77*, 31447–31467. [CrossRef]
12. Lan, X.; Ye, M.; Zhang, S.; Zhou, H.; Yuen, P.C. Modality-correlation-aware sparse representation for RGB-infrared object tracking. *Pattern Recognit. Lett.* **2020**, *130*, 12–20. [CrossRef]
13. Wan, M.; Gu, G.; Qian, W.; Ren, K.; Maldague, X.; Chen, Q. School of Electronic and Optical Engineering, Nanjing University of Science and Technology, Nanjing, China. Unmanned aerial vehicle video-based target tracking algorithm using sparse representation. *IEEE Internet Things J.* **2019**, *6*, 9689–9706. [CrossRef]

14. Elfring, J.; Torta, E.; van de Molengraft, R. Particle filters: A hands-on tutorial. *Sensors* **2021**, *21*, 438. [CrossRef]
15. Zhang, T.; Xu, C.; Yang, M.H. Learning multi-task correlation particle filters for visual tracking. *IEEE Trans. Pattern Anal. Mach. Intell.* **2018**, *41*, 365–378. [CrossRef] [PubMed]
16. Nai, K.; Li, Z.; Gan, Y.; Wang, Q. Robust Visual Tracking via Multitask Sparse Correlation Filters Learning. *IEEE Trans. Neural Netw. Learn. Syst.* 2021, in press. [CrossRef] [PubMed]
17. Gurkan, F.; Gunsel, B. Integration of regularized l1 tracking and instance segmentation for video object tracking. *Neurocomputing* **2021**, *423*, 284–300. [CrossRef]
18. Javanmardi, M.; Qi, X. Robust structured multi-task multi-view sparse tracking. In Proceedings of the IEEE International Conference on Multimedia and Expo (ICME), San Diego, CA, USA, 23–27 July 2018; pp. 1–6.
19. Lu, R.; Liu, J.; Lian, S.; Zuo, X. Multi-view representation learning in multi-task scene. *Neural Comput. Appl.* **2020**, *32*, 10403–10422. [CrossRef]
20. Yan, P.; Yao, S.; Zhu, Q.; Zhang, T.; Cui, W. Real-time detection and tracking of infrared small targets based on grid fast density peaks searching and improved KCF. *Infrared Phys. Technol.* **2022**, *123*, 104181. [CrossRef]
21. Lukezic, A.; Vojir, T.; Zajc, L.C.; Matas, J.; Kristan, M. Discriminative correlation filter with channel and spatial reliability. In Proceedings of the IEEE Conference Computer Vision and Pattern Recognition (CVPR), Honolulu, HI, USA, 21–26 July 2017; pp. 4847–4856.
22. Yuan, D.; Kang, W.; He, Z. Robust visual tracking with correlation filters and metric learning. *Knowl. Based Syst.* **2020**, *195*, 105697. [CrossRef]
23. Cheng, G.; Li, R.; Lang, C.; Han, J. Task-wise attention guided part complementary learning for few-shot image classification. *Sci. China Inf. Sci.* **2021**, *64*, 120104. [CrossRef]
24. Yang, Y.; Xing, W.; Zhang, S.; Gao, L.; Yu, Q.; Che, X.; Lu, W. Visual tracking with long-short term based correlation filter. *IEEE Access* **2020**, *8*, 20257–20269. [CrossRef]
25. Kurani, A.; Doshi, P.; Vakharia, A.; Shah, M. A comprehensive comparative study of artificial neural network (ANN) and support vector machines (SVM) on stock forecasting. *Ann. Data Sci.* **2021**, 1–26. [CrossRef]
26. Abbass, M.Y.; Kwon, K.C.; Kim, N.; Abdelwahab, S.A.; El-Samie, F.E.A.; Khalaf, A.A.M. Visual tracking using convolutional features with sparse coding. *Artif. Intell. Rev.* **2021**, *54*, 3349–3360. [CrossRef]
27. Bardow, P.; Davison, A.J.; Leutenegger, S. Simultaneous optical flow and intensity estimation from an event camera. In Proceedings of the IEEE Conference on Computer Vision and Pattern Recognition, Las Vegas, NV, USA, 27–30 June 2016; pp. 884–892.
28. Tripathi, R.P.; Ghosh, S.; Chandle, J.O. Tracking of object using optimal adaptive Kalman filter. In Proceedings of the IEEE International Conference on Engineering and Technology, Coimbatore, India, 17–18 March 2016; pp. 1128–1131.
29. Zhang, S.; Yang, Y.; Zhang, M.; Mi, P. An Efficient Tracker via Multi-feature Adaptive Correlation Filter. *J. Syst. Simul.* **2022**, *34*, 1864.
30. Weng, X.; Ivanovic, B.; Pavone, M. Mtp: Multi-hypothesis tracking and prediction for reduced error propagation. In Proceedings of the 33rd IEEE Intelligent Vehicles Symposium (IV), Aachen, Germany, 5–9 June 2022; pp. 1218–1225.
31. Meshgi, K.; Oba, S.; Ishii, S. Efficient diverse ensemble for discriminative co-tracking. In Proceedings of the Conference Computer Vision and Pattern Recognition (CVPR), Salt Lake City, UT, USA, 18–23 June 2018; pp. 4814–4823.
32. Wang, N.; Zhou, W.; Tian, Q.; Hong, R.; Wang, M.; Li, H. Multi-cue correlation filters for robust visual tracking. In Proceedings of the IEEE Conference Computer Vision and Pattern Recognition (CVPR), Salt Lake City, UT, USA, 18–23 June 2018; pp. 4844–4853.
33. Jiao, L.; Wang, D.; Bai, Y.; Chen, P.; Liu, F. Deep learning in visual tracking: A review. *IEEE Trans. Neural Netw. Learn. Syst.* **2021**. [CrossRef] [PubMed]
34. Dai, K.; Wang, D.; Lu, H.; Sun, C.; Li, J. Visual tracking via adaptive spatially-regularized correlation filters. In Proceedings of the IEEE Conference Computer Vision and Pattern Recognition (CVPR), Long Beach, CA, USA, 15–20 June 2019; pp. 4670–4679.
35. He, A.; Luo, C.; Tian, X.; Zeng, W. A twofold siamese network for real-time object tracking. In Proceedings of the IEEE Conference Computer Vision and Pattern Recognition (CVPR), Salt Lake City, UT, USA, 18–23 June 2018; pp. 4834–4843.
36. Gao, J.; Zhang, T.; Xu, C. Graph convolutional tracking. In Proceedings of the IEEE Conference Computer Vision and Pattern Recognition (CVPR), Long Beach, CA, USA, 15–20 June 2019; pp. 4649–4659.
37. Dong, B.; Zhou, Y.; Hu, C.; Fu, K.; Chen, G. BCNet: Bidirectional collaboration network for edge-guided salient object detection. *Neurocomputing* **2021**, *437*, 58–71. [CrossRef]
38. Wu, L.; Fang, S.; Ma, Y.; Fan, F.; Huang, J. Infrared small target detection based on gray intensity descent and local gradient watershed. *Infrared Phys. Technol.* **2022**, *123*, 104171. [CrossRef]
39. Liu, J.; Zhong, X. An object tracking method based on Mean Shift algorithm with HSV color space and texture features. *Clust. Comput.* **2019**, *22*, 6079–6090. [CrossRef]
40. Humeau-Heurtier, A. Texture feature extraction methods: A. survey. *IEEE Access* **2019**, *7*, 8975–9000. [CrossRef]
41. Chen, X.; Chen, H.; Wu, H.; Huang, Y.; Yang, Y.; Zhang, W.; Xiong, P. Robust visual ship tracking with an ensemble framework via multi-view learning and wavelet filter. *Sensors* **2020**, *20*, 932. [CrossRef] [PubMed]
42. Mao, J.L. Adaptive multi-view learning and its application in image classification. *J. Comput. Appl.* **2013**, *33*, 1955–1959. [CrossRef]
43. Deng, J.; Czarnecki, K. MLOD: A multi-view 3D object detection based on robust feature fusion method. In Proceedings of the 2019 IEEE Intelligent Transportation Systems Conference (ITSC), Auckland, New Zealand, 27–30 October 2019; pp. 279–284.
44. Zhou, W.; Gao, S.; Zhang, L.; Lou, X. Histogram of oriented gradients feature extraction from raw bayer pattern images. *IEEE Trans. Circ. Syst. II Express Briefs* **2020**, *67*, 946–950. [CrossRef]

45. Hassaballah, M.; Alshazly, H.A.; Ali, A.A. Ear recognition using local binary patterns: A comparative experimental study. *Expert Syst. Appl.* **2019**, *118*, 182–200. [CrossRef]
46. Tareen, S.A.K.; Saleem, Z. A comparative analysis of sift, surf, kaze, akaze, orb, and brisk. In Proceedings of the IEEE International Conference on Computing, Mathematics and Engineering Technologies, Sukkur, Pakistan, 3–4 March 2018; pp. 1–10.
47. Li, G.; Feng, Y. Moving object detection based on SIFT feature matching and K-means clustering. *J. Comput. Appl.* **2012**, *32*, 2824–2826.
48. Hu, M.K. Visual pattern recognition by moment invariants. *IRE Trans. Inf. Theory* **1962**, *8*, 179–187.
49. Chen, X.; Wang, S.; Shi, C.; Wu, H.; Zhao, J.; Fu, J. Robust ship tracking via multi-view learning and sparse representation. *J. Navig.* **2019**, *72*, 176–192. [CrossRef]
50. Mei, X.; Hong, Z.; Prokhorov, D. Robust Multitask Multiview Tracking in Videos. *IEEE Trans. Neural Netw. Learn. Syst.* **2015**, *26*, 2874–2890. [CrossRef] [PubMed]
51. Hong, Z.; Mei, X.; Prokhorov, D.; Tao, D. Tracking via robust multi-task multi-view joint sparse representation. In Proceedings of the IEEE International Conference on Computer Vision, Sydney, NSW, Australia, 1–8 December 2013; pp. 649–656.

*applied sciences*

MDPI

*Article*

# Multi-Input Attention Network for Dehazing of Remote Sensing Images

**Zhijie He [1,2], Cailan Gong [1,*], Yong Hu [1], Fuqiang Zheng [1] and Lan Li [1]**

1 Key Laboratory of Infrared System Detection and Imaging Technologies, Shanghai Institute of Technical Physics, Chinese Academy of Sciences, Shanghai 200083, China
2 University of Chinese Academy of Sciences, Beijing 100049, China
* Correspondence: gcl@mail.sitp.ac.cn

**Abstract:** The non-uniform haze distribution in remote sensing images, together with the complexity of the ground information, brings many difficulties to the dehazing of remote sensing images. In this paper, we propose a multi-input convolutional neural network based on an encoder–decoder structure to effectively restore remote sensing hazy images. The proposed network can directly learn the mapping between hazy images and the corresponding haze-free images. It also effectively utilizes the strong haze penetration characteristic of the Infrared band. Our proposed network also includes the attention module and the global skip connection structure, which enables the network to effectively learn the haze-relevant features and better preserve the ground information. We build a dataset for training and testing our proposed method. The dataset consists of remote sensing images with two different resolutions and nine bands, which are captured by Sentinel-2. The experimental results demonstrate that our method outperforms traditional dehazing methods and other deep learning methods in terms of the final dehazing effect, peak signal-to-noise ratio (PSNR), structural similarity (SSIM) and feature similarity (FSIM).

**Keywords:** non-uniform haze; dehazing; deep learning; remote sensing images; Sentinel-2

## 1. Introduction

In recent years, the quality and quantity of satellite data have tremendously increased. However, the impact of haze has been a common issue with optical remote sensing data. Haze can severely interfere with the transmittance in all optical spectral bands, which impacts the reflected signal and hinders the observation of the underlying surface of the haze. This further results in huge data loss in both the spatial and temporal domains. Haze becomes serious interference for applications requiring time consistency (such as agricultural monitoring) and applications requiring observation of a scene at a specific time (such as disaster monitoring). Therefore, effective recovery from haze will greatly increase the usability of remote sensing data.

Early studies on the dehazing problem of remote sensing images use different methods to eliminate the influence of haze [1–4]. They use multi-source or multi-temporal images in the same area as auxiliary data, the complementary relationship between images, image fusion, pixel replacement, etc. All these methods have achieved good results. However, the need to obtain multiple sets of data from the same area as auxiliary data leads to poor applicability; especially for some remote sensing data with long collection interval, it will be more difficult to obtain available auxiliary data. Therefore, the single image dehazing method has attracted more and more attention. Some studies on single image dehazing use image enhancement methods, including processing the histogram of the image [5], and enhancing the contrast [6] and saturation [7] of the image. Additionally, some dehazing methods are based on homomorphic filtering [8] and the retinex color constancy theory [9].

Image enhancement lacks the hazy imaging mechanism, which could lead to a certain degree of distortion in the restored image. Researchers build an image dehazing method based on the Atmospheric Scattering Model (ASM). The most popular ASM model was proposed by McCartney and further improved by Narasimhan [10] and Nayar [6]. The model can usually be written as in Formula (1):

$$I(x) = J(x)t(x) + A[1 - t(x)] \tag{1}$$

where $I(x)$ is the image disturbed by haze; $J(x)$ is the haze-free image that needs to be restored; $t(x)$ is the transmittance of light passing through the atmospheric medium; $x$ represents the image pixel; $A$ represents the global constant: atmospheric light. To obtain the haze-free image $J(x)$, we first need to obtain the transmittance $t(x)$ and the global atmospheric light $A$. However, using hazy images to estimate transmittance requires prior information. At present, studies on prior information are mainly based on statistical properties of hazy images, such as contrast prior [11], dark channel prior (DCP) [12], color attenuation prior [13], etc. However, this prior information will very likely become less applicable in images of different scenes, which affects the dehazing results.

The development of deep learning brings new ideas to the dehazing research: convolutional neural network (CNN). Some earlier studies use neural networks to replace prior information to estimate the parameters of Formula (1) [14–16] and to obtain the dehazed image. Since the real transmittance value of the hazy image cannot be obtained, the training data can be achieved through simulated parameters, which could impact the accuracy of the estimated transmittance. Meanwhile, the transmittance model is a simplified expression for hazy imaging. The advantage of feature extraction capability of neural networks cannot be fully utilized.

Some research uses end-to-end networks to directly explore the mapping relationship from hazy images to haze-free images and, finally, generate dehazed images [17–20]. This research obtained prominent dehazing effects. However, there are limitations. First, the dataset used in the dehazing models is usually less disturbed by haze, which is relatively uniformly distributed. In remote sensing images, the distribution of haze is often uneven, and thin clouds often exist, which makes the image more disturbed than the images studied in their research. Second, remote sensing images usually have more than three channels (RGB) as in ordinary natural images. For example, the images obtained by the Operational Land Imager (OLI) of landsat8 have nine bands. The images acquired by the Multispectral Imager (MSI) of Sentinel-2 have 13 bands. The bands beyond RGB also interfere with haze. Figure 1 shows an example of an image captured by Sentinel-2. It can be observed that the infrared bands, such as Band11 and Band12, have stronger penetrating power and are less impacted by haze.

Most of the previous dehazing methods cannot effectively remove non-uniformly distributed haze. They cannot deal with the impact on the bands beyond RGB as well. Inspired by the dehazing of natural images using convolutional neural networks (CNNs), some researchers apply CNNs to the dehazing of remote sensing images [21–24]. Meanwhile, the infrared bands and Synthetic Aperture Radar (SAR) microwave bands in multispectral remote sensing images can penetrate haze more easily compared to visible bands. They can better reserve the ground information in the area with haze. Therefore, some research uses infrared band images and SAR images as auxiliary information, which is used as input for CNNs to obtain dehazing models [25–27]. These methods can better handle the non-uniform distribution of haze. However, most of them focus on the RGB band or a few near-infrared bands, instead of the more abundant infrared bands. Furthermore, most of the training data are synthetic hazy images, which could be different from the actual hazy images with a lot more complexity.

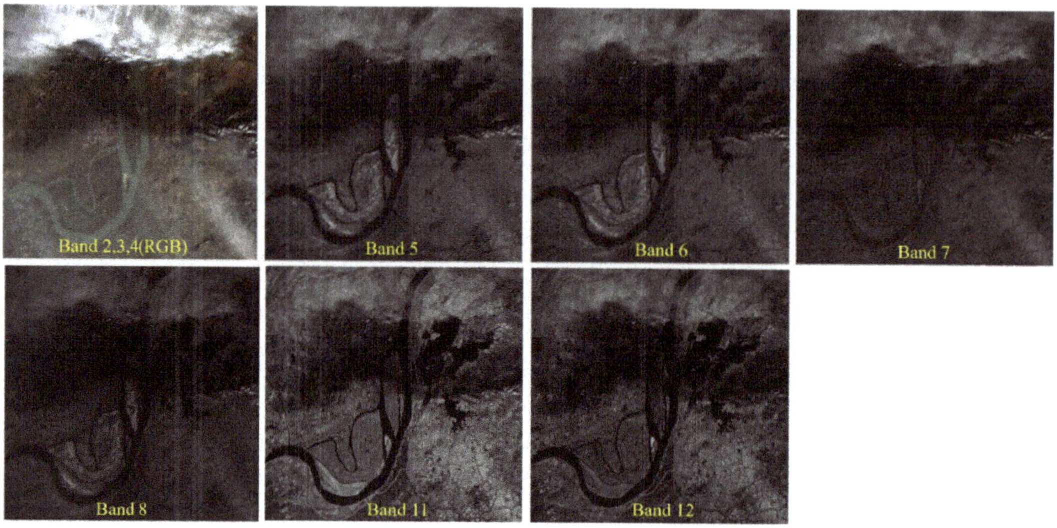


**Figure 1.** The visible (RGB) band and the infrared bands in the image captured by Sentinel-2.

To address these issues, we propose a multi-input attention network for the dehazing of a single multispectral remote sensing image. Since different bands in multispectral images have different features, we utilize the strong penetration capability of richer infrared bands and divide the multiple bands into three groups to extract features. Our proposed network consists of an encoder–decoder structure and uses head-to-tail connections and a multi-scale output structure, similar to the feature pyramid network. This structure enables the dehazing model to effectively remove haze while maintaining ground details. It can directly process bands of different resolutions. Furthermore, improved channel attention and spatial attention structure are added for extracting features from different inputs, which improves the efficiency and adaptability of training. In this research, we use real haze and haze-free multispectral remote sensing image pairs as datasets. Our dehazing model achieves very good results in restructuring a variety of cloud-contaminated multispectral images.

The main contributions of this study are as follows:

- We propose a multi-input attention network to dehaze multispectral remote sensing images. This method does not require upsampling/downsampling on the training data. It can dehaze the images captured by Sentinel-2 with different resolutions in nine bands, effectively avoiding information loss due to upsampling/downsampling. To obtain the best recovery effect, the visible light band and the features of the infrared band are fused. It takes the advantage of the strong penetration capability of the infrared band.
- We build an end-to-end dehazing network with an encoder–decoder structure, which directly obtains haze-free images from learning hazy images. To improve training efficiency, the structure of weighted multiplication and residual connection between different input lines are used to adjust the feature extraction.
- We use skip connections and a multi-layer output structure in the network, which can produce multi-spectral dehazing images with different resolutions. Connecting the shallow part with the tail of the network preserves the ground details and allows the network to fully extract deep features. Meanwhile, adding an improved attention module to the connection part further improves feature extraction. Finally, the network can effectively remove the disturbances, including clouds and cloud shadows under non-uniform distribution.

## 2. Related Work

At present, the research on single-image dehazing falls into two categories: traditional methods and deep learning methods.

### 2.1. Traditional Dehazing Methods

Traditional methods can be further divided into methods based on image enhancement and methods based on atmospheric scattering models. Methods based on image enhancement restore hazy images by enhancing contrast [5,6] and suppressing low-frequency information [8,9]. Chaudhry et al. [28] combined mixed median filtering with Laplacian to dehaze images and apply it to remote sensing images. Huang et al. [29] combined the phase-consistency feature of remote sensing images with multi-scale retinex theory and used it to dehaze urban remote sensing images.

The method based on image enhancement is considered to be unstable in many cases because it lacks a foundation in physics. Therefore, the method based on the atmospheric scattering model eventually became the mainstream of traditional dehazing methods. ASM-based methods mainly use prior knowledge to estimate the parameters in Formula (1) and then obtain a haze-free image.

He et al. [12] proposed a dark channel prior (DCP) method through statistics and research on a large number of haze-free images. Studies have shown that in the non-sky area of the image without haze, there are always pixels with very low values close to 0 in the RGB bands. Therefore, the DCP-based dehazing method achieves outstanding dehazing effects and wide applicability. Many research efforts have been dedicated to the improvement of the DCP-based dehazing method. Zhu et al. [13] proposed a dehazing method based on the color decay prior. This method obtains the depth of field by modeling the relationship between scene depth and color, obtains the parameters through the supervised learning method to obtain the transmittance, and, finally, restores the hazy image effectively according to Formula (1).

Berman et. al. [30] proposed a global-based transmittance estimation method, which is different from the previous local-based transmittance estimation. The method estimates the transmittance and restores the image based on the prior knowledge that the color distribution of pixels in a hazy image will generate haze lines. The global-based estimation is more efficient and robust. Long et al. [31] refined the atmospheric veil through a low-pass filter and redefined the transmittance to reduce color distortion. The experimental results demonstrate well the preservation of ground details and effective dehazing of remote sensing images. Shen et al. [32] proposed a spatial–spectral adaptive dehazing method to effectively remove the haze effect from visible light remote sensing images. This method establishes the relationship between the image gradient and the transmittance between different wavelength bands.

### 2.2. Neural Networks

In recent years, increasing efforts have been dedicated to data-driven methods using deep learning. The end-to-end learning of deep neural networks can potentially solve many problems in traditional algorithms. Researchers first estimate the transmittance in the atmospheric scattering model of Equation (1) by building a neural network and then restore the hazy image according to the model.

Cai et al. [14] used a network based on multi-scale feature extraction to restore images according to the degradation model. It takes the hazy image as input and outputs the transmittance map. Ren et al. [15] used a coarse-scale network that took the hazy image as an input to estimate the rough transmittance map, then fed into the fine-scale network to obtain an optimized transmittance map, and finally obtained a more refined dehazing image. Li et al. [16] took the transmittance and atmospheric light in Equation (1) as one variable and used a neural network to estimate it. Different from the previous estimation of atmospheric light by experience, this method uses the learning ability of the network to perform the estimation. Neural networks demonstrate powerful feature extraction capability, which greatly advocates the research on end-to-end direct dehazing networks.

Chen et al. [18] proposed an end-to-end gated context aggregation network to improve the finesse of dehazing results. It combines smooth hole convolution and multi-level feature fusion techniques. Liu et al. [19] proposed an attention-based grid dehazing network (GridDehazeNet), adding a densely connected grid network to effectively alleviate the bottleneck problem of traditional multi-scale networks. The attention module enables the network to better estimate model parameters. In [20], a Domain Adaptation framework was proposed. It employs a bidirectional transformer network to bridge the gap between synthetic and real domains by transforming images from one domain to another. This method effectively reduces the gap between the synthetic hazy image and the real hazy image. In [23], a spatial attention-based adversarial generative network was proposed to dehaze remote sensing images. The model is separately trained based on haze and small-scale cloud, and, finally, effectively removes both interferences. In [15], SkyGAN was proposed for haze removal in aerial images. The network reconstructs multispectral data from RGB bands in aerial images and then uses a conditional generative network to train these reconstructed data. This method can effectively expand multispectral datasets.

Overall, there are many issues in applying the natural image dehazing methods to remote sensing images. Most research on the dehazing of remote sensing images focuses on the visible light band. The recovery from hazy images is also limited. In this paper, we propose a multi-input multi-spectral remote sensing image dehazing network. It can effectively remove haze in multi-spectral remote sensing images.

## 3. Materials and Methods

### 3.1. Dataset

For natural image dehazing, there are some datasets for training the dehazing network models, for example, FRIDA [33], Hazy Cityscapes datasets [34], D-Hazy [35] and RESIDE [36]. The RESIDE dataset includes a large amount of indoor and outdoor clear images and synthetic hazy images. It has been widely accepted as a benchmark dataset for natural image dehazing research in recent years. Due to the huge difference between remote sensing images and natural images, these datasets cannot be directly applied to dehazing remote sensing images. At present, datasets for remote sensing image dehazing mainly include the haze detection and removal dataset [37], Haze1K [27] and RICE [38]. These datasets include some non-uniform haze but are limited to the visible light band and have no multispectral data. In this research, we build our dataset by collecting Sentinel-2 images from January 2021 to January 2022, from 112° E, 36° N in central China to 120° E and 29° N in eastern China. The band information of Sentinel-2 is shown in Table 1.

**Table 1.** Sentinel-2 bands.

| Sentinel-2 Bands | Central Wavelength (μm) | Resolution (m) |
|---|---|---|
| Band 1—Coastal Aerosol | 0.443 | 60 |
| Band 2—Blue | 0.490 | 10 |
| Band 3—Green | 0.560 | 10 |
| Band 4—Red | 0.665 | 10 |
| Band 5—Vegetable Red Edge | 0.705 | 20 |
| Band 6—Vegetable Red Edge | 0.740 | 20 |
| Band 7—Vegetable Red Edge | 0.783 | 20 |
| Band 8—Near Infrared | 0.842 | 10 |
| Band 8A—Vegetable Red Edge | 0.865 | 20 |
| Band 9—Water Vapor | 0.945 | 60 |
| Band 10—Shortwave Infrared—Cirrus | 1.375 | 60 |
| Band 11—Shortwave Infrared | 1.610 | 20 |
| Band 12—Shortwave Infrared | 2.190 | 20 |

We chose 9 bands of 10 m and 20 m, which are commonly used in earth observation, including Band 2, Band 3, Band 4, Band 5, Band 6, Band 7, Band 8, Band 11 and Band 12 for research. We also selected 30 sets of hazy and haze-free image pairs as experimental data. Figure 2 shows an example of those image pairs.

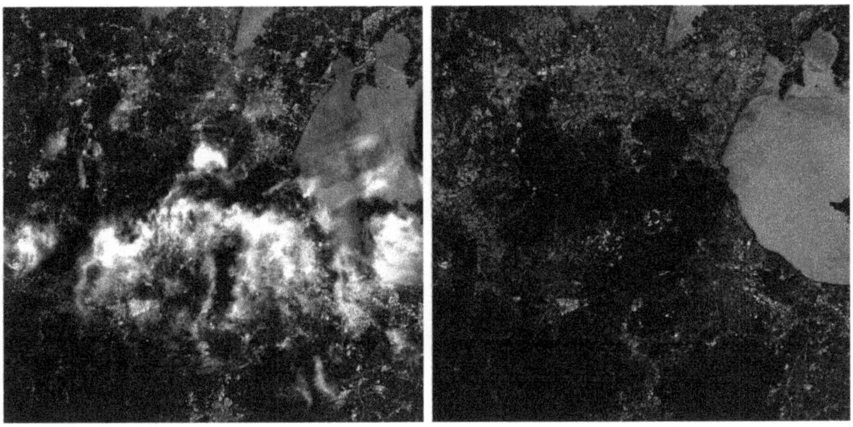

(a) Hazy Image (b) Haze-free Image

**Figure 2.** A Sentinel-2 foggy and fog-free image pair.

The image size of 10 m resolution images is 10,980 × 10,980, while the image size of 20 m resolution is 5490 × 5490. We first chose the hazy areas of the collected images and the corresponding areas of the haze-free images. We then resized the image size of the 10 m resolution training data to be 1024 × 1024 and resized the 20 m resolution training data to be 512 × 512. The selected area was randomly cropped. We finally obtained 1500 sets of 9-band hazy and haze-free data pairs as the training data set for this research.

### 3.2. Network Architecture

Figure 3 shows the network architecture in this paper. Inspired by U-Net [39] and Feature Pyramid [40], we designed a multi-input network based on an encoder–decoder structure. It takes multi-spectral remote sensing images of two resolutions as the input and outputs the corresponding haze-free image.

#### 3.2.1. Encoder

As shown in Figure 3, the encoder consists of three inputs, two double convolution layers, two channel attention modules and seven downsampling layers. The numbers in the figure are the number of channels after the feature map passes through the network layer.

On the input side, the 9 bands with different features are categorized into three parts as the input to the network. The visible light bands (Band 2, Band 3 and Band 4) with a resolution of 10 m and the near-infrared band (Band 8) have richer ground detail information. They are the main parts for feature extraction. All other five bands have a resolution of 20 m. The ground detail information in clear cases is relatively poor.

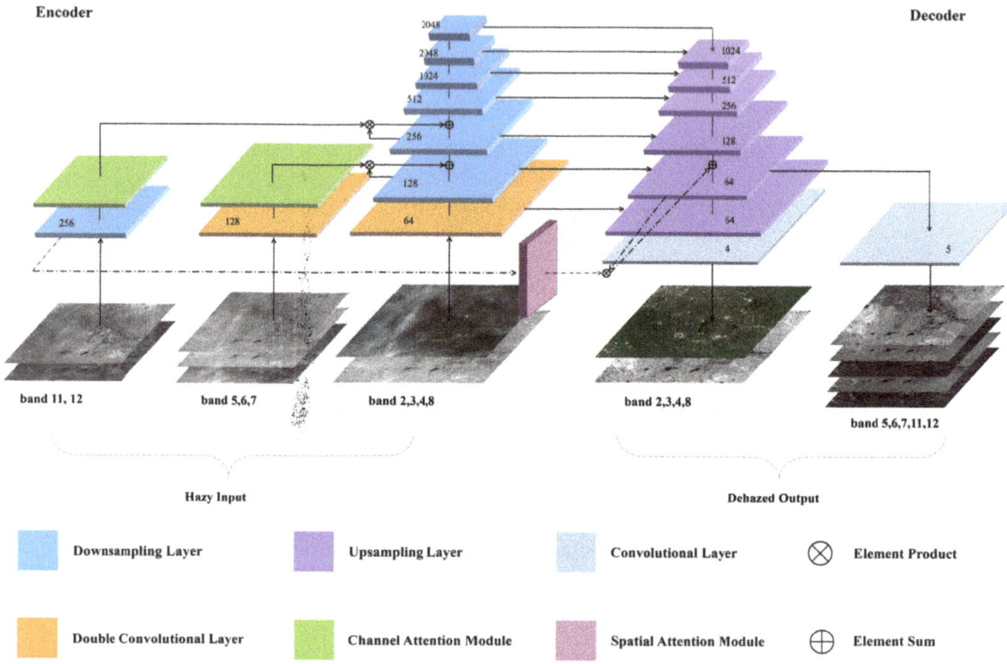


**Figure 3.** The architecture of the proposed network.

Band 5, Band 6 and Band 7 are Red Edge bands, which are mainly used to observe the mutation properties of vegetation reflectance in remote sensing applications. From Table 1, it can be observed that the wavelengths of Band 5, Band 6 and Band 7 have a small difference from the wavelengths of Band 8, as well as a small difference in the penetrating capability. Therefore, the features extracted from Band 5, Band 6 and Band 7 are fused with the shallow features of the main network to standardize and correct the features extracted by the main network.

Band 11 and Band 12 have relatively longer wavelengths and stronger penetration power (as shown in Figure 1). The features extracted from Band 11 and Band 12 are fused with the deeper downsampling features. Meanwhile, the features extracted from Band 11 and Band 12 are fused with the features of the upsampling stage in the decoder. The feature extracted from the infrared band (less interfered with haze), together with the learning of deep high-order features and the learning of shallow spatial detail features, make the final restored results closer to the real situation.

For feature extraction, the double convolution network group is used for initial feature extraction for the input image. As illustrated in Figure 4a, it includes two sets of 3 × 3 convolutions. The double convolution layer can be formulated as Formula (2)

$$F_c = \delta(BN(Conv(\delta(BN(Conv(F_i)))))) \tag{2}$$

where $F_i$ is the input image data or feature map. $F_c$ is the feature map after feature extraction from double convolution. $Conv$ and $BN$ are 3 × 3 convolution and Batch Normalization. $\delta$ is the Leaky ReLU function. In the downsampling module, as shown in Figure 4b, the input features are subjected to maximum pooling of stride = 2. Then, downsampling and feature extraction are implemented through the double convolutional network.

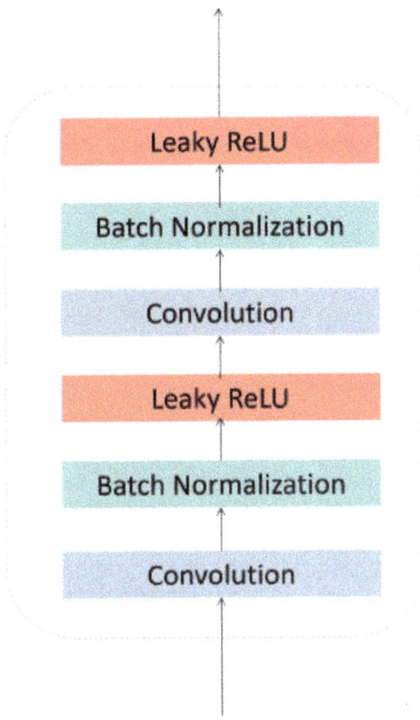


(**a**) Double Convolution Layer.

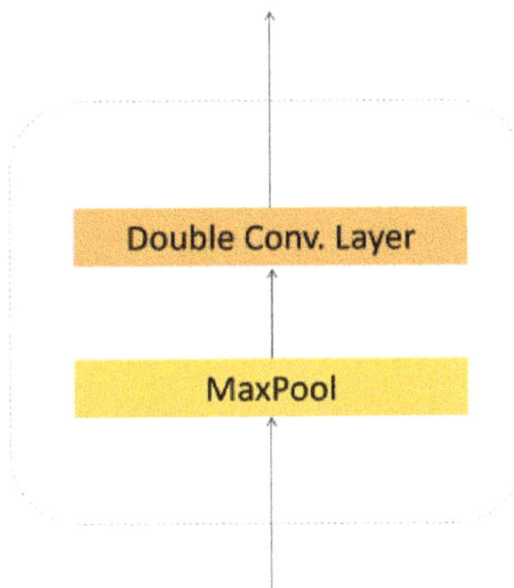


(**b**) Downsampling Layer.

**Figure 4.** The architecture of the convolution layer in encoder.

3.2.2. Decoder

The decoder consists of six upsampling layers, one spatial attention layer, two convolutional layers and an output. The decoder upsamples high-level features from the encoder and finally restores the image to a dehazed image. It uses different upsampling layers to output a multi-band image with the same two resolutions as the input. Figure 5 explains the structure of the upsampling layer. It starts with the upsampling of the upper layer features by Deconvolution with scale = 2, which, next, will be concatenated with the same-size downsampling feature map. After that, the double convolutional layer will restore the channel layer by layer. The "Concat" operation on the upsampling feature and downsampling feature enriches the information in the network, which contains both detailed information in shallow feature maps and the haze feature information in deep feature maps.

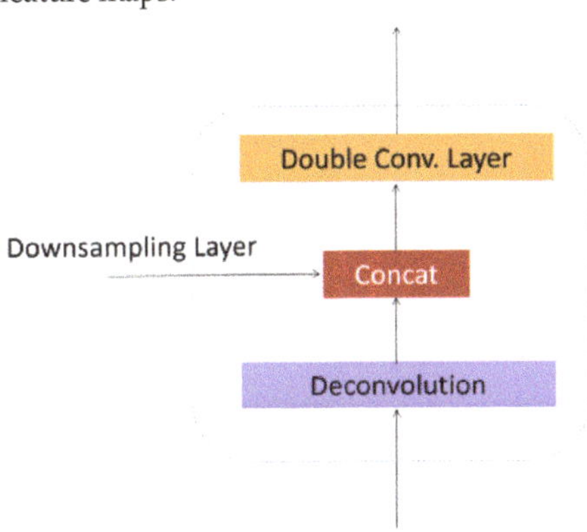


**Figure 5.** The architecture of the upsampling layer.

3.2.3. Attention Module

Inspired by the attention mechanisms widely applied in the field of image vision [41,42], we introduce the attention module in the feature extraction of the input data. The attention module can reinforce the focusing capability of the model. It enables the system to emphasize important information and suppress relatively irrelevant information so that the system can effectively extract non-uniform haze features. For the feature maps of Band 11 and Band 12 after Double Convolutional Layer and Band 5, Band 6 and Band 7 after the Downsampling Layer, the attention module infers the attention of the image along two independent dimensions of channel and space in turn. Then, the attention map is multiplied with the feature map after Downsampling and Upsampling in the backbone network, which is then fused with the feature map of the backbone network. The process is formulated in Formula (3).

$$F^* = \mu_m(A_i \bigotimes F) + F \tag{3}$$

where $A_i$ is the attention map generated after different attention modules. $F$ is the feature map of the backbone network with size $C \times H \times W$. $F^*$ is the output feature map after the fusion with the attention modules, and $\mu_m$ is the correction coefficient. Different bands impact the final haze removal in different ways. Therefore, different bands have different correction coefficients for feature fusion. The feature extraction correction coefficient for Band 11 and Band 12 is $\mu_m = 0.5$, while the correction coefficient for Band 5, Band 6 and Band 7 is $\mu_m = 0.3$.

The channel attention module generates channel attention maps using the channel relationship between features. Each feature map is achieved by a feature detector, which focuses on meaningful features. The structure of the channel attention module is shown in Figure 6a. Global max pooling and global average pooling are used to compress the spatial dimension of feature maps. The global max pooling and global average pooling can be expressed by Formulas (4) and (5).

$$g_m = H_{mp}(F_C) = \max_{(i,j)\in X_C} X_C(i,j) \tag{4}$$

$$g_a = H_{ap}(F_C)\frac{1}{H \times W}\sum_{i=1}^{H}\sum_{j=1}^{W} X_C(i,j) \tag{5}$$

$H_{mp}$ and $H_{ap}$ perform global max pooling and global average pooling for input feature map $F_C$ of size $H \times W$. $X_C(i,j)$ is the value of channel C at $(i,j)$. The size of the feature maps after compression is $C \times 1 \times 1$. Since the feature map contains rich information, we chose Leaky ReLU as the activation function instead of Sigmoid, which is widely used in the attention structure to suppress gradient vanishing. The process of channel attention module is formulated in Formula (6).

$$A_C = \delta(Conv(\delta(Conv(g_m))) + Conv(\delta(Conv(g_a)))) \tag{6}$$

where $A_C$ is the channel weight of the output. $\delta$ is Leaky ReLU function, and Conv is $1 \times 1$ convolution.

The spatial attention module generates the spatial attention feature map by using the internal spatial relationship between features. It focuses on different spatial information within a feature map. The structure of the spatial attention module is shown in Figure 6b. In the channel dimension, MaxPool and AveragePool are used to aggregate the channel information of the feature map. The processes of MaxPool and AveragePool are formulated in Formulas (7) and (8).

$$F_{max} = \max_{C} X(i,j) \tag{7}$$

$$F_{avg} = \underset{C}{avg} X(i,j) \tag{8}$$

After obtaining the maximum and average values of each pixel of the input feature map $X(i,j)$ on the C channels, two cross-channel feature maps of size $1 \times H \times W$ ($F_{max}$ and $F_{avg}$) are generated. Finally, they are concatenated through a convolutional layer into a 2D spatial attention feature map output. The process of spatial attention module is formulated in Formula (9).

$$A_S = \delta(Conv([MaxPool(F_S); AvgPool(F_S)])) \tag{9}$$

where $A_S$ is the spatial weight of the output. $\delta$ is Leaky ReLU function, and Conv is $7 \times 7$ convolution.

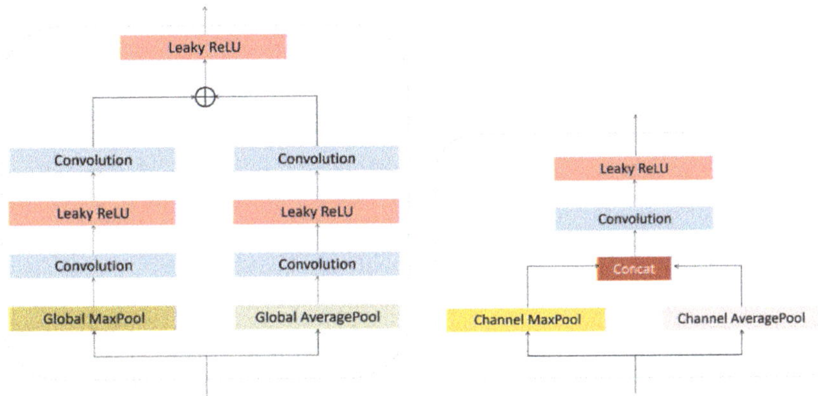


(**a**) Channel attention module. (**b**) Spatial attention module.

**Figure 6.** The architecture of the attention module.

### 3.3. Loss Function

In this research, there are many parameters of the haze restoration network model. We chose the loss $L_2$ to effectively train the network. The loss $L_2$ is the mean square loss, which has a relatively stable solution and can converge more effectively. Formula (10) explains the calculation.

$$L_2 = \frac{1}{N} \sum_{x=1}^{N} \sum_{i} (\hat{J}_i(x) - J_i(x))^2, (i = 2,3,4,5,6,7,8,11,12) \tag{10}$$

$\hat{J}_i(x)$ and $J_i(x)$ represent the dehazing image and the real haze-free image, respectively, $i$ represents the band and N represents the number of image pixels.

## 4. Experiment and Results

### 4.1. Model Training

Our experiments were performed on PyTorch. The training of the model was conducted on NVIDIA A100 GPU. The batch size was 4 and ADAMW was the optimizer for model training, where the range of betas was set to be (0.5, 0.999). The initial learning rate of the model was 0.0001. At the same time, CosineAnnealingLR with $T_{max} = 60$ and $eta_{min} = 1 \times 10^{-7}$ were used to adjust the learning rate. During training, 80% of the dataset was used as the training set and 20% as the test set. In addition, we took the two sets of haze and haze-free Sentinel II images outside the data set and used them to produce 150 data pairs as a validation set using the approach in Section 3.1.

Training was performed for a total of 200 Epochs, and the model preservation rule was to preserve only the best model of the validation set results. The loss was recorded during training, as shown in Figure 7. It can be seen that after 200 epochs of training, the loss curve tends to a lower and flatter value, and the training curve of the training set is clearly split from the validation set and the test set, indicating that the model has been adequately trained.

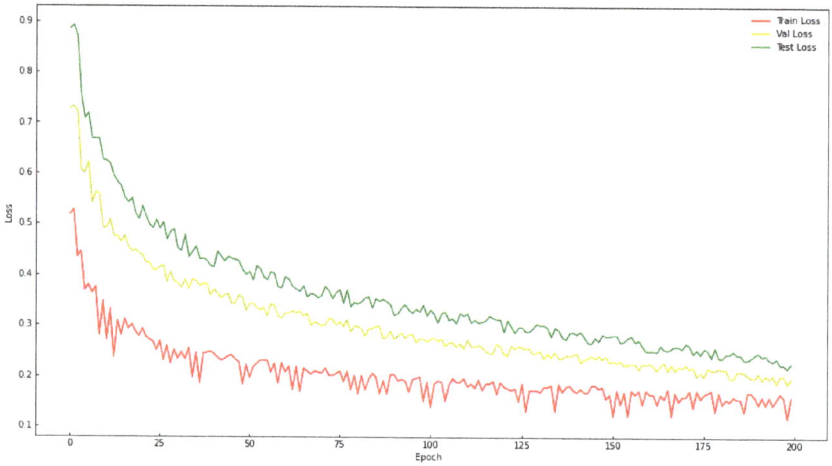


**Figure 7.** Training loss curve.

### 4.2. Metrics

We used peak signal-to-noise ratio (PSNR), structural similarity (SSIM) and feature similarity (FSIM) as the metrics to evaluate the performance of our models. PSNR is commonly used in image fusion tasks. It measures the ratio between the effective information of the image and the noise, which can reflect whether the image is distorted. The larger the value, the better the quality of the final dehazed image. SSIM describes structural similarity. The closer it is to 1, the higher the similarity with the haze-free image, which indicates a better dehazing effect. FSIM is a variant of SSIM that uses phase coherence to focus more on the contribution of different local features to the overall structure of the image. Still, the closer the value is to 1, the higher the similarity is.

### 4.3. Experimental Results

In this section, we apply the dehazing model to the hazy multispectral images captured by Sentinel-2. We also compare our model with the traditional method DCP [12] and four neural network methods, DehazeNet [14], AOD-Net [16], GridDehazeNet [19] and MSBDN [43]. For each hazy image used in the experiment, the corresponding clear sky image within a week was collected and used as the haze-free image for reference.

Figure 8 demonstrates the dehazing results of nine bands. It can be observed that the visible light as well as Band 5, Band 6 and Band 7 have highly interfered with haze. The restoration results of DCP, DehazeNet and AOD-Net have haze residues. GridDehazeNet, MSBDN and our proposed method have no haze residue after dehazing. MSBDN has a certain color distortion. Infrared bands (Band 8, Band 11 and Band 12) are less interfered with haze. All of the methods achieve great recovery effects visually. For the visible light bands, the most interfered bands, the dehazing effect of DCP, DehazeNet and AOD-Net is relatively poor. MSBDN has a certain color distortion, while GridDehazeNet and our proposed method have better fidelity.

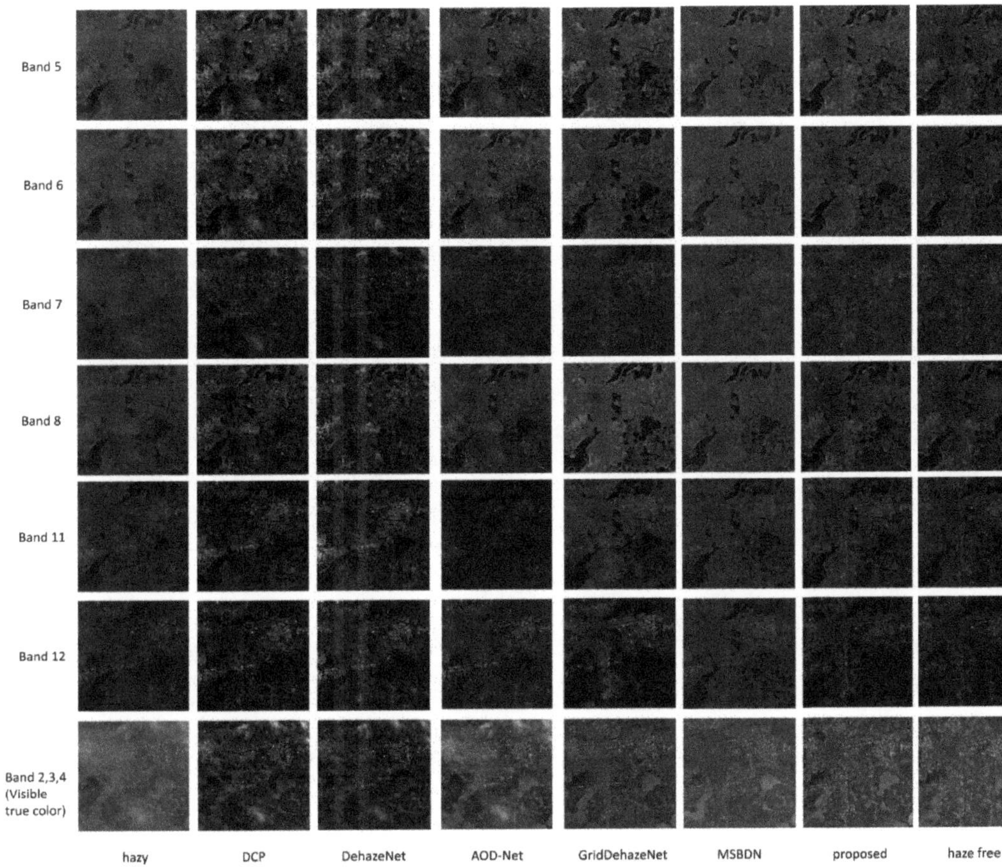


**Figure 8.** The dehazing effect on the hazy multi-spectral images captured by Sentinel-2.

Tables 2–4 demonstrate the performance evaluation using PSNR, SSIM and FSIM. It can be observed that the results are basically consistent with the visual effects. Meanwhile, our proposed method significantly outperforms GridDehazeNet and MSBDN, which have better dehazing effects. It indicates that the method in this paper can better maintain ground details and color fidelity.

**Table 2.** Experimental results on different bands: PSNR.

| Image | DCP | DehazeNet | AOD-Net | GridDehazeNet | MSBDN | Proposed |
|---|---|---|---|---|---|---|
| Band 5 | 17.316 | 17.283 | 19.233 | 25.311 | 20.103 | 29.651 |
| Band 6 | 18.673 | 18.074 | 20.206 | 25.682 | 22.572 | 28.304 |
| Band 7 | 19.325 | 18.792 | 21.461 | 26.104 | 24.305 | 27.342 |
| Band 8 | 21.353 | 20.386 | 22.718 | 25.933 | 26.438 | 31.541 |
| Band 11 | 22.176 | 20.850 | 21.648 | 26.452 | 24.349 | 29.072 |
| Band 12 | 21.981 | 21.098 | 20.052 | 26.939 | 25.406 | 27.508 |
| Visible true color | 15.683 | 14.532 | 19.797 | 24.261 | 23.821 | 30.106 |

**Table 3.** Experimental results on different bands: SSIM.

| Image | DCP | DehazeNet | AOD-Net | GridDehazeNet | MSBDN | Proposed |
|---|---|---|---|---|---|---|
| Band 5 | 0.436 | 0.474 | 0.637 | 0.737 | 0.677 | 0.853 |
| Band 6 | 0.474 | 0.468 | 0.623 | 0.722 | 0.665 | 0.878 |
| Band 7 | 0.351 | 0.332 | 0.593 | 0.693 | 0.621 | 0.817 |
| Band 8 | 0.452 | 0.422 | 0.685 | 0.749 | 0.734 | 0.906 |
| Band 11 | 0.576 | 0.532 | 0.723 | 0.805 | 0.756 | 0.874 |
| Band 12 | 0.509 | 0.502 | 0.629 | 0.723 | 0.692 | 0.825 |
| Visible true color | 0.379 | 0.461 | 0.582 | 0.718 | 0.665 | 0.867 |

**Table 4.** Experimental results on different bands: FSIM.

| Image | DCP | DehazeNet | AOD-Net | GridDehazeNet | MSBDN | Proposed |
|---|---|---|---|---|---|---|
| Band 5 | 0.744 | 0.732 | 0.792 | 0.856 | 0.827 | 0.922 |
| Band 6 | 0.766 | 0.782 | 0.812 | 0.866 | 0.834 | 0.903 |
| Band 7 | 0.713 | 0.722 | 0.775 | 0.823 | 0.802 | 0.887 |
| Band 8 | 0.813 | 0.821 | 0.843 | 0.897 | 0.857 | 0.944 |
| Band 11 | 0.778 | 0.791 | 0.813 | 0.885 | 0.866 | 0.939 |
| Band 12 | 0.771 | 0.782 | 0.829 | 0.863 | 0.855 | 0.911 |
| Visible true color | 0.755 | 0.762 | 0.835 | 0.865 | 0.847 | 0.896 |

In this study, we conducted dehazing experiments on different cases of hazy images. Figure 9 shows the case of slightly hazy images. The results are the dehazing of visible true colors (Band 2, Band 3 and Band 4). It can be observed that haze is unevenly distributed, while haze shadows also exist in Figure 9A(a). The traditional method DCP works effectively in the area with a uniform haze distribution but not in the non-uniform part nor the haze shadow area. DehazeNet and AOD-NET also have the same problem with DCP. GridDehazeNet and MSBDN can effectively remove the haze effect, but MSBDN has a certain color distortion after restoration. Compared with these two methods, our proposed method shows outstanding performance in maintaining color and ground details.

Figure 10 shows the visible true color dehazing result for an image with moderate haze interference. It can be observed that the results are similar to Figure 8. GridDehazeNet and MSBDN have removed haze, but they have also lost ground details to a certain extent. The method we propose maintains the ground details in a better way.

Figure 11 shows the visible true color dehazing results of an image with heavy haze. DCP, DehazeNet and AOD-NET demonstrate poor restoration. MSBDN has a poor effect on maintaining the details of ground objects in Figure 11A(f), and there is still residual haze in the upper part in Figure 11B(f). GridDehazeNet shows great restoration, but there is still a certain loss in the ground details in Figure 11A(e). There are also some haze residues in Figure 11A(e), and there is some color distortion. Our proposed method has demonstrated outstanding performance in haze removal and ground detail preservation.

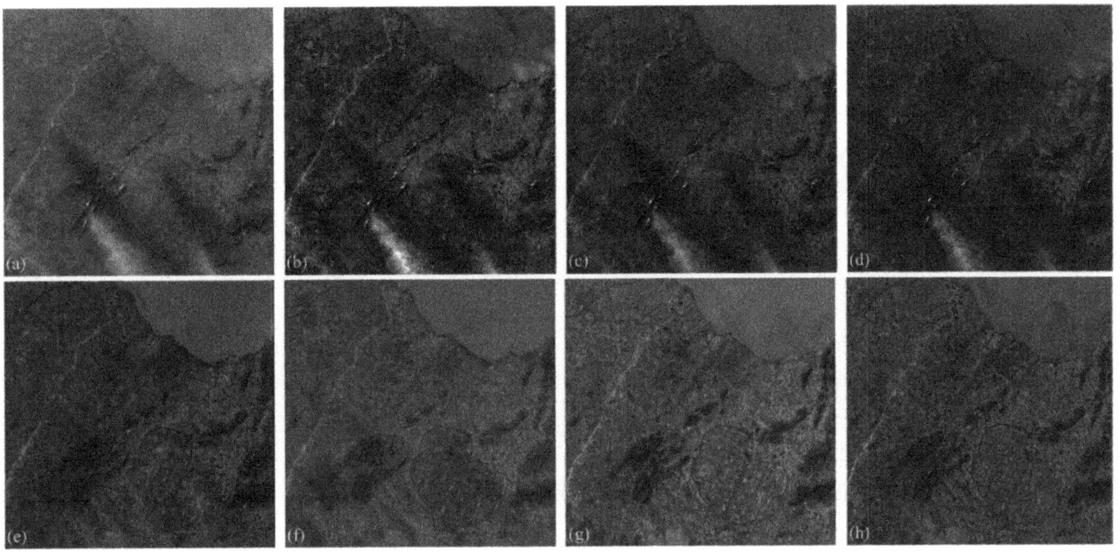


(**A**) Example Image 1.

(**B**) Example Image 2.

**Figure 9.** Dehazing results on images with slight haze interference. (**a**) Hazy image; (**b**) DCP; (**c**) DehazeNet; (**d**) AOD-NET; (**e**) GridDehazeNet; (**f**) MSBDN; (**g**) proposed method; (**h**) haze-free image.

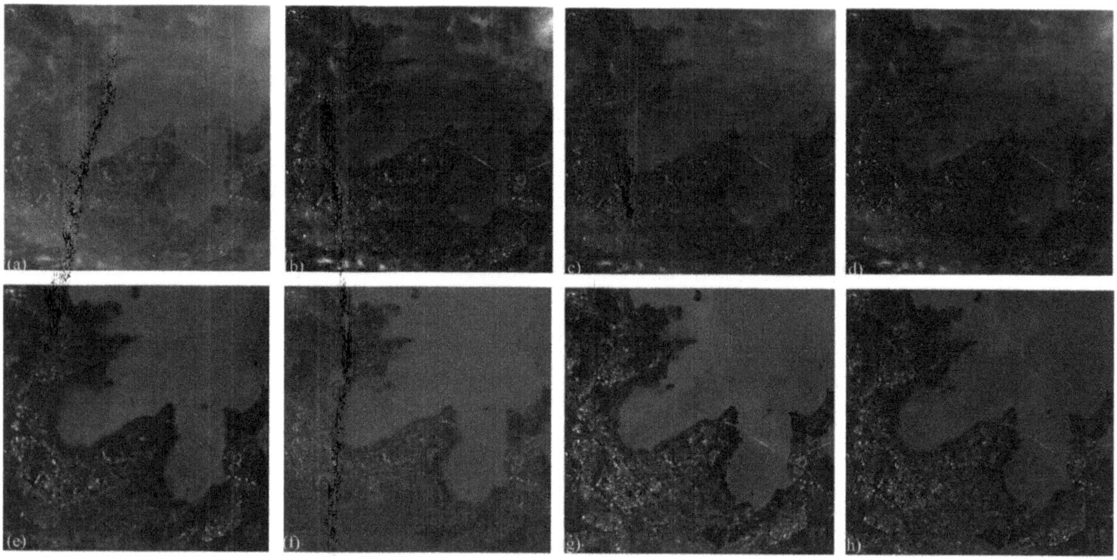


(**A**) Example Image 1.

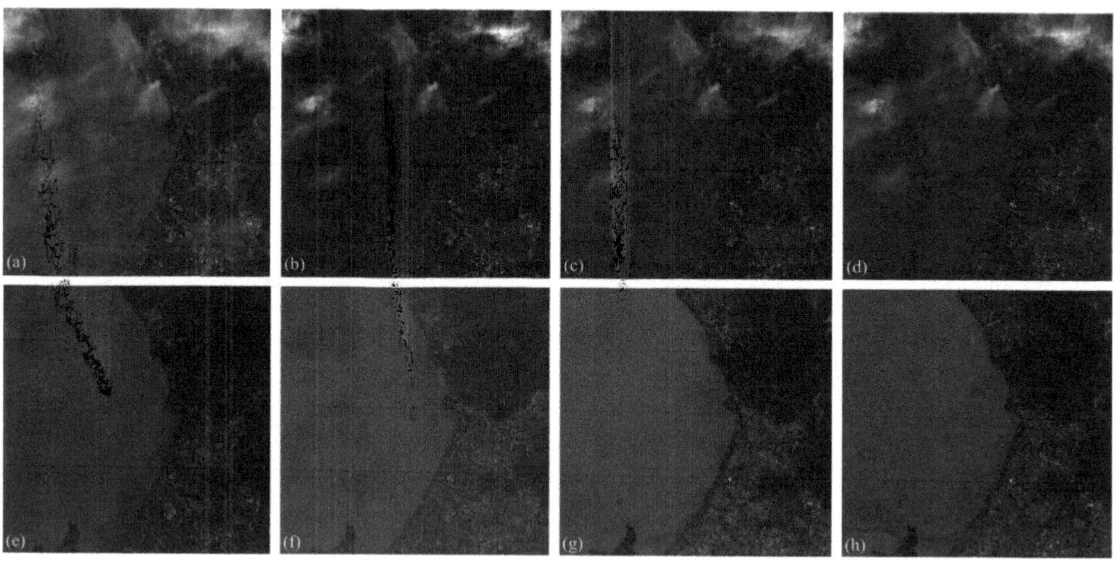


(**B**) Example Image 2.

**Figure 10.** Dehazing results on images with moderate haze interference. (**a**) Hazy image; (**b**) DCP; (**c**) DehazeNet; (**d**) AOD-NET; (**e**) GridDehazeNet; (**f**) MSBDN; (**g**) proposed method; (**h**) haze-free image.

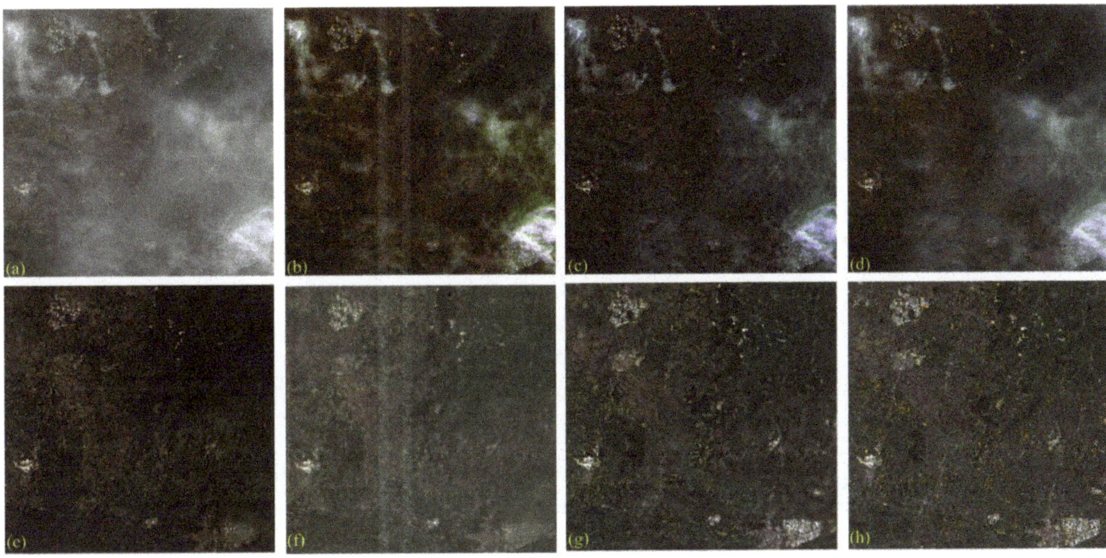


(**A**) Example Image 1.

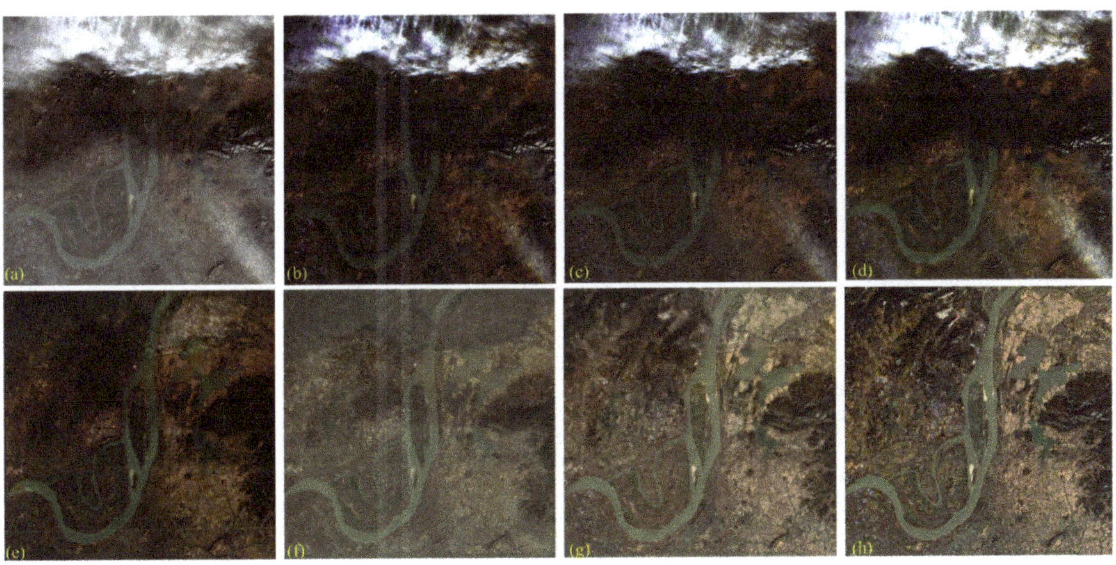


(**B**) Example Image 2.

**Figure 11.** Dehazing results on images with heavy haze interference. (**a**) Hazy image; (**b**) DCP; (**c**) DehazeNet; (**d**) AOD-NET; (**e**) GridDehazeNet; (**f**) MSBDN; (**g**) proposed method; (**h**) haze-free image.

Tables 5–7 list the PSNR, SSIM and FSIM of the above experiments. It can be seen that for the images with slight haze interference, the PSNR and FSIM values are close. DCP, DehazeNet and AOD-Net achieve relatively close results. MSBDN is slightly better and DehazeNet is even better. Our proposed method outperforms all these methods.

For moderate and heavy hazy images, the dehazing effect of the first three methods drops sharply. The results of GridDehazeNet and MSBDN have a slight drop. Our proposed method maintains a stable and good recovery effect, which, again, outperforms all other methods.

**Table 5.** Experimental results on different images: PSNR.

| Image | DCP | DehazeNet | AOD-Net | GridDehazeNet | MSBDN | Proposed |
|---|---|---|---|---|---|---|
| Slight hazy Image 1 | 21.600 | 22.492 | 22.255 | 25.516 | 23.811 | 26.541 |
| Slight hazy Image 2 | 21.002 | 22.199 | 22.871 | 23.212 | 22.906 | 27.840 |
| Moderate hazy Image 1 | 18.973 | 20.612 | 21.291 | 24.228 | 23.387 | 27.838 |
| Moderate hazy Image 2 | 15.739 | 17.614 | 18.044 | 23.649 | 22.299 | 28.591 |
| Heavy hazy Image 1 | 16.843 | 17.051 | 18.411 | 22.353 | 22.645 | 25.596 |
| Heavy hazy Image 2 | 11.566 | 12.497 | 11.505 | 21.376 | 20.694 | 26.100 |

**Table 6.** Experimental results on different images: SSIM.

| Image | DCP | DehazeNet | AOD-Net | GridDehazeNet | MSBDN | Proposed |
|---|---|---|---|---|---|---|
| Slight hazy Image 1 | 0.567 | 0.533 | 0.673 | 0.755 | 0.699 | 0.824 |
| Slight hazy Image 2 | 0.598 | 0.547 | 0.578 | 0.683 | 0.642 | 0.818 |
| Moderate hazy Image 1 | 0.506 | 0.621 | 0.714 | 0.750 | 0.742 | 0.836 |
| Moderate hazy Image 2 | 0.469 | 0.553 | 0.643 | 0.728 | 0.734 | 0.821 |
| Heavy hazy Image 1 | 0.405 | 0.411 | 0.535 | 0.697 | 0.653 | 0.787 |
| Heavy hazy Image 2 | 0.292 | 0.319 | 0.478 | 0.632 | 0.604 | 0.778 |

**Table 7.** Experimental results on different images: FSIM.

| Image | DCP | DehazeNet | AOD-Net | GridDehazeNet | MSBDN | Proposed |
|---|---|---|---|---|---|---|
| Slight hazy Image 1 | 0.833 | 0.828 | 0.844 | 0.878 | 0.854 | 0.893 |
| Slight hazy Image 2 | 0.842 | 0.851 | 0.866 | 0.892 | 0.873 | 0.932 |
| Moderate hazy Image 1 | 0.756 | 0.764 | 0.778 | 0.859 | 0.838 | 0.910 |
| Moderate hazy Image 2 | 0.744 | 0.761 | 0.791 | 0.843 | 0.822 | 0.898 |
| Heavy hazy Image 1 | 0.702 | 0.711 | 0.767 | 0.813 | 0.824 | 0.877 |
| Heavy hazy Image 2 | 0.681 | 0.693 | 0.755 | 0.833 | 0.791 | 0.874 |

### *4.4. Ablation Experiment*

In order to validate the role of different structures in our network, we performed ablation experiments. We are concerned with the multi-input feature fusion structure and the attention module. First, we only kept the main input/output structures corresponding to Band 2, Band 3, Band 4 and Band 8 in the network. We then upsampled the remaining five bands of the training data with 20 m resolution and set the size of input/output images to be 1024 × 1024. At the same time, we removed the spatial attention and channel attention modules. We used this version as the baseline. After that, we added the structural modules one by one as Model-1 to Model-6. We used the hazy images (outside the training set) as the verification dataset and calculated the average PSNR, SSIM and FSIM of 100 verification images for quantitative evaluation purposes. Table 8 lists the results. It can be observed that with the multi-input structure, the dehazing effect is greatly improved. The results are also improved by adding the attention module.

**Table 8.** Experimental results of different structural models.

| Method | Components | | | | | |
|---|---|---|---|---|---|---|
| | Multi-Input | SA | CA | PSNR | SSIM | FSIM |
| Baseline | | | | 24.573 | 0.725 | 0.818 |
| Model-1 | | ✓ | | 24.861 | 0.733 | 0.806 |
| Model-2 | | | ✓ | 25.061 | 0.744 | 0.832 |
| Model-3 | | ✓ | ✓ | 25.224 | 0.761 | 0.840 |
| Model-4 | ✓ | | | 26.641 | 0.809 | 0.858 |
| Model-5 | ✓ | ✓ | | 27.021 | 0.831 | 0.889 |
| Model-6 | ✓ | | ✓ | 26.891 | 0.813 | 0.894 |
| Proposed | ✓ | ✓ | ✓ | 27.866 | 0.858 | 0.908 |

## 5. Conclusions

Traditional dehazing methods rely on prior features and are less versatile, which makes them not applicable, especially for remote sensing images with widespread non-uniform haze. In recent years, deep learning methods have been applied for automatic feature extraction. However, the structure of remote sensing images with haze is relatively complex. It is difficult for general neural networks to effectively extract features. Meanwhile, there are very few network models targeting muti-band remote sensing images. In this research, we propose a multi-input, multi-spectral remote sensing image dehazing network, which effectively utilizes the penetrating capability of the infrared band to haze. We used global skip connections and attention modules to achieve effective feature extraction and maintain ground details in the meantime. Finally, we designed experiments to test the performance of the proposed method on multispectral images captured by Sentinel-2, which have different degrees of haze effects. Our method can effectively restore the images. It outperforms the traditional dark channel method and several neural network methods, such as DehazeNet, AOD-Net, MSBDN and GridDehazeNet, in terms of haze residues and quantitative evaluation metrics.

Meanwhile, there are some limits in this research. First, the training dataset is not categorized based on the types of haze, which could impact the effectiveness of the proposed model. Second, even though the ground details are well maintained in the restored images, there is still some loss compared to haze-free images. As for our future work, we will formulate an indicator to describe the degree of the haze effect, which will be used to classify the images in the training dataset. At the same time, we will improve the model by referring to the method that can effectively improve the detail resolution in super-resolution research.

**Author Contributions:** Conceptualization, Z.H.; Data curation, Z.H.; Software, Z.H. and F.Z.; Writing—original draft preparation, Z.H.; Supervision, C.G.; Methodology, Y.H.; Resources, Y.H.; Validation, F.Z. and L.L.; Writing—review and editing, C.G. and L.L. All authors have read and agreed to the published version of the manuscript.

**Funding:** This research was funded by National Natural Science Foundation of China (Grant No. 31970378), Science and Technology Commission of Shanghai, Shanghai 2021 "Science and Technology Innovation Action Plan" social development science and technology research project (Grant No. 21DZ1202500). Shanghai Water Authority Science and Technology Project (Grant No. 2021-10). Jiangsu provincial water resources department, Jiangsu Province Water Conservancy Science and Technology Project (Grant No. 2020068).

**Institutional Review Board Statement:** Not applicable.

**Informed Consent Statement:** Not applicable.

**Data Availability Statement:** Not applicable.

**Conflicts of Interest:** The authors declare no conflict of interest.

## References


1. Qi, Q.; Zhang, C.; Yuan, Q.; Li, H.; Shen, H.; Cheng, Q. An Adaptive Haze Removal Method for Single Remotely Sensed Image Considering the Spatial and Spectral Varieties. *Geomat. Inf. Sci. Wuhan Univ.* **2019**, *44*, 1369–1376.
2. Pyongsop, R.I.; Zhangbao, M.A.; Qingwen, Q.I.; Gaohuan, L. Cloud and shadow removal from Landsat TM data. *J. Remote Sens.* **2010**, *14*, 534–545.
3. Melgani, F. Contextual reconstruction of cloud-contaminated multitemporal multispectral images. *IEEE Trans. Geofence Remote Sens.* **2006**, *44*, 442–455. [CrossRef]
4. Li, X.; Shen, H.; Zhang, L.; Zhang, H.; Yuan, Q.; Yang, G. Recovering Quantitative Remote Sensing Products Contaminated by Thick Clouds and Shadows Using Multitemporal Dictionary Learning. *IEEE Trans. Geoence Remote Sens.* **2014**, *52*, 7086–7098.
5. Xu, Z.; Liu, X.; Ji, N. Fog Removal from Color Images using Contrast Limited Adaptive Histogram Equalization. In Proceedings of the 2009 2nd International Congress on Image and Signal Processing, Tianjin, China, 17–19 October 2009; pp. 1–5. [CrossRef]
6. Narasimhan, S.G.; Nayar, S.K. Contrast restoration of weather degraded images. *IEEE Trans. Pattern Anal. Mach. Intell.* **2003**, *25*, 713–724. [CrossRef]
7. McDonald, J.E. The Saturation Adjustment in Numerical Modelling of Fog. *J. Atmos. Ences* **2010**, *20*, 476–478. [CrossRef]
8. Yu, L.; Liu, X.; Liu, G. A new dehazing algorithm based on overlapped sub-block homomorphic filtering. In Proceedings of the Eighth International Conference on Machine Vision, Barcelona, Spain, 19–20 November 2015.
9. Jobson, D.J.; Rahman, Z.; Woodell, G.A. A multiscale retinex for bridging the gap between color images and the human observation of scenes. *IEEE Trans. Image Process.* **2002**, *6*, 965–976. [CrossRef]
10. Nayar, S.K.; Narasimhan, S.G. Vision in bad weather. In Proceedings of the Seventh IEEE International Conference on Computer Vision, Kerkyra, Greece, 20–27 September 1999.
11. Tan, R.T. Visibility in bad weather from a single image. In Proceedings of the IEEE Conference on Computer Vision and Pattern Recognition, Anchorage, AK, USA, 24–26 June 2008; pp. 1–8.
12. He, K.; Sun, J.; Tang, X. Single image haze removal using dark channel prior. *IEEE Trans. Pattern Anal. Mach. Intell.* **2010**, *33*, 2341–2353.
13. Zhu, Q.; Mai, J.; Shao, L. A fast single image haze removal algorithm using color attenuation prior. *IEEE Trans. Image Process.* **2015**, *24*, 3522–3533.
14. Cai, B.; Xu, X.; Jia, K.; Qing, C.; Tao, D. Dehazenet: An end-to-end system for single image haze removal. *IEEE Trans. Image Process.* **2016**, *25*, 5187–5198. [CrossRef]
15. Ren, W.; Liu, S.; Zhang, H.; Pan, J.; Cao, X.; Yang, M. Single image dehazing via multi-scale convolutional neural networks. In Proceedings of the European Conference on Computer Vision, Amsterdam, The Netherlands, 11–14 October 2016; Springer: Cham, Switzerland, 2016; pp. 154–169.
16. Li, B.; Peng, X.; Wang, Z.; Xu, J.; Feng, D. Aod-net: All-in-one dehazing network. In Proceedings of the IEEE International Conference on Computer Vision, Venice, Italy, 22–29 October 2017; pp. 4770–4778.
17. Li, R.; Pan, J.; Li, Z.; Tang, J. Single image dehazing via conditional generative adversarial network. In Proceedings of the IEEE Conference on Computer Vision and Pattern Recognition, Salt Lake City, UT, USA, 18–22 June 2018; pp. 8202–8211.
18. Chen, D.; He, M.; Fan, Q.; Liao, J.; Zhang, L.; Hou, D.; Yuan, L.; Hua, G. Gated context aggregation network for image dehazing and deraining. In Proceedings of the 2019 IEEE Winter Conference on Applications of Computer Vision (WACV), Waikoloa Village, HI, USA, 7–11 January 2019; pp. 1375–1383.
19. Liu, X.; Ma, Y.; Shi, Z.; Chen, J. Griddehazenet: Attention-based multi-scale network for image dehazing. In Proceedings of the IEEE/CVF International Conference on Computer Vision, Seoul, Korea, 27 October–2 November 2019; pp. 7314–7323.
20. Shao, Y.; Li, L.; Ren, W.; Gao, C.; Sang, N. Domain adaptation for image dehazing. In Proceedings of the IEEE/CVF Conference on Computer Vision and Pattern Recognition, Seattle, WA, USA, 13–19 June 2020; pp. 2805–2814.
21. Jiang, H.; Lu, N. Multi-Scale Residual Convolutional Neural Network for Haze Removal of Remote Sensing Images. *Remote Sens.* **2018**, *10*, 945. [CrossRef]
22. Qin, M.; Xie, F.; Li, W.; Shi, Z.; Zhang, H. Dehazing for multispectral remote sensing images based on a convolutional neural network with the residual architecture. *IEEE J. Sel. Top. Appl. Earth Obs. Remote Sens.* **2018**, *11*, 1645–1655. [CrossRef]
23. Pan, H. Cloud removal for remote sensing imagery via spatial attention generative adversarial network. *arXiv* **2020**, arXiv:2009.13015.
24. Mehta, A.; Sinha, H.; Mandal, M.; Narang, P. Domain-aware unsupervised hyperspectral reconstruction for aerial image dehazing. In Proceedings of the IEEE/CVF Winter Conference on Applications of Computer Vision, Waikoloa, HI, USA, 3–8 January 2021; pp. 413–422.
25. Enomoto, K.; Sakurada, K.; Wang, W.; Fukui, H.; Matsuoka, M.; Nakamura, R.; Kawaguchi, N. Filmy cloud removal on satellite imagery with multispectral conditional generative adversarial nets. In Proceedings of the IEEE Conference on Computer Vision and Pattern Recognition Workshops, Honolulu, HI, USA, 21–26 July 2017; pp. 48–56.
26. Grohnfeldt, C.; Schmitt, M.; Zhu, X. A conditional generative adversarial network to fuse SAR and multispectral optical data for cloud removal from Sentinel-2 images. In Proceedings of the 2018 IEEE International Geoscience and Remote Sensing Symposium, Valencia, Spain, 22–27 July 2018; pp. 1726–1729.

27. Huang, B.; Li, Z.; Yang, C.; Sun, F.; Song, Y. Single satellite optical imagery dehazing using sar image prior based on conditional generative adversarial networks. In Proceedings of the IEEE/CVF Winter Conference on Applications of Computer Vision, Snowmass Village, CO, USA, 1–5 March 2020; pp. 1806–1813.
28. Chaudhry, A.M.; Riaz, M.M.; Ghafoor, A. A framework for outdoor RGB image enhancement and dehazing. *IEEE Geosci. Remote Sens. Lett.* **2018**, *15*, 932–936. [CrossRef]
29. Huang, S.; Liu, Y.; Wang, Y.; Wang, Z.; Guo, J. A New Haze Removal Algorithm for Single Urban Remote Sensing Image. *IEEE Access* **2020**, *8*, 100870–100889. [CrossRef]
30. Berman, D.; Avidan, S. Non-local image dehazing. In Proceedings of the IEEE Conference on Computer Vision and Pattern Recognition, Las Vegas, NV, USA, 27–30 June 2016; pp. 1674–1682.
31. Long, J.; Shi, Z.; Tang, W.; Zhang, C. Single remote sensing image dehazing. *IEEE Geosci. Remote Sens. Lett.* **2013**, *11*, 59–63. [CrossRef]
32. Shen, H.; Zhang, C.; Li, H.; Yuan, Q.; Zhang, L. A Spatial–Spectral Adaptive Haze Removal Method for Visible Remote Sensing Images. *IEEE Trans. Geosci. Remote Sens.* **2020**, *58*, 6168–6180. [CrossRef]
33. Tarel, J.; Hautière, N.; Cord, A.; Gruyer, D.; Halmaoui, H. Improved visibility of road scene images under heterogeneous fog. In Proceedings of the IEEE Intelligent Vehicles Symposium, La Jolla, CA, USA, 21–24 June 2010; pp. 478–485.
34. Sakaridis, C.; Dai, D.; Van Gool, L. Semantic foggy scene understanding with synthetic data. *Int. J. Comput. Vis.* **2018**, *126*, 973–992. [CrossRef]
35. Ancuti, C.; Ancuti, C.O.; De Vleeschouwer, C. D-hazy: A dataset to evaluate quantitatively dehazing algorithms. In Proceedings of the IEEE International Conference on Image Processing (ICIP), Phoenix, AZ, USA, 25–28 September 2016; pp. 2226–2230.
36. Li, B.; Ren, W.; Fu, D.; Tao, D.; Feng, D.; Zeng, W.; Wang, Z. Benchmarking single-image dehazing and beyond. *IEEE Trans. Image Process.* **2018**, *28*, 492–505. [CrossRef]
37. Ji, S.; Dai, P.; Lu, M.; Zhang, Y. Simultaneous cloud detection and removal from bitemporal remote sensing images using cascade convolutional neural networks. *IEEE Trans. Geosci. Remote Sens.* **2020**, *59*, 732–748. [CrossRef]
38. Lin, D.; Xu, G.; Wang, X.; Wang, Y.; Sun, X.; Fu, K. A remote sensing image dataset for cloud removal. *arXiv* **2019**, arXiv:1901.00600.
39. Ronneberger, O.; Fischer, P.; Brox, T. U-net: Convolutional networks for biomedical image segmentation. In Proceedings of the International Conference on Medical Image Computing and Computer-Assisted Intervention, Munich, Germany, 5–9 October 2015; Springer: Cham, Switzerland, 2015; pp. 234–241.
40. Lin, T.; Dollár, P.; Girshick, R.B.; He, K.; Hariharan, B.; Belongie, S.J. Feature pyramid networks for object detection. In Proceedings of the IEEE Conference on Computer Vision and Pattern Recognition, Honolulu, HI, USA, 21–26 July 2017; pp. 2117–2125.
41. Hu, J.; Shen, L.; Sun, G. Squeeze-and-excitation networks. In Proceedings of the IEEE Conference on Computer Vision and Pattern Recognition, Salt Lake City, UT, USA, 18–22 June 2018; pp. 7132–7141.
42. Woo, S.; Park, J.; Lee, J.; Kweon, I.S. Cbam: Convolutional block attention module. In Proceedings of the European Conference on Computer Vision (ECCV), Munich, Germany, 8–14 September 2018; pp. 3–19.
43. Dong, H.; Pan, J.; Xiang, L.; Hu, Z.; Zhang, X.; Wang, F.; Yang, M.H. Multi-scale boosted dehazing network with dense feature fusion. In Proceedings of the IEEE/CVF Conference on Computer Vision and Pattern Recognition, Seattle, WA, USA, 13–19 June 2020; pp. 2157–2167.

*applied sciences*

MDPI

*Article*

# YOLO-DSD: A YOLO-Based Detector Optimized for Better Balance between Accuracy, Deployability and Inference Time in Optical Remote Sensing Object Detection

**Hengxu Chen, Hong Jin and Shengping Lv ***

College of Engineering, South China Agricultural University, Guangzhou 510642, China; hengxuchen@stu.scau.edu.cn (H.C.); hjin@scau.edu.cn (H.J.)
* Correspondence: lvshengping@scau.edu.cn; Tel.: +86-187-1937-3880

**Abstract:** Many deep learning (DL)-based detectors have been developed for optical remote sensing object detection in recent years. However, most of the recent detectors are developed toward the pursuit of a higher accuracy, but little toward a balance between accuracy, deployability and inference time, which hinders the practical application for these detectors, especially in embedded devices. In order to achieve a higher detection accuracy and reduce the computational consumption and inference time simultaneously, a novel convolutional network named YOLO-DSD was developed based on YOLOv4. Firstly, a new feature extraction module, a dense residual (DenseRes) block, was proposed in a backbone network by utilizing a series-connected residual structure with the same topology for improving feature extraction while reducing the computational consumption and inference time. Secondly, convolution layer–batch normalization layer–leaky ReLu (CBL) $\times 5$ modules in the neck, named S-CBL$\times 5$, were improved with a short-cut connection in order to mitigate feature loss. Finally, a low-cost novel attention mechanism called a dual channel attention (DCA) block was introduced to each S-CBL$\times 5$ for a better representation of features. The experimental results in the DIOR dataset indicate that YOLO-DSD outperforms YOLOv4 by increasing mAP$^{0.5}$ from 71.3% to 73.0%, with a 23.9% and 29.7% reduction in Params and Flops, respectively, but a 50.2% improvement in FPS. In the RSOD dataset, the mAP$^{0.5}$ of YOLO-DSD is increased from 90.0~94.0% to 92.6~95.5% under different input sizes. Compared with the SOTA detectors, YOLO-DSD achieves a better balance between the accuracy, deployability and inference time.

**Keywords:** optical remote sensing; object detection; feature extraction; attention mechanism

## 1. Introduction

Object detection in optical remote sensing images (ORSIs) is a crucial but challenging task for remote sensing technology and has been widely applied in many fields, such as military, natural resources exploration, urban construction, agriculture and mapping [1,2]. The development of a cost-effective detector considering the characteristic of ORSIs is the persistently pursued direction, and has attracted a large amount of attention from scholars and practitioners.

The approaches for object detection can be roughly divided into traditional detectors and deep learning (DL)-based detectors. DL-based detectors, especially convolutional neural network (CNN) detectors, have gradually replaced traditional detectors since they possess better adaptability and generalization in different application scenarios. There are two categories of DL-based detectors: one-stage [3–9] and two-stage [10–13]. The one-stage detectors directly regress bounding boxes and probabilities for each object simultaneously without region proposals; thus, they perform well regarding inference speed. Two-stage detectors employ the region proposals to improve the location and detection accuracy, with the sacrifice of the inference speed. With the emergence of large-scale natural scene images (NSIs) datasets for object detection tasks such as Pascal VOC [14] and MS COCO [15],

DL-based detectors have been further developed for a better tradeoff between accuracy and cost, including Faster-RCNN [12], single shot multibox detector (SSD) [3], the series of You Only Look Once (YOLO) [4–6,8], CenterNet [7], EfficientDet [9] and RetinaNet [16]. These detectors with continuous improvement have been widely applied in various natural scene visual detection tasks.

Since ORSIs are photographed from an overhead perspective at different heights, whereas NSIs are shot from a horizontal perspective at relatively close distance, three main differences have emerged as follows: first, the available feature of most detected objects in ORSIs is less obvious than that in NSIs and leads to greater inter-class similarity. Second, the intra-class difference is more prominent since object scales of the same category in ORSIs usually vary greater. Third, the background in ORSIs is more complex and abundant than that in NSIs. Differences between ORSIs and NSIs with instances are shown in Figure 1. These differences make object detection in ORSIs more difficult, and most of the well-designed detectors for NSIs are not elaborately optimized for ORSIs. For the problems of a greater intra-class difference and inter-class similarity caused by the characteristic of objects in ORSIs, the detector needs to extract more abundant object features with high-level semantics to overcome it. However, the feature of objects in ORSIs are easily submerged by the redundant and complex background information and thus will decrease or even disappear when transmitted in the detector. Thus, DL-based detectors also require a stronger feature extraction and transmission ability.

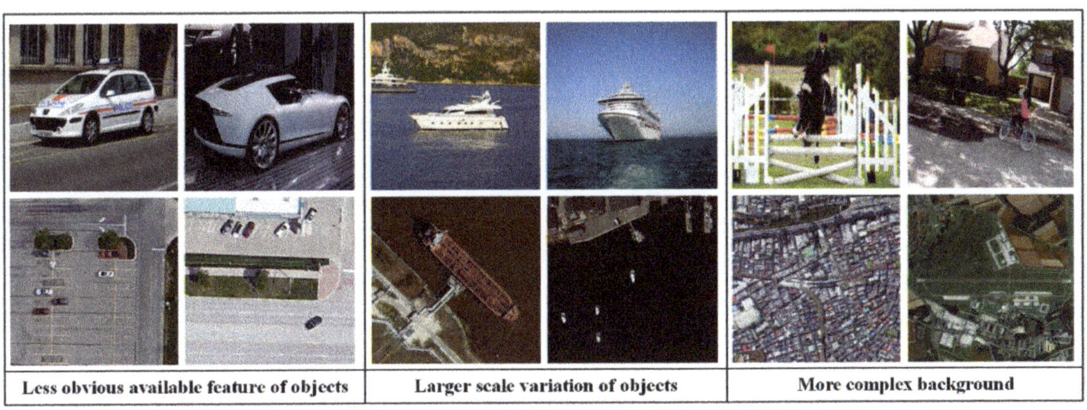


**Figure 1.** Three main differences between RSIs and NSIs. The first and second lines show instances from NSIs and RSIs, respectively.

With the popularity and wide application of embedded devices such as unmanned aerial vehicles (UAVs), the demand for real-time optical remote sensing object detection deployed on edge devices has increased rapidly. UAVs with far less computing resource and storage space than computers involve wide application scenarios such as rescue, military and surveying tasks, which require a high detection accuracy, flexible equipment deployment and less inference time for detectors [17].

In recent years, several outstanding achievements have been made by researchers in fields related to ORSIs, and can be roughly divided into heavyweight [18–21] and lightweight detectors [22–25]. Most of the heavyweight detectors usually have a high accuracy but require a large computational cost, and thus hinder their real-time response and the deployment on UAVs, whereas lightweight detectors have practical deployability and a fast inference speed but it is difficult for them to achieve as high a competitive accuracy as heavyweight detectors, especially for large multi-category object detection tasks [23,24,26]. Therefore, optimizing the structure of heavyweight detectors toward a better balance between accuracy, deployability and inference time is an issue well worth investigating. To establish a detector with a better balance between accuracy, deployability and inference time, a novel detector called YOLO-DSD for real-time optical remote sensing object detection based on YOLOv4 was developed in this study. The main contributions are as follows: (1) a new feature extraction module named a dense residual (DenseRes) Block was designed for better feature extraction and to reduce the computational cost and inference time in the backbone network. (2) Convolution layer–batch normalization layer–leaky ReLu (CBL) $\times$5 modules in the neck were improved with a short-cut connection and named S-CBL$\times$5 to strengthen the transmission of object features. (3) A novel low-cost attention mechanism called a dual channel attention (DCA) Block was proposed to enhance the representation of the object feature. The experimental results in the DIOR dataset indicate that YOLO-DSD outperforms YOLOv4 by increasing $mAP^{0.5}$ from 71.3% to 73.0%, with a 23.9% and 29.7% reduction in Params and Flops, respectively, but a 50.2% improvement in FPS. In the RSOD dataset, the $mAP^{0.5}$ of YOLO-DSD is increased from 90.0~94.0% to 92.6~95.5% under different input sizes. Compared with the SOTA detectors, YOLO-DSD achieves a better balance between accuracy, deployability and inference time.

## 2. Related Works

### 2.1. DL-Based Detectors for Optical Remote Sensing Object Detection

DL-based detectors have been widely applied in natural sense visual tasks. However, detectors established on NSIs need to further improve their feature extraction ability for optical remote sensing object detection tasks due to the problems of a greater intra-class difference, inter-class similarity and feature loss in ORSIs. Therefore, some heavyweight detectors have been improved and applied in ORSIs by many scholars. Xu et al. [18] modified YOLOv3 with a multi-receptive field to take full advantage of the feature information and to detect optical remote sensing objects effectively. Cheng et al. [19] designed an end-to-end cross-scale feature fusion framework for ORSIs object detection based on Faster R-CNN with a feature pyramid network (FPN) [16]. Yin et al. [20] proposed a multi-scale feature extraction network based on RetinaNet, which strengthens the detection performance of irregular objects in ORSIs. Yuan et al. [21] established a multi-FPN that performs well in object detection with a complex background. The above research has successfully made obvious improvements in detection accuracy, but come with the non-ignorable sacrifice of the deployability or inference speed, and thus further hinder the application of detectors in edge devices. As a consequence, some lightweight DL-based detectors have been elaborately designed and improved to facilitate the application in edge devices. Li et al. [22] designed a lightweight detector by taking advantage of YOLOv3 and DenseNet [27]. Lang et al. [23] employed the backbone network of ThunerNet [28] and constructed a six-layer feature fusion pyramid to enhance the detection performance. The improved YOLOv4-tiny proposed by Lei et al. [24] was constructed with an efficient channel attention mechanism to enhance the information sensitivity in each channel. Li et al. [25] established a lightweight detector for vehicle and ship detection through using a semantic transfer block and the distillation loss. Although these lightweight detectors have a better accuracy after improvement, there is still an obvious gap in the detection accuracy compared with heavyweight detectors.

Our motivation is to propose an end-to-end detector that can achieve a higher detection accuracy, better deployability and less inference time in order to meet the requirements of edge device real-time detection. YOLOv4 [8] is one of the widely used one-stage detectors, with an impressive performance in accuracy, deployability and inference time. It has been improved and applied in various fields, such as agriculture, industry and transportation [29–32], which verify its excellent generalization. In this study, YOLOv4 was utilized as the basic framework, while it was optimized from feature extraction modules, structures of the neck and the attention mechanism for a better application in optical remote sensing object detection.

### 2.2. Feature Extraction Modules in Backbone

The backbone that is utilized to extract high-level semantic features of images is the first part of the DL-based detector. It comprises several feature extraction modules. VGG [33] is one of the earliest backbones for object detection and utilizes 3 × 3 convolution layers as the feature extraction module. However, its heavy computation burden and shallow depth hinder the deployability and performance of detectors.

To solve this problem, He et al. [34] introduced a new feature extraction module named a Res Block to deepen the depth of backbones by adding short-cut connections. ResNet based on the Res Block achieves a better accuracy than VGG in the natural scene dataset, with a lower computation burden and deeper depth. The backbone DarkNet53 of YOLOv3 [6] also uses the Res Block as the main feature extraction module. Since then, many feature extraction modules based on the Res Block, such as a ResNeXt Block [35], Res2 Block [36], Dense Block [27] and CSP Block [37], have been improved and developed. The trunk of the ResNeXt Block is split into 32 paths that transform the input from high to low dimensions and back to high dimensions using the same topology, and aggregates them through element-wise addition. Although the ResNeXt Block outperforms the Res Block with fewer parameters and a higher detection accuracy in the natural scene dataset, since the semantic relevance between background and detected objects in ORSIs is stronger than that in NSIs [38], the operation of the ResNeXt Block easily breaks this relevance, and is thus not conducive to the detection performance in ORSIs. The Res2 Block can generate multi-scale features through a hierarchical short-cut connection and increase in receptive fields, thus improving the detection accuracy and reducing the computational consumption. However, its structure with parallel convolution and interactive operations significantly increases the inference time. The Dense Block contains several dense layers. The output of the dense layer is concatenated with its input, and the concatenated feature map serves as the input of the next dense layer. This structure takes full advantage of the short-cut that can better retain the feature and reduce the computation burden. However, the Dense Block will deteriorate in the situation where the background submerges features of detected objects in ORSIs, since the background information is more redundant and complex. Meanwhile, the structure of the Dense Block will reduce its inference speed due to the asymmetry of the input channels number and output channels number for a convolution operation. The CSP Block is the feature extraction module of the backbone CSP DarkNet in YOLOv4. It is mainly composed of several Res units based on a short-cut and a cross-stage part containing a 1 × 1 convolution layer. Although this structure can double the number of gradient paths and improve the detection accuracy through a splitting and merging strategy, there is parallel convolution and the problem of the trunk of the CSP Block being stacked alternately by an excessive convolution layer, which significantly increase the degree of network fragmentation and thus decrease the inference speed [39].

In order to alleviate the shortcomings of the above Blocks, a novel feature extraction module DenseRes Block is proposed in this study to improve the backbone in YOLOv4. Firstly, the input feature map of the DenseRes Block was compressed in order to increase the proportion of object feature information. Then, the series-connected residual structure with the same topology was utilized not only to obtain the high-level semantics of the object feature but also to reduce the computational consumption and inference time. Finally, the feature map output from the residual structure was combined with the input of the DenseRes Block to enhance the semantic relevance between background and detected objects.

### 2.3. Structure of the Neck

In the neck, feature maps output from the backbone will be processed and transmitted to the prediction part of the detector. The neck of the early DL-based detectors only directly transmits the last feature map of the backbone to the prediction part. The shallow feature map contains rich location information but low-level semantic information, whereas the deep feature map is the opposite; thus, this structure is not conducive to object detection, especially for small objects. In order to improve the detection performance of detectors for small objects, Liu et al. [3] proposed a neck structure that directly transfers the feature maps of different levels from the backbone to the prediction part of the detector for multi-scale detection, and proves that the utilization of a shallow feature map is beneficial for small object detection. However, shallow feature maps still lack high-level semantic information, while deep feature maps are still short of location information. FPN [16] is designed to transfer the high-level semantic information to the shallow feature map through the bottom-up structure to further improve the detection performance of the detector for small objects. In order to make the deep feature map possess rich location information and high-level semantic information, BFPN [40] has been developed to fuse the penultimate feature map and the last feature map based on FPN, while PANet [41] adds a top-down structure based on FPN to transmit location information to the deep feature map. Both BFPN and PANet can improve the detection performance for middle and large objects while maintaining a high detection accuracy for small objects.

YOLOv4 adopts the PANet as the framework in the neck. However, YOLOv4 suffers from the problem of feature loss in ORSIs due to many convolution operations in the neck. Therefore, a short-cut connection based on a residual is introduced to each CBL×5 in the neck for strengthening the transmission of object features without an increase in the computational burden and inference time.

### *2.4. Attention Mechanism*

The attention mechanism assigns different weights to the pixel according to the spatial or channel relationship between the pixels in the feature map to enhance the representation of the feature, and it mainly includes three categories: a channel attention mechanism (e.g., an SE Block [42] and ECA Block [43]), spatial attention mechanism (e.g., a CA Block [44]) and hybrid attention mechanism (e.g., a CBAM Block [45]). The attention mechanism can improve the detection accuracy in NSIs with a few parameters and computation burden increase for detectors. The SE Block squeezes and then extends channel information through two full connection layers in order to learn the relationship of global channel information and effectively improve the detection performance, but the relationship between local channel information is not considered. The ECA Block learns the relationship between local channels through 1-D convolution with an adaptive convolution kernel, which promotes the detection performance but ignores the relationship of global channel information. In the CA Block, the information is extracted by average pooling in horizontal and vertical directions, respectively, and then concatenated and fused by 2-D convolution. The fused information is split into two parts and each part is further extracted by the convolution layer, respectively. The hybrid attention mechanism CBAM Block combines the channel and spatial attention mechanism. Both the CA Block and CBAM Block bring an obvious improvement in the detection accuracy in the natural scene dataset, but their complex structure increase the inference time. Meanwhile, it is difficult for them to use a few parameters to extract the spatial information of ORSIs due to their more complex background and less spatial feature information for detected objects in ORSIs.

In order to more efficiently highlight the feature related to the detection task in ORSIs with a better robustness, a novel channel attention mechanism named a DCA Block was proposed to enhance the representation of the object feature in ORSIs through combing global and local channel information with a slight inference time increase.

## 3. Proposed Methods

### *3.1. Method Overview*

The structure of YOLOv4 is given in Figure 2. YOLOv4 consists of a backbone, neck and prediction. YOLOv4 is established for NSIs and not practical enough to be adopted in ORSIs directly. Specifically, the backbone CSP DarkNet in YOLOv4 utilizes the CSP Block [37] as the feature extraction module and performs well in detection accuracy, but its model complexity and computational burden can be further reduced to improve its deployability and inference speed for ORSIs. The neck PANet [41] employed in YOLOv4 can strengthen the integration of a shallow and deep feature map, but its CBL×5 modules will easily cause the problem of feature loss, which is not conducive to information transmission for objects in ORSIs. Moreover, attention mechanisms that can enhance the feature representation are not utilized in YOLOv4.

The proposed detector YOLO-DSD based on YOLOv4 is shown in Figure 3. Three new modules are presented to improve the performance of YOLOv4. In the backbone, we developed a DenseRes Block as the main module for a better feature extraction and reduction in computational cost. In the neck, S-CBL×5 was proposed to handle the information loss problem, and the proposed attention mechanism, the DCA Block, was added after each S-CBL×5 module to enhance the representation of features.

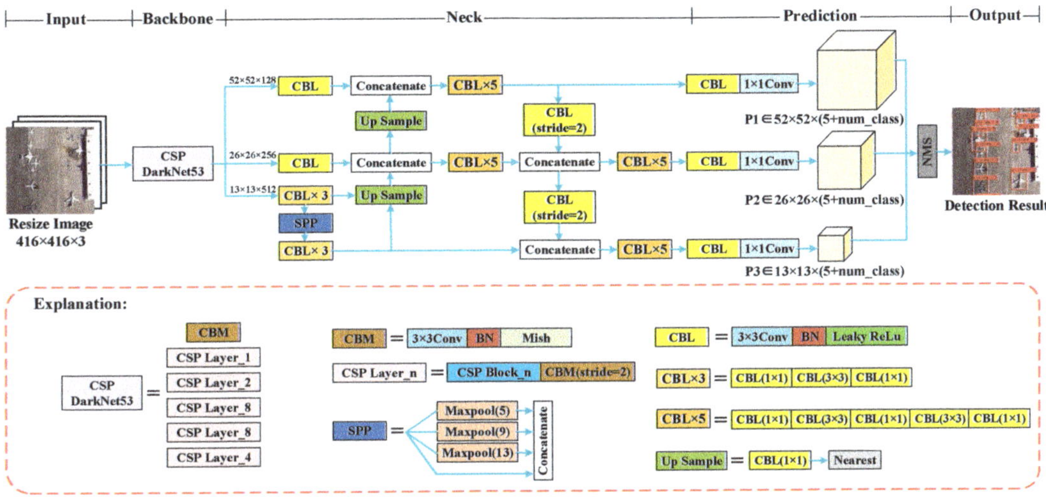


**Figure 2.** The architecture of YOLOv4.

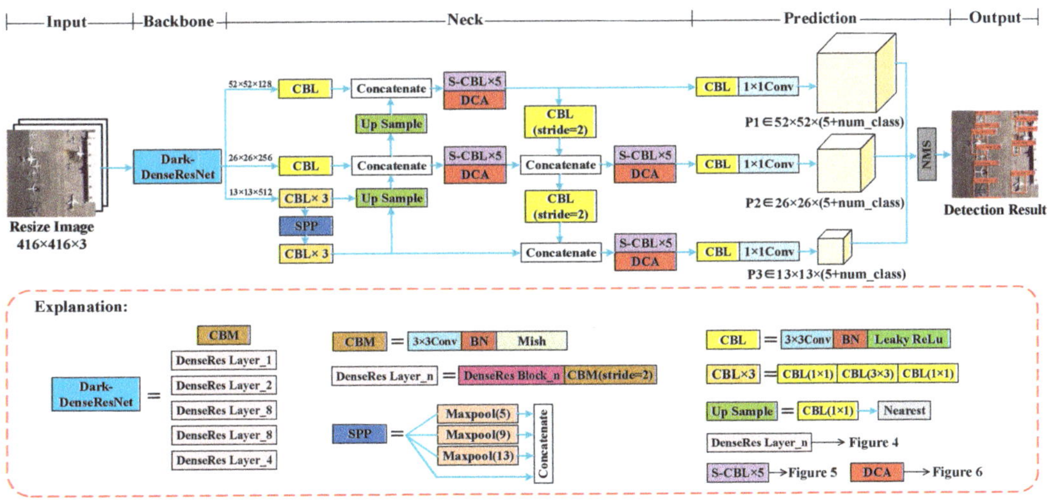


**Figure 3.** The architecture of YOLO-DSD.

### 3.2. Improvement in the Backbone

YOLOv4 adopts a CSP Block, shown in Figure 4a, to extract features of images in the backbone. Although the CSP Block performs well in detection accuracy, the structure of the CSP Block containing a parallel convolution operation for reusing the feature of the 'Input' and excessive convolution layers caused by 'Res Unit' takes up a large amount of computing resources and inference time [39]. Aiming at this problem of the CSP Block, we proposed a DenseRes Block, shown in Figure 4b, and employed it in the backbone for feature extraction.

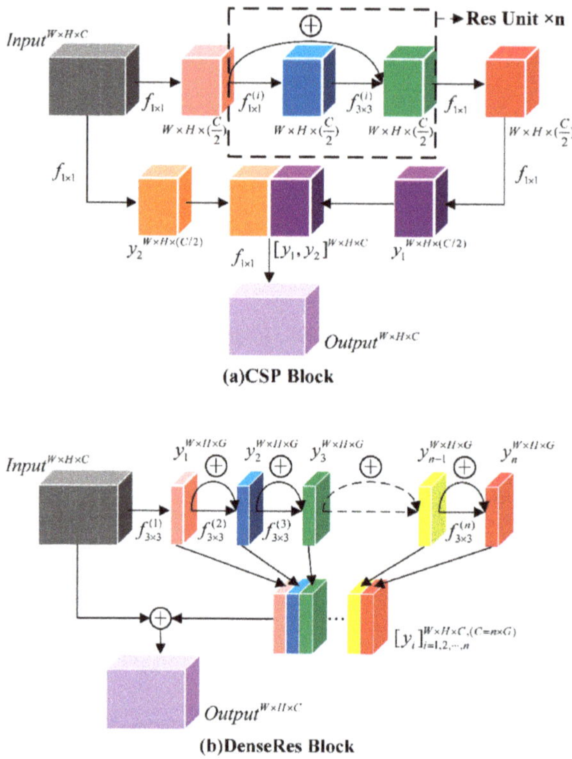


**Figure 4.** The structure comparison between CSP Block (**a**) and DenseRes Block (**b**).

The DenseRes Block is only composed of several series-connected 3 × 3 convolution operations $f_{3\times3}^{(i)}$ $(i = 1, 2, \cdots, n)$ and short-cut connections based on residual learning. $y_i$ is the output feature map of $f_{3\times3}^{(i)}$. For the feature map $Input \in \mathbb{R}^{W\times H\times C}$, $W$, $H$ and C indicate the height, width and channel number of the map, respectively. Since the feature of detected objects in ORSIs is easily overwhelmed by that of the background when transmitted, we utilized a feature map with fewer channels as the output of the first convolution operation to compress the '*Input*' to focus on object features and reduce the proportion of background information. Therefore, the feature map $y_1 \in \mathbb{R}^{W\times H\times G}$ was computed by

$$y_1^{W\times H\times G} = f_{3\times3}^{(1)}\left(Input^{W\times H\times C}\right) \quad (1)$$

where C = $n$ × $G$, $f_{3\times3}^{(1)}$ contains the 3 × 3 convolution layer that compacts the number of channels from C to G, the BN layer and the leaky ReLu activation function. If $n$ = 1, the DenseRes Block is the same as the Res Block. When $n$ > 1, the DenseRes Block will compress the '*Input*' and make a feature extraction. It was proven in Ref. [39] that the following operations can effectively reduce the memory access cost and the inference time of the model: (1) the input channel and output channel of the convolution layer should be equal as much as possible; (2) the number of fragmented operators (i.e., the number of individual convolution or parallel operations in one building block) should be reduced. Therefore, $y_j (1 < j \leq n) \in \mathbb{R}^{W \times H \times G}$ could be designed as

$$y_{j(1<j\leq n)}^{W\times H\times G} = y_{j-1}^{W\times H\times G} \oplus f_{3\times 3}^{(j)}\left(y_{j-1}^{W\times H\times G}\right) \tag{2}$$

where $f_{3\times 3}^{(j)} (1 < j \leq n)$ contains the $3 \times 3$ convolution layer with the same number $G$ of input and output channels, the BN layer and the leaky ReLu activation function. $\oplus$ indicates the element-wise addition. From the comparison between the CSP Block and the DenseRes Block shown in Figure 4, the output of each 'Res Unit' in the CSP Block will go through two convolution layers with different kernel sizes, whereas that of each '$y_i$' in the DenseRes Block only goes through one $3 \times 3$ convolution layer. Therefore, the fragment degree can be decreased. Moreover, we used a short-cut based on residual learning to connect $y_j (1 < j \leq n)$ and $y_{j-1} (1 < j \leq n)$ for the problem of feature loss in the process of feature extraction.

In ORSIs, there will be potential semantic relevance between the object and the background [21,38]. For example, cars and airplanes tend to park on land whereas ships tend to sail on the sea, and bridges are built over water whereas overpasses are built over land. In order to make the network better learn high-level semantic relevance, the $Output \in \mathbb{R}^{W \times H \times C}$ was designed as

$$Output^{W\times H\times C} = Input^{W\times H\times C} \oplus [y_i]_{i=1,2,\cdots,n}^{W\times H\times C(C=n\times G)} \tag{3}$$

where $[y_i]_{i=1,2,\cdots,n}^{W\times H\times C(C=n\times G)}$ concatenates $y_1, y_2, \ldots, y_n$ in the channel dimension to a feature map with the same size as $Input \in \mathbb{R}^{W \times H \times C}$. $[y_i]_{i=1,2,\cdots,n}^{W\times H\times C(C=n\times G)}$ possessing more object information was combined with the $Input \in \mathbb{R}^{W \times H \times C}$ holding more background information by element-wise addition directly to improve the detection accuracy. Compared with the CSP Block, such a designed structure in the DenseRes Block not only reuses the feature of '*Input*' but also omits a parallel convolution operation, which can further reduce the degree of the fragment in the backbone.

The DenseRes Block was utilized in order to replace the original module, the CSP Block, in the backbone. The architecture and complexity of the restructured backbone, named DarkNet-DenseRes, is shown in Table A1, Appendix A.

### 3.3. Improvement in the Neck

YOLOv4 uses the feature pyramid structure of PANet in the neck to fuse feature maps of different levels and extract a feature, which performs well in object detection in natural scenes. However, the feature information of objects in ORSIs is usually far less obvious than that of objects in natural scenes, and information loss caused by excessive convolutional operations in PANet limits the detection performance of the network for the objects in ORSIs. In order to solve this problem, S-CBL×5 was utilized to replace each CBL×5 in the original neck as shown in Figure 3. The structure comparison between CBL×5 and S-CBL×5 is given in Figure 5. S-CBL×5 adds two short-cuts based on CBL×5 and does not add additional parameters and inference time.

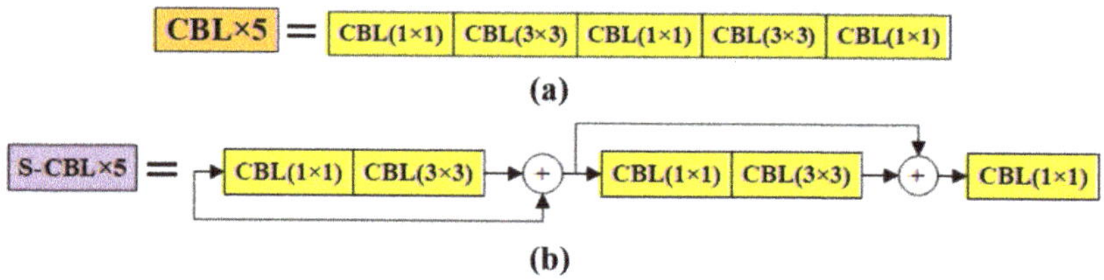


**Figure 5.** The structure comparison between CBL×5 (**a**) and the proposed S-CBL×5 (**b**).

To highlight significant features related to the detection task, the DCA Block was proposed to optimize the weight distribution of each feature map in the channel dimension by combining the local and global relationship between channels with a slight increase in computations cost and inference time. The structure of the DCA Block is shown in Figure 6.

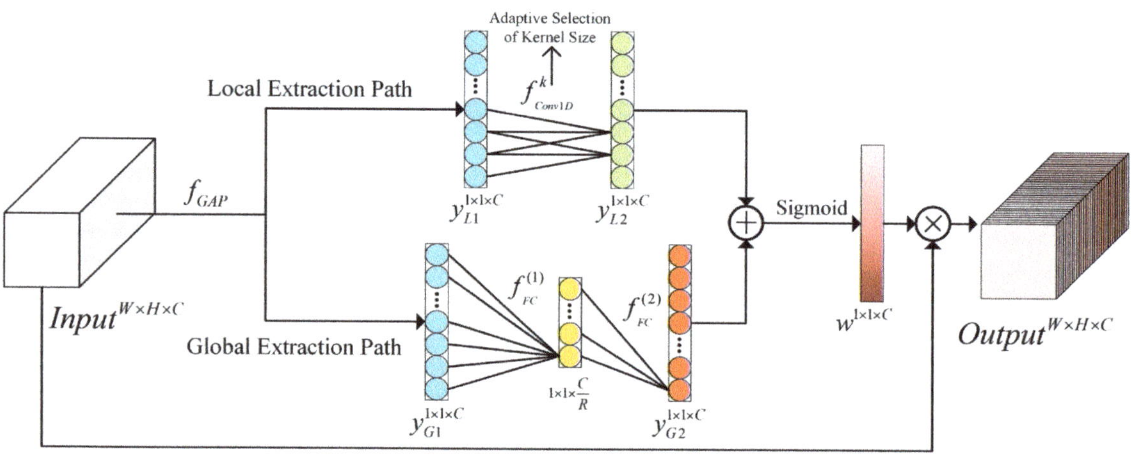


**Figure 6.** The structure of the proposed DCA Block.

The DCA Block is composed of a 'Local Extraction Path' and 'Global Extraction Path' in parallel. The 'Global Extraction Path' is used to learn the global relationship between channels, whereas the 'Local Extraction Path' is employed to extract the local channel relationship.

Firstly, global average pooling was employed to obtain the integrated information in the space dimension of each channel $y_{L1}^{1\times1\times C}$ and $y_{G1}^{1\times1\times C}$, where $y_{L1}^{1\times1\times C}$ and $y_{G1}^{1\times1\times C}$ indicate the input of the 'Local Extraction Path' and 'Global Extraction Path', respectively, and $y_{L1}^{1\times1\times C} = y_{G1}^{1\times1\times C}$.

Secondly, $y_{L2}^{1\times1\times C}$ in the 'Local Extraction Path' could be computed by

$$y_{L2}^{1\times1\times C} = f_{Conv1D}^{k}\left(y_{L1}^{1\times1\times C}\right) \quad (4)$$

$$k = \frac{\log_2 C + 1}{2} \quad (5)$$

where $f_{Conv1D}^{k}$ represents the 1-dimension convolution layer. Since each feature map has a different number of channels and the kernel size of the convolution layer is proportional to the number of the channels [43], the mapping between its kernel size ($k$) and the number of input channels ($C$) is given in Equation (5). $f_{Conv1D}^{k}$ could adaptively select the kernel size according to non-linearly mapping Equation (5); thus, it can extract the local relationship between covered channels more effectively than the convolution layer with a hand-given convolution kernel size.

At the same time, two full connection layers were used as a bottleneck in the 'Global Extraction Path' to build the global relationship of each channel:

$$y_{G2}^{1\times1\times C} = f_{FC}^{(2)}\left(f_{FC}^{(1)}\left(y_{G1}^{1\times1\times C}\right)\right) \tag{6}$$

where $f_{FC}^{(1)}$ is the first full connection layer that compresses the channel number from $C$ to $C/R$, and $f_{FC}^{(2)}$ is the second full connection layer that extends the channel number from $C/R$ to C. The value of the zoom factor R that could reduce the complexity of the structure was set to 32 according to the experimental results in Section 4.4.1. The structure of the 'Global Extraction Path' with two full connection layers has a stronger non-linearity and can fit better with the complex global relationship between each channel.

Thirdly, the output of the 'Global Extraction Path' and 'Local Extraction Path' were combined by element-wise addition, and the sigmoid function was applied to generate the weight $w \in \mathbb{R}^{1\times1\times C}$. Finally, the output of the DCA Block was calculated as:

$$w^{1\times1\times C} = Sigmoid\left(y_{G3}^{1\times1\times C} \oplus y_{L2}^{1\times1\times C}\right) \tag{7}$$

$$Output^{W\times H\times C} = w^{1\times1\times C} \otimes Input^{W\times H\times C} \tag{8}$$

where $\otimes$ represents the operation of the element-wise product. As shown in Figure 3, we added the proposed DCA Block after each S-CBL×5 to generate an improved PANet (shown in Figure 7) with a structure that is more suitable for optical remote sensing object detection and has a nearly equal computational cost compared to the original structure.

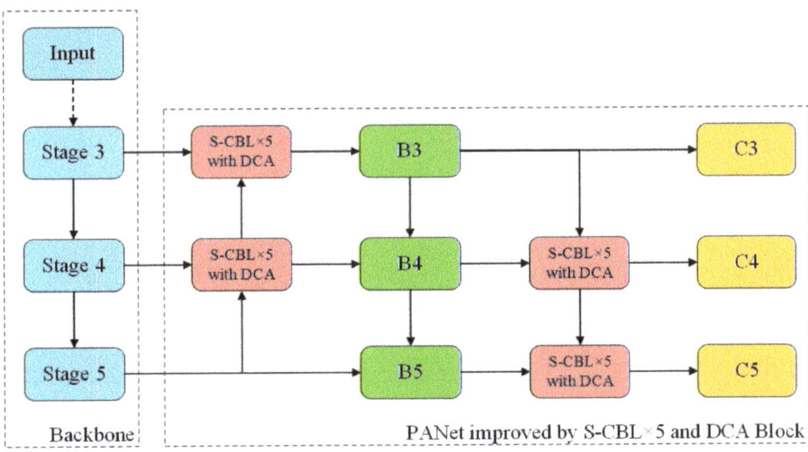


**Figure 7.** The structure of the improved PANet.

### 3.4. Prediction

Decoding and obtaining the detection result were processed in the prediction. As shown in Figure 3, each output of the neck went through with a CBL module and a 1 × 1 convolution layer, and three feature maps, $P_1 \in \mathbb{R}^{52\times52\times num\_class}$, $P_2 \in \mathbb{R}^{26\times26\times num\_class}$ and $P_3 \in \mathbb{R}^{13\times13\times num\_class}$, were generated. Then, as shown in Figure 8, $P_1$, $P_2$ and $P_3$ were mapped back to the original image and the image was divided into 52 × 52, 26 × 26 and 13 × 13 sizes of grids. Each grid corresponding to a feature map contains the information of three anchors. In each anchor, $(x, y)$ and $(w, h)$ are the offset coefficient and size coefficient, respectively, $Cf$ is the confidence of the grid containing the object and $C_1, C_2, C_3, \ldots, C_n$ are the confidence of each object class, respectively.

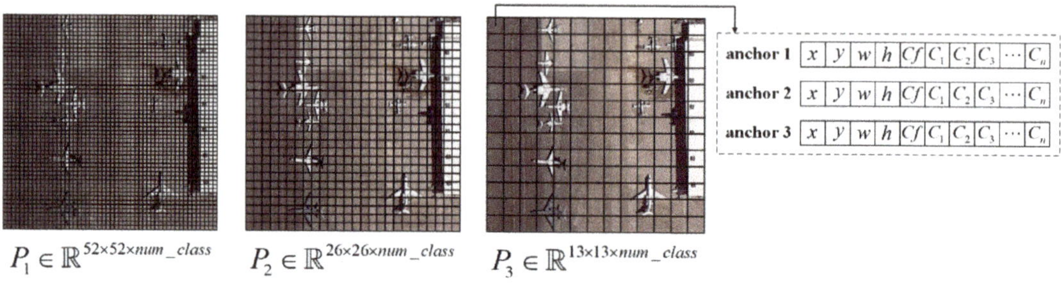


**Figure 8.** The image is divided into 52 × 52, 26 × 26 and 13 × 13 by $P_1$, $P_2$ and $P_3$, respectively.

Then, each grid generated three bounding boxes according to the information combined with anchors, and the process of converting the anchor to the bounding box is illustrated in Figure 9. $(C_x, C_y)$ are the upper left corner position of the current grid and the center of each grid anchor. $(\sigma(x), \sigma(y))$ is the offset of the bounding box relative to the anchor. The width $b_w$ and height $b_h$ of the bounding box were obtained through multiplying the width $p_w$ and height $p_h$ of the anchor by scaling factors $e^w$ and $e^h$, respectively. Finally, the detection results were obtained after redundant bounding boxes were removed through NMS.

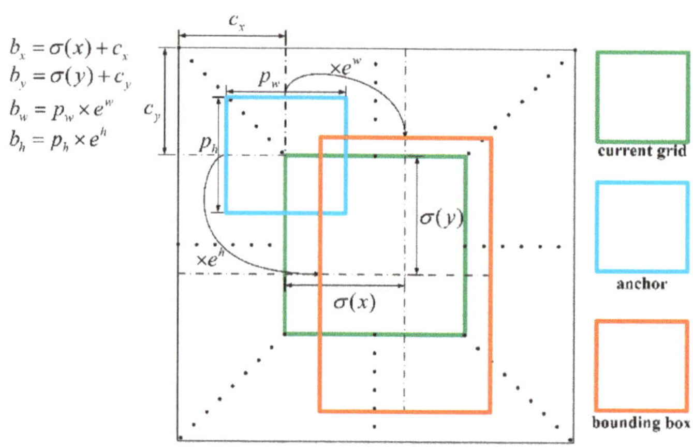


**Figure 9.** The process of converting anchor to bounding box.

### 3.5. Loss Function

The loss function of YOLOv4 includes three parts: confidence, classification and bounding box regression loss. YOLOv4 employs the complete intersection over union ($IoU$) loss ($CIoU$) [46], replacing the mean squared error loss adopted in YOLOv3 with the bounding box regression loss. $CIoU$ takes the overlap area, center point distance and aspect ratio into consideration simultaneously, and the convergence speed and detection accuracy were improved. $CIoU$ introduces a penalty item $\alpha v$ based on the distance $IoU$ loss to impose the consistency of the aspect ratio for the ground truth ($bb^{gt}$) and bounding box ($bb^{b}$). The loss of *CIoU* can be defined as Equation (9).

$$Loss_{CIoU} = 1 - \left( IoU - \frac{\rho^2(b^{gt}, b^b)}{c^2} - \alpha v \right)$$
$$\alpha = \frac{v}{1 - IoU + v}, \; v = \frac{4}{\pi^2}\left(\arctan\frac{w^{gt}}{h^{gt}} - \arctan\frac{w^b}{h^b}\right)^2 \tag{9}$$

where $b^{gt}$, $b^b$ are the center of $bb^{gt}$, $bb^b$, respectively, $\rho$ denotes the Euclidean distance, $c$ represents the diagonal length of the smallest enclosing rectangle covering $bb^{gt}$, $bb^b$, $\alpha$ is a positive trade-off value and $v$ means the consistency of the aspect ratio. $w^{gt}$, $w^b$ are the width of the $bb^{gt}$, $bb^b$, respectively. $h^{gt}$, $h^b$ are the height of the $bb^{gt}$, $bb^b$, respectively.

*CIOU* can directly minimize the distance between the bounding box and ground truth and accelerate the model convergence. Previous works [47–49] have proved that *CIOU* can perform better in detecting objects with diverse sizes, which can match well with the characteristics of remote sensing object detection tasks.

## 4. Experiments and Discussion

In this section, we conduct ablation and comparative experiments on a public optical remote sensing dataset DIOR [2] with 20 categories to validate the proposed YOLO-DSD, considering the accuracy, deployability and speed indictors. Another optical remote sensing dataset RSOD [50] with 4 categories was utilized to further verify the effectiveness of the proposed YOLO-DSD compared with YOLOv4.

### *4.1. Datasets*

#### 4.1.1. DIOR Dataset

DIOR [2] is a large ORSIs dataset that was established in 2020 to develop and validate data-driven methods. It contains 23,463 images and 192,472 objects in total, covering 20 categories in optical remote sensing field. Images in this benchmark dataset have been clipped into 800 × 800 pixels. There are vast scale variations across objects in DIOR because it contains images with spatial resolutions ranging from 0.5 m to 30 m. According to the definition of COCO [15], objects with area of ground truth less than 32 × 32 pixels, between 32 × 32 pixels and 96 × 96 pixels and larger than 96 × 96 pixels are taken as small, middle and large-sized objects, respectively. Each category and the size distribution of objects in DIOR is shown in Figure 10. It can be seen that objects in DIOR possess great size difference and are concentrated in small and middle-sized.

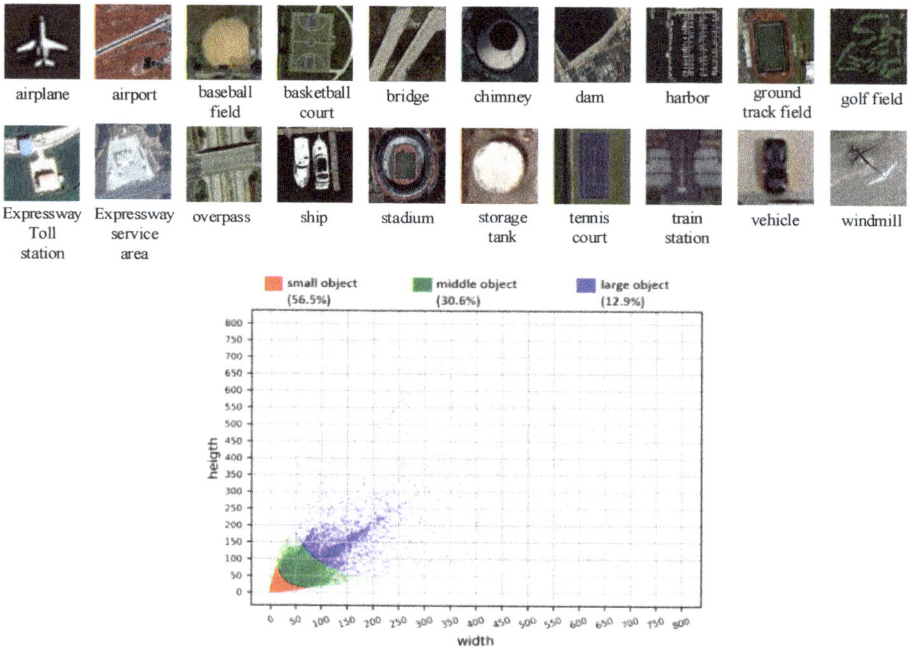


**Figure 10.** Each category and the size distributions of objects in DIOR.

Moreover, since images in DIOR are carefully collected under various environment conditions, such as different weathers and seasons, these images possess richer variations in viewpoint, background, occlusion, etc. Problems of intra-class diversity and intra-class similarity are more laborious due to the above characteristics. The main difficulties in real-world tasks can be well reflected by DIOR; thus, ablation experiments of YOLO-DSD and comparative experiments with SOTA detectors were conducted in DIOR dataset.

4.1.2. RSOD Dataset

RSOD [50] contains 976 images that have been clipped into approximately 1000 × 1000 pixels, and the spatial resolution of these images ranges from 0.3 m to 3 m. There are 6950 object instances in this dataset in total, covered by 4 common classes in ORSIs, including 4993 aircraft, 1586 oil tanks, 180 overpasses and 191 playgrounds. Each instance of classes is shown in Figure 11.

In addition, instances in RSOD dataset are under various scenes, including urban, grasslands, mountains, lakes, airport, etc. Although the scale of RSOD is not as large as that of DIOR, the characteristics of images in optical remote sensing object detection task can also be reflected by RSOD dataset. Therefore, we further analyzed the effectiveness of YOLO-DSD compared with YOLOv4 in RSOD dataset.

**Figure 11.** Each category of objects in RSOD.

## 4.2. Evaluation Indicator

Detectors in this study were analyzed from three perspectives, including detection accuracy, deployability and speed. The evaluation indicators of each performance are shown in Table 1. The higher the mAP and FPS, but the lower Params and Flops, the better the detector.

**Table 1.** The evaluation indicators.

| Indicator Class | Indicator | Description |
|---|---|---|
| Accuracy | $mAP^{0.5}$ (%) | Average precision when IOU = 0.5. It is the most used indicator in remote sensing object detection. |
| | $mAP^{0.5:0.95}$ (%) | Mean values of mAPs under each IOU, which are taken at an interval of 0.05 between 0.5 and 0.95. |
| | $mAP^{S}$, $mAP^{M}$, $mAP^{L}$ (%) | The $mAP^{0.5:0.95}$ of small, middle and large-sized object defined in MS COCO. |
| Deployability | Params | Number of detector parameters. |
| | Flops | Floating point operations. |
| Speed | FPS (img/s) | Frames transmitted per second. |

## 4.3. Experiment Setting

In this study, the deep learning framework PyTorch1.7.1 was utilized to implement all of the detectors in this study. The experimental environment was ubuntu18.04, CUDA11.1, CUDNN8.0.5 and NVIDIA GeForce RTX 3080. In order to ensure enough training samples and to make the test set reflect the characteristics of each dataset well, training and test sets in DIOR were split by 1:1, whereas those in RSOD were split by 4:1 randomly. A total of 90% of the training set was utilized for training detectors, and 10% was used for monitoring to avoid overfitting. The input size and batch size of detectors was set to $416 \times 416$ and 7, respectively. Adam optimizer was employed to update the parameters, with a weight decay of $2 \times 10^{-4}$. The relationship between learning rate and epoch is shown in Figure 12. For anchor-based detectors, K-means was utilized to optimize the size of anchors before training.

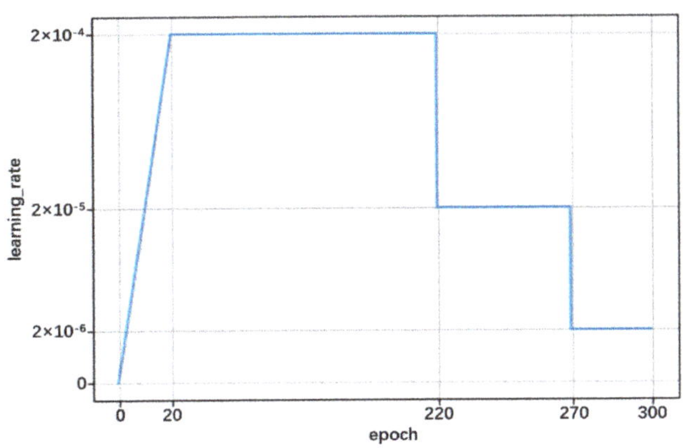


**Figure 12.** The relationship between learning rate and epoch.

## 4.4. Experiment Results and Discussion in DIOR Dataset

### 4.4.1. Ablation Experiment

Ablation experiments were conducted to verify the effectiveness of each improved module in YOLO-DSD, and the results are shown in Table 2. The detector improved with the DenseRes Block reduces Params by 23.9% ($\frac{48.81-64.17}{64.17} \times 100\%$) and Flops by 29.7%

($\frac{21.12-30.07}{30.07} \times 100\%$), and increases FPS by 63.4% ($\frac{65.7-40.2}{40.2} \times 100\%$), while achieving a 0.2% higher $mAP^{0.5}$ and almost the same $mAP^{0.5:0.95}$ compared with YOLOv4 as the baseline. The detector improved by S-CBL×5 in the neck based on "+DenseRes Block" is beneficial for $mAP^{0.5}$ and $mAP^{0.5:0.95}$, which are brought about by the increase in $mAP^{M}$ and $mAP^{L}$ without affecting the deployability and inference speed. However, the $mAP^{S}$ slightly decreased by 0.3% because the short-cut utilized in S-CBL×5 strengthened the transmitting of the feature, and thus introduced background features additionally, which attenuated the representation of the feature for small-sized objects. The detector further improved by the DCA Block achieved a significant increase in mAP due to the enhancement of feature expression, and made up for the loss of $mAP^{S}$ caused by the short-cut with the same Params and Flops, while the FPS was only slightly reduced by 5.3 img/s.

In summary, YOLO-DSD outperforms YOLOv4 both in the detection accuracy, deployability and speed evaluation indicator. YOLO-DSD based on YOLOv4 increases the commonly used indicator $mAP^{0.5}$ by 1.7% and the more rigorous indicator $mAP^{0.5:0.95}$ by 0.9%. Specifically, YOLO-DSD has a greater advantage in $mAP^{M}$ and $mAP^{L}$, while it achieves a similar and competitive $mAP^{S}$ compared with YOLOv4. In terms of deployability performance, the Params and Flops of YOLO-DSD decreased by 23.9% and 29.7% more than those of YOLOv4, respectively. YOLO-DSD also performs well in inference speed: it is 50.2% faster than YOLOv4 in FPS.

**Table 2.** The ablation results of YOLO-DSD.

| Detectors | Params | Flops | FPS | $mAP^{0.5}$ | $mAP^{0.5:0.95}$ | $mAP^{S}$ | $mAP^{M}$ | $mAP^{L}$ |
|---|---|---|---|---|---|---|---|---|
| YOLOv4(Baseline) | 64.17 M | 30.07 G | 40.2 | 71.3 | 39.1 | 10.1 | 30.2 | 55.1 |
| +DenseRes Block | 48.81 M | 21.12 G | 65.7 | 71.5 | 38.8 | 9.4 | 30.4 | 54.9 |
| +S-CBL×5 | 48.81 M | 21.12 G | 65.7 | 71.9 | 39.2 | 9.1 | 30.9 | 55.7 |
| +DCA(YOLO-DSD) | 48.81 M | 21.12 G | 60.4 | 73.0 | 40.0 | 9.6 | 31.6 | 56.4 |

We further analyzed the performance of the DenseRes Block. The ablation results of the DenseRes Block are shown in Table 3. The structure of the DenseRes Block in each detector is shown in Figure 13. Model 1 is the detector improved by the DenseRes Block without the structure of the 'Short-cut' and 'Combine'. 'Short-cut' and 'Combine' are introduced to the DenseRes Block in Model 2 and Model 3, respectively. Model 4 utilizes the complete DenseRes Block to improve the backbone of YOLOv4. From the comparison between Model 1 and Model 2, the 'Short-cut' introduced to DenseRes Block for the mitigation of feature loss can improve the mAP of objects in each size. After adding the 'Combine' to DenseRes Block, Model 3 performs better on the middle and large-sized object, while the $mAP^{S}$ decreases slightly by 0.1%. The possible reason for this is that the feature of the middle and large-sized object is obvious enough to build high-level semantic relevance with the background feature, while the feature of the small object is not obvious enough and thus it is easy for it to be overwhelmed. Model 4 improved by the complete DenseRes Block achieves the highest mAP and a significant increase in $mAP^{S}$, $mAP^{M}$ and $mAP^{L}$. It is probable that, on the basis of the 'Short-cut', the feature of each size object can be better retained when transmitting in the DenseRes Block, and can thus benefit the building of high-level semantic relevance with a background feature through 'Combine'.

**Table 3.** The ablation results of DenseRes Block.

| Detectors | DenseRes Block | | FPS | mAP$^{0.5}$ | mAP$^{0.5:0.95}$ | mAP$^{S}$ | mAP$^{M}$ | mAP$^{L}$ |
|---|---|---|---|---|---|---|---|---|
| | +Short-Cut* | +Combine* | | | | | | |
| Model 1 | | | 65.7 | 70.8 | 38.2 | 9.1 | 29.9 | 54.4 |
| Model 2 | √ | | 65.7 | 71.3 | 38.7 | 9.5 | 30.3 | 54.7 |
| Model 3 | | √ | 65.7 | 71.2 | 38.5 | 9.0 | 30.2 | 54.9 |
| Model 4 | √ | √ | 65.7 | 71.5 | 38.8 | 9.4 | 30.4 | 54.9 |

Note: 'Short-cut*' indicates a short-cut to connect $y_i$ $(1 < i \leq n)$ and $y_{(i-1)}$ $(1 < i \leq n)$ in DenseRes Block; 'Combine*' means $[y_i]$ $(1 \leq i \leq n) \oplus$ *Input* in DenseRes Block.

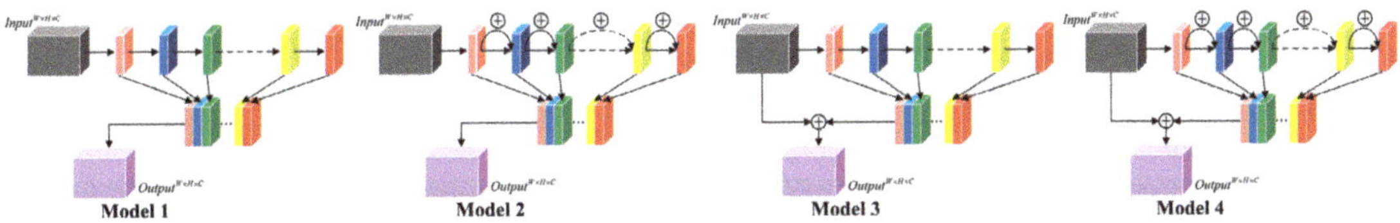


**Figure 13.** The structure of DenseRes Block in each detector in Table 3.

The experimental results of DCA module ablation are shown in Tables 4 and 5. Table 4 shows the influence of scaling factor R on the performance of the DCA Block. The results show that, when R = 32, DCA can achieve the best performance. Table 5 exhibits the influence of three different fusion methods shown in Figure 14 on the performance of the DCA Block. The results show that the DCA Block with a different fusion method can effectively improve the detection accuracy. Specifically, compared with DCA in series, DCA in parallel has a more obvious advantage in small and middle-sized objects, while the FPS is slightly reduced by 0.7 img/s. This may be due to the fact that, when employing the same number of operation layers in one building block, although the structure designed in parallel has a higher fragment, it can keep the integrity of the feature better compared with that in series. For the proposed DCA Block, which has a small structure complexity, utilizing the structure in parallel makes it perform better in the enhancement of feature expression without an obvious sacrifice of inference time.

**Table 4.** Results of different zoom factor 'R's in DCA Block.

| Detectors | DCA Block | FPS | mAP$^{0.5}$ | mAP$^{0.5:0.95}$ | mAP$^{S}$ | mAP$^{M}$ | mAP$^{L}$ |
|---|---|---|---|---|---|---|---|
| | Zoom Factor 'R' | | | | | | |
| Model 1 | R = 1 | 60.3 | 72.4 | 39.3 | 9.2 | 31.1 | 55.8 |
| Model 2 | R = 2 | 60.3 | 72.2 | 39.2 | 9.1 | 30.3 | 55.9 |
| Model 3 | R = 4 | 60.3 | 71.8 | 39.0 | 9.2 | 30.1 | 55.9 |
| Model 4 | R = 8 | 60.4 | 72.8 | 39.8 | 9.7 | 31.1 | 56.4 |
| Model 5 | R = 16 | 60.4 | 72.3 | 39.5 | 9.4 | 31.2 | 56.2 |
| Model 6 | R = 32 | 60.4 | 73.0 | 40.0 | 9.6 | 31.6 | 56.4 |
| Model 7 | R = 64 | 60.4 | 72.1 | 39.6 | 9.2 | 31.4 | 56.1 |

**Table 5.** Results of different fusion forms in DCA Block.

| Detectors | DCA Block | FPS | mAP$^{0.5}$ | mAP$^{0.5:0.95}$ | mAP$^{S}$ | mAP$^{M}$ | mAP$^{L}$ |
|---|---|---|---|---|---|---|---|
| | Fusion Form | | | | | | |
| Model 1 | 'Global Path' + 'Local Path' (In series) | 61.1 | 72.2 | 39.7 | 9.3 | 31.2 | 56.5 |
| Model 2 | 'Local Path' + 'Global Path' (In series) | 61.1 | 72.0 | 39.5 | 9.3 | 30.5 | 56.6 |
| Model 3 | 'Global Path' + 'Local Path' (In parallel) | 60.4 | 73.0 | 40.0 | 9.6 | 31.6 | 56.4 |

Note: 'Global Path' and 'Local Path' refer to 'Global Extraction Path' and 'Local Extraction Path', respectively.

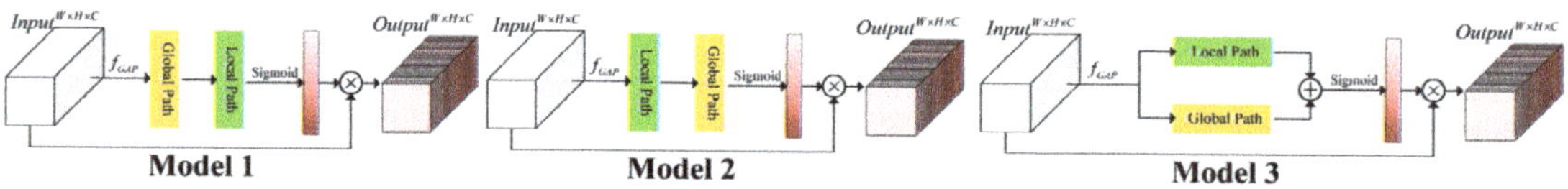


**Figure 14.** The structure of DCA Block in each detector in Table 5.

#### 4.4.2. Comparative Experiment

Four experiments were conducted in this study to verify the superiority of the proposed method. (1) ResNet50 [34], VGG16 [33] and the backbones that were established based on the CSP DarkNet framework with different feature extraction modules, including the Res Block [34], ResNeXt Block [35], Res2 Block [36], Dense Block [27], CSP Block [37] and DenseRes Block, were compared. (2) A comparison of different neck structures, including FPN [16], BFPN [40], PANet [41], S-PANet (PANet improved with the proposed S-CBL×5) and none (without feature pyramid structure), was conducted. (3) The performance of different attention mechanisms, including the SE Block [42], ECA Block [43], CA Block [44], CBAM Block [45] and DCA Block(R = 32), was compared and analyzed. (4) YOLO-DSD was compared with eight SOTA detectors, including Faster-RCNN, SSD, RetinaNet, YOLOv3, YOLOv4, YOLO-Lite (MobileNetV2 [51]—YOLOv4), CenterNet [7] and EfficientDet [9], which have been widely applied in various natural scene visual detection tasks due to their acceptable tradeoff between accuracy, deployability and inference time.

Comparative experiment for different backbones: The performances of the CSP DarkNet, which is improved by the proposed DenseRes Block (DarkNet-DenseRes) and other backbones, are demonstrated in Table 6. Based on the CSP DarkNet framework, the proposed DenseRes Block outperforms the ResNeXt Block and Dense Block in all indicators. Although the $mAP^{0.5}$ and $mAP^{0.5:0.95}$ of DarkNet-DenseRes are slightly lower than those of DarkNet-Res by 0.1% and 1.3%, the Params and Flops of DarkNet-DenseRes are only approximately 1/3 and 1/4 of DarkNet-Res, while the FPS of DarkNet-DenseRes is approximately 1/4 higher than that of DarkNet-Res. Similarly, the $mAP^{0.5}$ and $mAP^{0.5:0.95}$ of DarkNet-DenseRes are 0.9% and 1.1% lower than those of DarkNet-Res2; however, the Params and Flops of DarkNet-DenseRes are only approximately 1/3 and 1/2 of those of DarkNet-Res2, while the interference speed is 2.3 times that of DarkNet-Res2 according to FPS. The superiority of DarkNet-DenseRes compared with CSP DarkNet was analyzed and proved in ablation experiments. DarkNet-DenseRes also has obvious advantage in all indicators compared with ResNet50. Although DarkNet-DenseRes has a similar accuracy and speed to VGG16, VGG16 has seven times as much Flops than that of DarkNet-DenseRes. Therefore, DarkNet-DenseRes achieves the optimal balance of accuracy, deployability and speed.

**Table 6.** Results of comparative experiment for different backbones.

| Backbone | Params | Flops | FPS | $mAP^{0.5}$ | $mAP^{0.5:0.95}$ | $mAP^{S}$ | $mAP^{M}$ | $mAP^{L}$ |
|---|---|---|---|---|---|---|---|---|
| CSP DarkNet (Baseline) | 26.61 M | 17.34 G | 40.2 | 71.3 | 39.1 | 10.1 | 30.2 | 55.1 |
| DarkNet-Res [1] | 40.58 M | 24.61 G | 52.8 | 71.6 | 40.1 | 9.9 | 31.7 | 56.1 |
| DarkNet-ResNeXt [2] | 20.55 M | 12.71 G | 39.1 | 68.4 | 36.4 | 8.2 | 28.3 | 52.6 |
| DarkNet-Res2 [3] | 31.65 M | 19.33 G | 28.4 | 72.4 | 39.9 | 10.2 | 31.6 | 55.4 |
| DarkNet-Dense [4] | 14.06 M | 8.16 G | 50.9 | 69.6 | 37.5 | 7.8 | 29.7 | 54.5 |
| DarkNet-DenseRes [5](Ours) | 11.26 M | 8.42 G | 65.7 | 71.5 | 38.8 | 9.4 | 30.4 | 54.9 |
| ResNet50 | 23.51 M | 13.41 G | 47.4 | 68.5 | 36.5 | 7.8 | 28.7 | 53.2 |
| VGG16 | 17.07 M | 54.64 G | 70.1 | 71.1 | 38.9 | 9.9 | 29.8 | 55.2 |

Note: [1, 2, 3, 4, 5] means CSP DarkNet utilizes Res Block, ResNeXt Block, Res2 Block, Dense Block and DenseRes Block as the main feature extraction module, respectively.

Comparative experiment for different necks: Table 7 shows the performance of each neck structure that was tested by applying a no-feature pyramid structure (None), FPN, BFPN, PANet (Baseline) and S-PANet to the modified YOLOv4, with the DenseRes Block in the backbone. 'None' has the lowest Params (18.83 M) and Flops (4.89 G) and the highest FPS (85.5 img/s), but it does not perform well in detection accuracy, and, in particular, its $mAP^S$ is only 8.1%, whereas that of the other four necks ranges from 9.1% to 9.5%. Therefore, the feature pyramid structure is vital for detection accuracy and, in particular, for small size objects, which occupy more than 50% in DIOR. Although FPN and BFPN are slightly better than PANet in deployability and inference speed, they have more than a 2.6% inferiority in mAP of middle and large-sized objects, which, in total, account for approximately 50% of objects in DIOR. It was proven that the structure of PANet is important to the detection accuracy in YOLOv4 for ORSIs. PANet and S-PANet have almost the same Params, Flops and FPS, but our S-PANet performs better than PANet in $mAP^{0.5}$ and $mAP^{0.5:0.95}$. In conclusion, S-PANet is more suitable for optical remote sensing object detection than other necks.

**Table 7.** Results of comparative experiment for different necks.

| Neck | Params | Flops | FPS | $mAP^{0.5}$ | $mAP^{0.5:0.95}$ | $mAP^S$ | $mAP^M$ | $mAP^L$ |
|---|---|---|---|---|---|---|---|---|
| None | 18.83 M | 4.89 G | 85.5 | 68.3 | 35.5 | 8.1 | 27.4 | 51.6 |
| FPN | 27.22 M | 8.50 G | 71.8 | 69.1 | 36.0 | 9.2 | 27.2 | 51.4 |
| BFPN | 35.68 M | 10.84 G | 68.1 | 69.9 | 36.8 | 9.5 | 27.8 | 52.1 |
| PANet (Baseline) | 37.55 M | 12.73 G | 65.7 | 71.5 | 38.8 | 9.4 | 30.4 | 54.9 |
| S-PANet(ours) | 37.55 M | 12.73 G | 65.7 | 71.9 | 39.2 | 9.1 | 30.9 | 55.7 |

Comparative experiments for different attention mechanisms: Taking modified YOLOv4 with the DenseRes Block in the backbone and S-PANet in the neck as the baseline (None), the indicator values of different attention mechanisms are exhibited and compared in Table 8. The CA Block and CBAM Block containing the spatial attention mechanism fail to improve the detection accuracy, and the FPS decreases significantly due to those complex structures. Most channel attention mechanisms, including the SE Block, ECA Block and DCA Block, can improve the detection accuracy. The DCA Block improves the detection accuracy for small, medium and large sizes of objects, and achieves the highest $mAP^{0.5}$ = 73.0% and $mAP^{0.5:0.95}$ = 40.0%, with an increase of 1.1% and 0.8% compared with 'None', respectively, when R = 32, and the FPS only decreases by 5.3 img/s. In the case of the SE Block, $mAP^{0.5}$ and $mAP^{0.5:0.95}$ increases by 0.2% and 0.1%, and the FPS decreases by 3.4 img/s. The ECA Block improves both $mAP^{0.5}$ and $mAP^{0.5:0.95}$ by 0.1%, and decreases the FPS by 2.8 img/s. Therefore, the proposed DCA Block can achieve the best balance between accuracy and speed.

**Table 8.** Results of comparative experiment for different attention mechanisms.

| Attention Mechanism | Params | Flops | FPS | $mAP^{0.5}$ | $mAP^{0.5:0.95}$ | $mAP^S$ | $mAP^M$ | $mAP^L$ |
|---|---|---|---|---|---|---|---|---|
| None (Baseline) | 0 | 0 | 65.7 | 71.9 | 39.2 | 9.1 | 30.9 | 55.7 |
| CA | 42.36 K | 1126.41 K | 57.2 | 71.6 | 39.0 | 9.1 | 30.4 | 55.9 |
| CBAM | 102.79 K | 516.59 K | 52.8 | 70.6 | 38.1 | 8.4 | 29.3 | 55.7 |
| SE | 51.20 K | 51.272 K | 62.3 | 72.1 | 39.3 | 9.0 | 30.8 | 56.5 |
| ECA | 0.02 K | 0.02 K | 62.9 | 72.0 | 39.3 | 9.2 | 30.7 | 56.1 |
| DCA (R = 32) | 51.22 K | 830.24 K | 60.4 | 73.0 | 40.0 | 9.6 | 31.6 | 56.4 |

Comparative experiments for different detectors: The performances of the proposed YOLO-DSD and eight SOTA detectors are demonstrated in Table 9. RetinaNet and Efficient-Det have a better deployability than YOLO-DSD, but their detection accuracy, especially for small-sized objects and speed, are far behind that of YOLO-DSD, so this hinders the application of these detectors in optical remote sensing object detection. The large Flops of SSD and Faster-RCNN require a huge amount of computing resources, which greatly increases the difficulty in deploying them on edge devices. Although the Params and Flops of CenterNet are 67% and 69% that of YOLO-DSD, and the FPS is 46% faster, the detection accuracy of CenterNet is significantly lower than that of YOLO-DSD ($mAP^{0.5:0.95}$:35.8%

vs. 40.0%), and the $mAP^S$ is only 62.5% that of YOLO-DSD. YOLO-Lite has an obvious disadvantage in detection accuracy for small and large-sized objects, even though it has a better deployability compared with YOLO-DSD. The inference speed of YOLOv3 is nearly the same as that of YOLO-DSD, but the deployability and detection accuracy of YOLOv3 are obviously inferior to that of YOLO-DSD. The superiority of YOLO-DSD compared with YOLOv4 was analyzed and proved in ablation experiments. Therefore, YOLO-DSD outperforms other SOTA detectors in the balance of accuracy, deployability and speed.

**Table 9.** Results of comparative experiment for different detectors.

| Detector | Params | Flops | FPS | $mAP^{0.5}$ | $mAP^{0.5:0.95}$ | $mAP^S$ | $mAP^M$ | $mAP^L$ |
|---|---|---|---|---|---|---|---|---|
| RetinaNet | 36.72 M | 17.24 G | 44.6 | 62.7 | 37.6 | 4.8 | 30.9 | 57.5 |
| EfficientDet | 3.60 M | 1.30 G | 18.1 | 50.4 | 29.4 | 2.4 | 24.9 | 46.0 |
| SSD | 26.15 M | 59.59 G | 87.3 | 61.9 | 37.8 | 4.6 | 31.0 | 58.2 |
| CenterNet | 32.67 M | 14.62 G | 88.5 | 61.4 | 35.8 | 6.0 | 27.3 | 55.3 |
| Faster-RCNN | 28.47 M | 364.14 G | 21.9 | 56.1 | 31.8 | 2.8 | 23.7 | 53.2 |
| YOLO-Lite | 10.48 M | 3.89 G | 54.1 | 64.5 | 33.1 | 6.5 | 26.1 | 48.7 |
| YOLOv3 | 61.63 M | 32.83 G | 61.4 | 69.2 | 34.7 | 7.8 | 27.4 | 49.8 |
| YOLOv4 | 64.17 M | 30.07 G | 40.2 | 71.3 | 39.1 | 10.1 | 30.2 | 55.1 |
| YOLO-DSD (ours) | 48.81 M | 21.12 G | 60.4 | 73.0 | 40.0 | 9.6 | 31.6 | 56.4 |

Figures 15–17 exhibit the detection performance of Faster-RCNN, CenterNet, YOLOv4 and YOLO-DSD on DIOR. The detection result of the small-sized instance in Figure 15 indicates that both Faster-RCNN and CenterNet obviously miss detection. Although YOLOv4 could completely detect airplanes, it incorrectly detected a storage tank. Our YOLO-DSD can correctly detect all airplanes without any false detection. Figure 16 presents the detection results of an instance in the complex urban background. We can see that Faster-RCNN only detects one ground track field, and that CenterNet misses two bridges and two ground track fields and misdetects an overpass. YOLOv4 misses one bridge and one ground track field, whereas YOLO-DSD detects all objects correctly. The detection results of instances in a complex suburban background are given in Figure 17. It can be seen that Faster-RCNN detects only one Expressway-Service-Area, CenterNet has two false detections of an overpass and windmill, YOLOv4 detects two Expressway-Service-Areas as one, and YOLO-DSD correctly detects all objects. The above instances verify that YOLO-DSD can handle object detection under different complex backgrounds well.

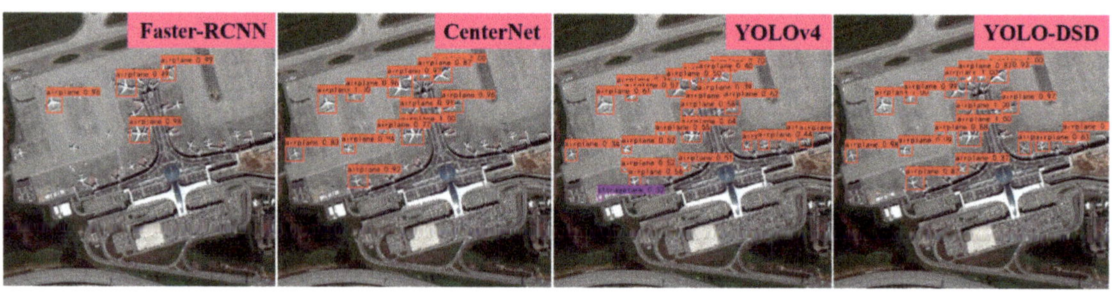


**Figure 15.** The detection result of small-sized instance.

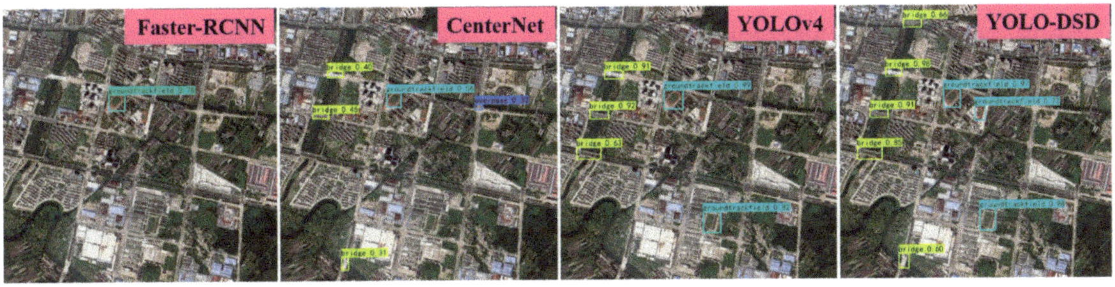


**Figure 16.** The detection result of instance in complex suburban background.

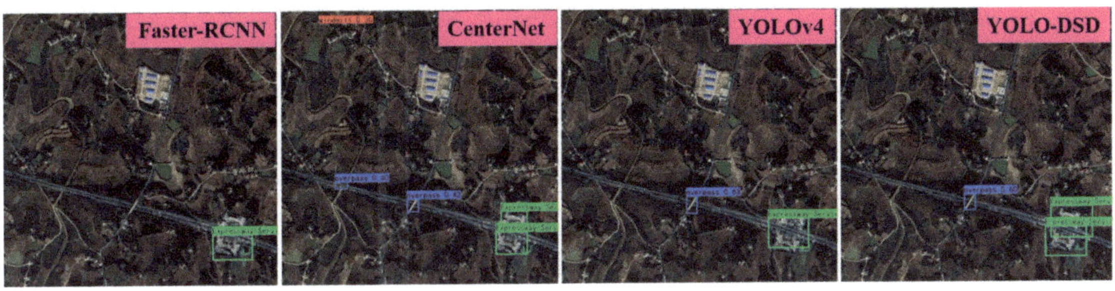


**Figure 17.** The detection result of instance in complex urban background.

The precision–recall curves and AP (IOU = 0.5) of YOLOv4 and YOLO-DSD in each category are given in Figure 18 for a better illustration of the difference in detection accuracy. It can be seen that YOLO-DSD detects better than YOLOv4 in 11 categories, including airplane, airport, baseball field, chimney, dam, Expressway-Service-Area, golffield, groundtrackfield, stadium, storagetank and transtation. In particular, the AP of YOLO-DSD in airport, baseballfield, Expressway-Service-Area and groundtrackfield is over 2% higher than that of YOLOv4. The AP of YOLO-DSD in airplane, transtation and stadium significantly increase by 6.63%, 5.21% and 17.02%, respectively. For the other nine categories, YOLO-DSD only slightly decreases by 0.35~1.78% compared with YOLOv4 in AP, but still has a competitive accuracy. Therefore, YOLO-DSD has a better accuracy performance than YOLOv4 in the large-scale ORSIs dataset DIOR in total.

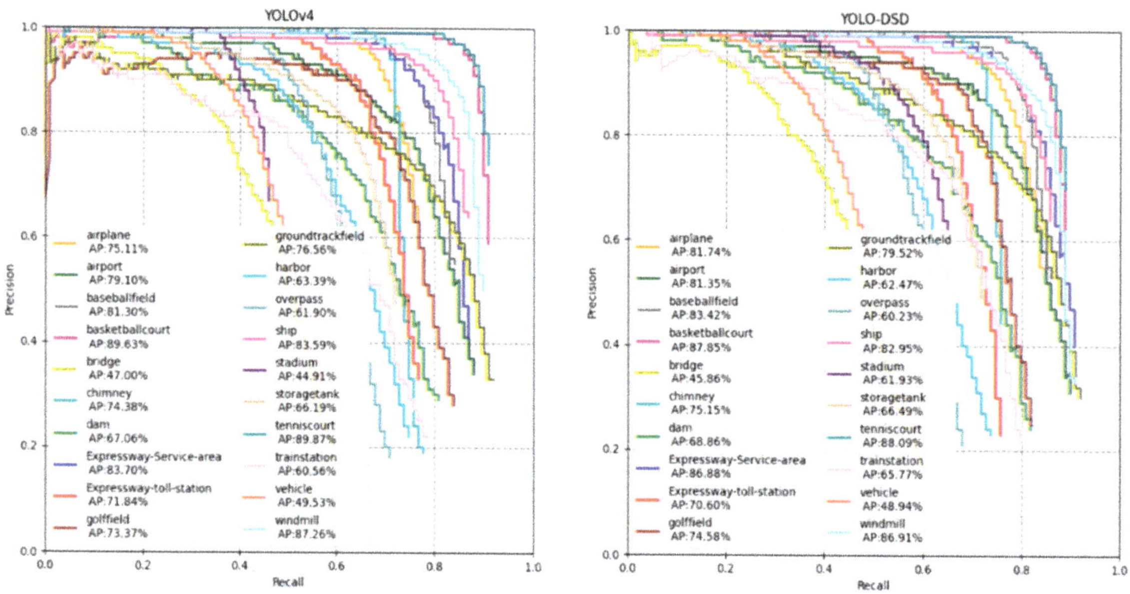


**Figure 18.** The precision–recall curves and AP (IOU = 0.5) of YOLOv4 and YOLO-DSD in each category.

### 4.5. Experiment Results and Discussion in RSOD Dataset

In order to further exhibit the superiority of the proposed YOLO-DSD based on YOLOv4 in optical remote sensing object detection application, another comparison experiment between YOLO-DSD and YOLOv4 was conducted in a four-category dataset, RSOD [50], which contained aircraft, oil tank, playground and overpass. The experiment result is shown in Table 10. It can be seen that YOLO-DSD outperforms in accuracy and inference time compared with YOLOv4 under different input sizes, including 416 × 416, 512 × 512 and 608 × 608. Specifically, YOLO-DSD increases $mAP^{0.5}$, $mAP^{0.5:0.95}$ and FPS by 2.6%, 0.8% and 50.2%, respectively, under the input size 416 × 416, while, under the input size 512 × 512, $mAP^{0.5}$, $mAP^{0.5:0.95}$ and FPS are improved by 2.1%, 1.9% and 54.9%, respectively. In terms of the input size 608 × 608, the $mAP^{0.5}$, $mAP^{0.5:0.95}$ and FPS of YOLO-DSD are 1.5%, 1.2% and 59.3% higher than those of YOLOv4.

However, it is noteworthy that the overpass AP of YOLO-DSD is higher than that of YOLOv4 in RSOD, whereas it is the opposite in DIOR. One possible reason for this is that 'bridge' and 'overpass' possess a significant inter-class similarity and thus interfere with the detection performance of YOLO-DSD in these two categories in DIOR. Therefore, how to overcome the inter-class similarity between 'bridge' and 'overpass' for a better detection accuracy while keeping its deployability and inference speed is one of our future works.

**Table 10.** Results of comparative experiment for YOLOv4 and YOLO-DSD in RSOD.

| Detector | Input Size | FPS | $AP^{0.5}$ | | | | $mAP^{0.5}$ | $mAP^{0.5:0.95}$ |
|---|---|---|---|---|---|---|---|---|
| | | | Aircraft | Oil Tank | Playground | Overpass | | |
| YOLOv4 | 416 × 416 | 40.2 | 97.8 | 95.7 | 99.4 | 67.2 | 90.0 | 52.1 |
| YOLO-DSD | | 60.7 | 98.0 | 98.2 | 99.6 | 74.4 | 92.6 | 52.9 |
| YOLOv4 | 512 × 512 | 37.1 | 98.1 | 97.5 | 99.5 | 73.5 | 92.2 | 55.1 |
| YOLO-DSD | | 57.5 | 98.5 | 98.6 | 99.8 | 80.1 | 94.3 | 57.0 |
| YOLOv4 | 608 × 608 | 34.9 | 98.2 | 98.5 | 99.9 | 79.2 | 94.0 | 58.5 |
| YOLO-DSD | | 55.6 | 99.1 | 98.9 | 99.9 | 84.2 | 95.5 | 59.7 |

## 5. Conclusions

In this study, a new detector, YOLO-DSD, based on YOLOv4, was proposed to balance the accuracy, deployability and inference time for remote sensing object detection. Three main improvements were utilized in YOLO-DSD, including the DenseRes Block, S-CBL×5 and DCA Block. Firstly, the DenseRes Block improves the backbone, which can better compress and extract the object feature with a high accuracy but less computational consumption. Secondly, S-CBL×5 introduced in the neck can mitigate feature loss without increasing the consumption and inference time. Finally, a new channel attention mechanism, the DCA Block, added to S-CBL×5 better highlights the important features in the channel dimension.

Experiments on a large dataset, DIOR, were conducted to analyze the detection performance from the accuracy (mAP), deployability (Params and Flops) and speed (FPS). The results of the experiments indicate that the proposed DenseRes Block is superior to other feature extraction modules, such as the Res Block, ResNeXt Block, Res2 Block, Dense Block and CSP Block. Moreover, S-CBL×5 performs better than currently widely used FPN, BFPN and PANet. In addition, the proposed DCA Block outperforms other attention mechanisms, including the SE Block, ECA Block, CA Block and CBAM Block. Compared with YOLOv4, YOLO-DSD reduces Params by 23.9% and Flops by 29.7%, but increases FPS by 50.2%, while $mAP^{0.5}$ and $mAP^{0.5:0.95}$ increased from 71.3% to 73.0% and 39.1% to 40.0%, respectively. Compared with other SOTA detectors, including Faster-RCNN, SSD, RetinaNet, YOLOv3, YOLOv4, CenterNet, YOLO-Lite and EfficientDet, YOLO-DSD achieves the optimal balance of accuracy, deployability and inference time. In terms of the RSOD dataset, compared with YOLOv4, YOLO-DSD achieves 1.5~2.6%, 0.8~1.2% and 50.2~59.3% increases in $mAP^{0.5}$, $mAP^{0.5:0.95}$ and FPS under different input sizes, including 416 × 416, 512 × 512 and 608 × 608.

However, YOLO-DSD has a limitation in processing a serious inter-class similarity, such as 'bridge' and 'overpass', compared with YOLOv4. In order to further improve the performance of the proposed detector, we will try to combine depthwise separable convolution [52] with the proposed DenseRes Block for a better feature extraction and deployability reduction. Moreover, other non-consumption methods, such as image preprocessing and anchor optimization, will be considered to improve the detector.

**Author Contributions:** Conceptualization, S.L. and H.C.; methodology, S.L. and H.C.; software, H.C.; validation, H.C. and H.J.; formal analysis, S.L. and H.C.; investigation, S.L. and H.C.; resources, S.L. and H.C.; data curation, H.C.; writing—original draft preparation, S.L. and H.C.; writing—review and editing, H.J.; visualization, H.C.; supervision, S.L.; project administration, S.L.; funding acquisition, S.L. All authors have read and agreed to the published version of the manuscript.

**Funding:** This research was funded by Natural Science Foundation of Guangdong, China with grant number 2021A1515012395, and was supported by earmarked fund for China Agriculture Research System, grant number CARS-17.

**Institutional Review Board Statement:** Not applicable.

**Informed Consent Statement:** Not applicable.

**Data Availability Statement:** All data used during the study have been uploaded at: https://gcheng-nwpu.github.io/#Datasets (last accessed on 27 July 2022) and https://github.com/RSIA-LIESMARS-WHU/RSOD-Dataset- (last accessed on 27 July 2022).

**Acknowledgments:** We gratefully appreciate the editor and anonymous reviewers for their efforts and constructive comments, which have greatly improved the technical quality and presentation of this study.

**Conflicts of Interest:** The authors declare no conflict of interest.

## Appendix A

**Table A1.** The architecture and complexity of DarkNet-DenseRes.

| Stage | Output Size | Operation | Number | Params | Flops |
|---|---|---|---|---|---|
| CBM | 416 × 416 × 32 | Conv-BN-Mish (k = 3 × 3, c = 32, s = 1) * | 1 | 928 | 166,133,760 |
| DenseRes Layer_1 | 208 × 208 × 64 | Conv-BN-Mish (k = 3 × 3, c = 64, s = 2) | 1 | 18,560 | 805,748,736 |
| | 208 × 208 × 64 | Conv-BN- Leaky ReLu (k = 3 × 3, c = 64, s = 1) | 1 | 36,992 | 1,603,190,784 |
| | 208 × 208 × 64 | Concatenation | 1 | / | / |
| DenseRes Layer_2 | 104 × 104 × 128 | Conv-BN-Mish (k = 3 × 3, c = 128, s = 2) | 1 | 73,984 | 801,595,392 |
| | 104 × 104 × 64 | Conv-BN- Leaky ReLu (k = 3×3, c = 64, s = 1) | 2 | 110,848 | 1,200,316,416 |
| | 104 × 104 × 128 | Concatenation | 1 | / | / |
| DenseRes Layer_3 | 52 × 52 × 256 | Conv-BN-Mish (k = 3 × 3, c = 256, s = 2) | 1 | 295,424 | 799,518,720 |
| | 52 × 52 × 32 | Conv-BN- Leaky ReLu (k = 3 × 3, c = 32, s = 1) | 8 | 138,752 | 375,861,632 |
| | 52 × 52 × 256 | Concatenation | 1 | / | / |
| DenseRes Layer_4 | 26 × 26 × 512 | Conv-BN-Mish (k = 3 × 3, c = 512, s = 2) | 1 | 1,180,672 | 798,480,384 |
| | 26 × 26 × 64 | Conv-BN- Leaky ReLu (k = 3 × 3, c = 64, s = 1) | 8 | 553,984 | 374,839,296 |
| | 26 × 26 × 512 | Concatenation | 1 | / | / |
| DenseRes Layer_5 | 13 × 13 × 1024 | Conv-BN-Mish (k = 3 × 3, c = 1024, s = 2) | 1 | 4,720,640 | 797,961,216 |
| | 13 × 13 × 256 | Conv-BN- Leaky ReLu (k = 3 × 3, c = 256, s = 1) | 4 | 4,130,816 | 698,280,960 |
| | 13 × 13 × 1024 | Concatenation | 1 | / | / |
| Total Params | | 11,261,600 | | | |
| Total Flops | | 8,421,927,296 | | | |

Note: * k, c, and s mean kernel size, output channels and stride of the convolution layer, respectively.

## References


1. Cheng, G.; Han, J.W. A survey on object detection in optical remote sensing images. *ISPRS J. Photogramm* **2016**, *117*, 11–28. [CrossRef]
2. Li, K.; Wan, G.; Cheng, G.; Meng, L.Q.; Han, J.W. Object detection in optical remote sensing images: A survey and a new benchmark. *ISPRS J. Photogramm* **2020**, *159*, 296–307. [CrossRef]
3. Liu, W.; Anguelov, D.; Erhan, D.; Szegedy, C.; Reed, S.; Fu, C.-Y.; Berg, A.C. Ssd: Single shot multibox detector. In Proceedings of the European Conference on Computer Vision, Amsterdam, The Netherlands, 11–14 October 2016; pp. 21–37.
4. Redmon, J.; Divvala, S.; Girshick, R.; Farhadi, A. You only look once: Unified, real-time object detection. In Proceedings of the IEEE Conference on Computer Vision and Pattern Recognition, Las Vegas, NV, USA, 27–30 June 2016; pp. 779–788.
5. Redmon, J.; Farhadi, A. YOLO9000: Better, faster, stronger. In Proceedings of the IEEE Conference on Computer Vision and Pattern Recognition, Honolulu, HI, USA, 21–26 July 2017; pp. 7263–7271.
6. Redmon, J.; Farhadi, A. Yolov3: An incremental improvement. *arXiv* **2018**, arXiv:1804.02767.
7. Zhou, X.; Wang, D.; Krähenbühl, P. Objects as points. *arXiv* **2019**, arXiv:1904.07850.
8. Bochkovskiy, A.; Wang, C.-Y.; Liao, H.-Y.M. Yolov4: Optimal speed and accuracy of object detection. *arXiv* **2020**, arXiv:2004.10934.
9. Tan, M.; Pang, R.; Le, Q.V. Efficientdet: Scalable and efficient object detection. In Proceedings of the IEEE/CVF Conference on Computer Vision and Pattern Recognition, Seattle, WA, USA, 13–19 June 2020; pp. 10781–10790.
10. Girshick, R.; Donahue, J.; Darrell, T.; Malik, J. Rich feature hierarchies for accurate object detection and semantic segmentation. In Proceedings of the IEEE Conference on Computer Vision and Pattern Recognition, Columbus, OH, USA, 23–28 June 2014; pp. 580–587.
11. Girshick, R. Fast r-cnn. In Proceedings of the IEEE International Conference on Computer Vision, Boston, MA, USA, 7–12 June 2015; pp. 1440–1448.
12. Ren, S.; He, K.; Girshick, R.; Sun, J. Faster r-cnn: Towards real-time object detection with region proposal networks. *Adv. Neural Inf. Processing Syst.* **2015**, *28*, 91–99. [CrossRef] [PubMed]
13. He, K.; Gkioxari, G.; Dollár, P.; Girshick, R. Mask r-cnn. In Proceedings of the IEEE International Conference on Computer Vision, Honolulu, HI, USA, 21–26 July 2017; pp. 2961–2969.
14. Everingham, M.; Van Gool, L.; Williams, C.K.I.; Winn, J.; Zisserman, A. The Pascal Visual Object Classes (VOC) Challenge. *Int. J. Comput. Vis.* **2010**, *88*, 303–338. [CrossRef]
15. Lin, T.-Y.; Maire, M.; Belongie, S.; Hays, J.; Perona, P.; Ramanan, D.; Dollár, P.; Zitnick, C.L. Microsoft coco: Common objects in context. In Proceedings of the European Conference on Computer Vision, Zurich, Switzerland, 6–12 September 2014; pp. 740–755.
16. Lin, T.-Y.; Goyal, P.; Girshick, R.; He, K.; Dollár, P. Focal loss for dense object detection. In Proceedings of the IEEE International Conference on Computer Vision, Honolulu, HI, USA, 21–26 July 2017; pp. 2980–2988.

17. Al Ridhawi, I.; Bouachir, O.; Aloqaily, M.; Boukerche, A. Design Guidelines for Cooperative UAV-supported Services and Applications. *ACM Comput. Surv.* **2022**, *54*, 1–35. [CrossRef]
18. Xu, D.Q.; Wu, Y.Q. MRFF-YOLO: A Multi-Receptive Fields Fusion Network for Remote Sensing Target Detection. *Remote Sens.* **2020**, *12*, 3118. [CrossRef]
19. Cheng, G.; Si, Y.J.; Hong, H.L.; Yao, X.W.; Guo, L. Cross-Scale Feature Fusion for Object Detection in Optical Remote Sensing Images. *IEEE Geosci. Remote Sens. Lett.* **2021**, *18*, 431–435. [CrossRef]
20. Yin, W.; Diao, W.; Wang, P.; Gao, X.; Li, Y.; Sun, X. PCAN—Part-based context attention network for thermal power plant detection in remote sensing imagery. *Remote Sens.* **2021**, *13*, 1243. [CrossRef]
21. Yuan, Z.C.; Liu, Z.M.; Zhu, C.B.; Qi, J.; Zhao, D.P. Object Detection in Remote Sensing Images via Multi-Feature Pyramid Network with Receptive Field Block. *Remote Sens.* **2021**, *13*, 862. [CrossRef]
22. Li, Z.L.; Zhao, L.N.; Han, X.; Pan, M.Y. Lightweight Ship Detection Methods Based on YOLOv3 and DenseNet. *Math. Probl. Eng.* **2020**, *2020*, 4813183. [CrossRef]
23. Huyan, L.; Bai, Y.P.; Li, Y.; Jiang, D.M.; Zhang, Y.N.; Zhou, Q.; Wei, J.Y.; Liu, J.N.; Zhang, Y.; Cui, T. A Lightweight Object Detection Framework for Remote Sensing Images. *Remote Sens.* **2021**, *13*, 683. [CrossRef]
24. Lang, L.; Xu, K.; Zhang, Q.; Wang, D. Fast and Accurate Object Detection in Remote Sensing Images Based on Lightweight Deep Neural Network. *Sensors* **2021**, *21*, 5460. [CrossRef]
25. Li, Y.Y.; Mao, H.T.; Liu, R.J.; Pei, X.; Jiao, L.C.; Shang, R.H. A Lightweight Keypoint-Based Oriented Object Detection of Remote Sensing Images. *Remote Sens.* **2021**, *13*, 2459. [CrossRef]
26. Huang, W.; Li, G.Y.; Chen, Q.Q.; Ju, M.; Qu, J.T. CF2PN: A Cross-Scale Feature Fusion Pyramid Network Based Remote Sensing Target Detection. *Remote Sens.* **2021**, *13*, 847. [CrossRef]
27. Huang, G.; Liu, Z.; Van Der Maaten, L.; Weinberger, K.Q. Densely connected convolutional networks. In Proceedings of the IEEE Conference on Computer Vision and Pattern Recognition, Honolulu, HI, USA, 21–26 July 2017; pp. 4700–4708.
28. Qin, Z.; Li, Z.; Zhang, Z.; Bao, Y.; Yu, G.; Peng, Y.; Sun, J. ThunderNet: Towards real-time generic object detection on mobile devices. In Proceedings of the IEEE/CVF International Conference on Computer Vision, Long Beach, CA, USA, 15–20 June 2019; pp. 6718–6727.
29. He, H.; Huang, X.; Song, Y.; Zhang, Z.; Wang, M.; Chen, B.; Yan, G. An insulator self-blast detection method based on YOLOv4 with aerial images. *Energy Rep.* **2022**, *8*, 448–454. [CrossRef]
30. Roy, A.M.; Bose, R.; Bhaduri, J. A fast accurate fine-grain object detection model based on YOLOv4 deep neural network. *Neural Comput. Appl.* **2022**, *34*, 3895–3921. [CrossRef]
31. Song, W.; Fu, C.; Zheng, Y.; Cao, L.; Tie, M.; Sham, C.W. Protection of image ROI using chaos-based encryption and DCNN-based object detection. *Neural Comput. Appl.* **2022**, *34*, 5743–5756. [CrossRef]
32. Gu, Y.; Si, B.J.E. A novel lightweight real-time traffic sign detection integration framework based on YOLOv4. *Entropy* **2022**, *24*, 487. [CrossRef]
33. Simonyan, K.; Zisserman, A. Very deep convolutional networks for large-scale image recognition. *arXiv* **2014**, arXiv:1409.1556.
34. He, K.; Zhang, X.; Ren, S.; Sun, J. Deep residual learning for image recognition. In Proceedings of the IEEE Conference on Computer Vision and Pattern Recognition, Las Vegas, NV, USA, 27–30 June 2016; pp. 770–778.
35. Xie, S.; Girshick, R.; Dollár, P.; Tu, Z.; He, K. Aggregated residual transformations for deep neural networks. In Proceedings of the IEEE Conference on Computer Vision and Pattern Recognition, Honolulu, HI, USA, 21–26 July 2017; pp. 1492–1500.
36. Gao, S.-H.; Cheng, M.-M.; Zhao, K.; Zhang, X.-Y.; Yang, M.-H.; Torr, P. Res2net: A new multi-scale backbone architecture. *IEEE Trans. Pattern Anal. Mach. Intell.* **2019**, *43*, 652–662. [CrossRef]
37. Wang, C.-Y.; Liao, H.-Y.M.; Wu, Y.-H.; Chen, P.-Y.; Hsieh, J.-W.; Yeh, I.-H. CSPNet: A new backbone that can enhance learning capability of CNN. In Proceedings of the IEEE/CVF Conference on Computer Vision and Pattern Recognition Workshops, Seattle, WA, USA, 13–19 June 2020; pp. 390–391.
38. Xu, C.; Li, C.; Cui, Z.; Zhang, T.; Yang, J. Hierarchical Semantic Propagation for Object Detection in Remote Sensing Imagery. *IEEE Trans. Geosci. Remote Sens.* **2020**, *58*, 4353–4364. [CrossRef]
39. Ma, N.; Zhang, X.; Zheng, H.-T.; Sun, J. Shufflenet v2: Practical guidelines for efficient cnn architecture design. In Proceedings of the European Conference on Computer Vision (ECCV), Salt Lake City, UT, USA, 18–23 June 2018; pp. 116–131.
40. Zhang, X.; Wan, T.; Wu, Z.; Du, B. Real-time detector design for small targets based on bi-channel feature fusion mechanism. *Appl. Intell.* **2022**, *52*, 2775–2784. [CrossRef]
41. Liu, S.; Qi, L.; Qin, H.; Shi, J.; Jia, J. Path aggregation network for instance segmentation. In Proceedings of the IEEE Conference on Computer Vision and Pattern Recognition, Salt Lake City, UT, USA, 18–23 June 2018; pp. 8759–8768.
42. Hu, J.; Shen, L.; Sun, G. Squeeze-and-excitation networks. In Proceedings of the IEEE Conference on Computer Vision and Pattern Recognition, Salt Lake City, UT, USA, 18–23 June 2018; pp. 7132–7141.
43. Wang, Q.; Wu, B.; Zhu, P.; Li, P.; Hu, Q. ECA-Net: Efficient Channel Attention for Deep Convolutional Neural Networks. In Proceedings of the 2020 IEEE/CVF Conference on Computer Vision and Pattern Recognition (CVPR), Seattle, WA, USA, 13–19 June 2020.
44. Hou, Q.; Zhou, D.; Feng, J. Coordinate attention for efficient mobile network design. In Proceedings of the IEEE/CVF Conference on Computer Vision and Pattern Recognition, Nashville, TN, USA, 20–25 June 2021; pp. 13713–13722.

45. Woo, S.; Park, J.; Lee, J.-Y.; Kweon, I.S. Cbam: Convolutional block attention module. In Proceedings of the European Conference on Computer Vision (ECCV), Salt Lake City, UT, USA, 18–23 June 2018; pp. 3–19.
46. Zheng, Z.; Wang, P.; Liu, W.; Li, J.; Ye, R.; Ren, D. Distance-IoU loss: Faster and better learning for bounding box regression. In Proceedings of the AAAI Conference on Artificial Intelligence, Seattle, WA, USA, 13–19 June 2020; pp. 12993–13000.
47. Dai, W.; Li, D.; Tang, D.; Jiang, Q.; Wang, D.; Wang, H.; Peng, Y. Deep learning assisted vision inspection of resistance spot welds. *J. Manuf. Processes* **2021**, *62*, 262–274. [CrossRef]
48. Tian, R.; Jia, M. DCC-CenterNet: A rapid detection method for steel surface defects. *Measurement* **2022**, *187*, 110211. [CrossRef]
49. Zheng, Z.; Wang, P.; Ren, D.; Liu, W.; Ye, R.; Hu, Q.; Zuo, W. Enhancing Geometric Factors in Model Learning and Inference for Object Detection and Instance Segmentation. *IEEE Trans. Cybern.* **2021**, *52*, 8574–8586. [CrossRef]
50. Long, Y.; Gong, Y.; Xiao, Z.; Liu, Q. Accurate object localization in remote sensing images based on convolutional neural networks. *IEEE Trans. Geosci. Remote Sens.* **2017**, *55*, 2486–2498. [CrossRef]
51. Sandler, M.; Howard, A.; Zhu, M.; Zhmoginov, A.; Chen, L.-C. Mobilenetv2: Inverted residuals and linear bottlenecks. In Proceedings of the IEEE Conference on Computer Vision and Pattern Recognition, Salt Lake City, UT, USA, 18–23 June 2018; pp. 4510–4520.
52. Howard, A.G.; Zhu, M.; Chen, B.; Kalenichenko, D.; Wang, W.; Weyand, T.; Andreetto, M.; Adam, H. Mobilenets: Efficient convolutional neural networks for mobile vision applications. *arXiv* **2017**, arXiv:1704.04861.

MDPI AG
Grosspeteranlage 5
4052 Basel
Switzerland
Tel.: +41 61 683 77 34

*Applied Sciences* Editorial Office
E-mail: applsci@mdpi.com
www.mdpi.com/journal/applsci

www.ingramcontent.com/pod-product-compliance
Lightning Source LLC
Chambersburg PA
CBHW061516210926
51850CB00008B/1935
*9783725872749*